INSIDE
BASIC GAMES

INSIDE BASIC GAMES

Richard Mateosian

Library of Congress Cataloging in Publication Data

Mateosian, Richard.
   Inside basic games.

   Includes index.
   1. Games—Data processing. 2. Basic (Computer program language) I. Title.
GV1469.2.M37   794   80-53281
ISBN 0-89588-055-5   AACR2

Cover Design by Daniel Le Noury
Technical illustrations by J. Trujillo Smith, Jeanne E. Tennant

NOTICES

Apple is a registered trademark of Apple Computer Inc.
TRS-80 is a registered trademark of Tandy Corporation.
PET/CBM are registered trademarks of Commodore Inc.
All uses of these terms throughout this work are to be construed as adjectives.

   Every effort has been made to supply complete and accurate information. However, Sybex assumes no responsibility for its use, nor for any infringements of patents or other rights of third parties which would result.
   Copyright © 1981 SYBEX Inc. World Rights reserved. No part of this publication may be stored in a retrieval system, transmitted, or reproduced in any way, including but not limited to photocopy, photograph, magnetic or other record, without the prior agreement and written permission of the publisher.

Library of Congress Card Number: 80-53281
ISBN 089588-055-5
First Edition 1981
Printed in the United States of America
10 9 8 7 6 5 4 3 2

Contents

PREFACE   *xv*

INTRODUCTION   *xvii*

1   ARITHMETIC GAMES   1

Addition Drill   1
The Addition Drill Program   1
Arithmetic Drill   9
The Arithmetic Drill Program   10
Possible Additions and Changes   19
Summary   19

2   GUESSING GAMES   23

General Form of Guessing Games   23
Four   24
A Sample Game   24
The Guessing Game Program   29
The Hangman Program   50
Possible Additions and Changes   59
Summary   60

3   TIME GAMES   63

The Pet Clock   63
Clock   64
The Clock Program   65
Card Memory   83
The Card Memory Program   84
Ten-Key Flicker   98
Timer   107
Summary   109

4   DATE GAMES   111

Birthday   111
The Birthday Program   112
Calendar   121
The Calendar Program   121
Summary   128

5 TAXMAN — 131
Instructions for Taxman   131
The Taxman Program   134
Suggestions for Improvements and Additions   142
Summary   142

6 PROGRAMMING WITH FREE BASIC — 145
Program Design Techniques   145
Free BASIC   146
Translating from Free BASIC into BASIC   149
Free BASIC, Structured Programming and Pascal   155
Summary   157

7 THE MATCH-UP GAME — 161
The Game-Building Phase   161
The Playing Phase   172
The Match-up Program   180
Changes and Improvements   229
Summary   231

8 CRAPS — 233
Instructions for Craps   233
The Craps Program   241
Suggested Additions and Improvements   271
Summary   272

9 ALIEN LIFE — 275
Alien Encounter   275
The Rules of Game of Life   286
The Alien Life Program   289
Improvements and Additions   321
Summary   321

Appendix A The ASCII Character Set   **322**
Appendix B TRS-80 BASIC Listings   **324**
Index   **346**

Illustrations

CHAPTER 1

- 1.1: Flowchart of Addition Drill   2
- 1.2: Addition Drill   3
- 1.3: Addition Drill in Free BASIC   9
- 1.4: Flowchart of Arithmetic Drill   12
- 1.5: Arithmetic Drill   13
- 1.6: Arithmetic Drill in Free BASIC   18

CHAPTER 2

- 2.1: Dialog for Four   25
- 2.2: Guessing Game   30
- 2.3: Think Subroutine for Guessing Game   36
- 2.4: Askword Subroutine for Guessing Game   37
- 2.5: Guess Subroutine for Guessing Game   42
- 2.6: Check Subroutine for Guessing Game   43
- 2.7: Hint Subroutine for Guessing Game   45
- 2.8: Stats Subroutine for Guessing Game   46
- 2.9: Setup Subroutine for Guessing Game   47
- 2.10: Two Small Subroutines for Guessing Game   49
- 2.11: Single-Character Input   51
- 2.12: Screen Clearing   52
- 2.13: Think Subroutine for Hangman   54
- 2.14: Guess Subroutine for Hangman   55
- 2.15: Check Subroutine for Hangman   57
- 2.16: Hint Subroutine for Hangman   58
- 2.17: Setup Subroutine for Hangman   59

CHAPTER 3

- 3.1: Pet Clock   64
- 3.2: Command Processing for Clock   67
- 3.3: Time Display for Clock   68
- 3.4: Event Checking for Clock   70
- 3.5: Speed Adjustment Commands for Clock   71
- 3.6: Time Setting for Clock   72

3.7:	Alarm Commands for Clock	73
3.8:	Alarm Display for Clock	74
3.9:	Inelegant Time Input for Clock	75
3.10:	Time Adjustment Routines for Clock	76
3.11:	Housekeeping for Clock Change	77
3.12:	Mechanics of Clock Change	78
3.13:	Julian Time Routines for Clock	79
3.14:	Utility Routines for Pet Clock	80
3.15:	TIME$ Routines for Clock	81
3.16:	Initialization for Clock	82
3.17:	Card Memory Test	85
3.18:	Shuffle Subroutine for Card Memory	86
3.19:	Card Display Routines for Card Memory	87
3.20:	Decode/Encode Routines for Card Memory	88
3.21:	Player Input for Card Memory	89
3.22:	Ask-for-Card Routine for Card Memory	90
3.23:	Parsing Player's Answer	90-91
3.24:	Give the Score for Card Memory	92-93
3.25:	Recall Record Performances for Card Memory	93
3.26:	Old Favorite Routines for Card Memory	94
3.27:	Delay Routine for Card Memory	95
3.28:	Initialization for Card Memory	96-97
3.29:	Pet Digit Arrangement	99
3.30:	Samples of Ten-Key Flicker Displays	100
3.31:	Ten-Key Flicker	101
3.32:	Pattern Display for Ten-Key Flicker	102
3.33:	Time-Limited Input for Ten-Key Flicker	103
3.34:	Delay Routine for Ten-Key Flicker	104
3.35:	Pattern Creation for Ten-Key Flicker	105
3.36:	Odds and Ends for Ten-Key Flicker	106
3.37:	Timer	107-108

CHAPTER 4

4.1:	Screen Display for Birthday	112
4.2:	Birthday	113
4.3:	Date Input	113
4.4:	Date String Examination	114
4.5:	Breaking Date Strings into Fields	115
4.6:	Determine Day of Week	116

4.7:	Gratuitous Remark about Day of Birth	117	
4.8:	Julian Date	118	
4.9:	Leap Year Checking	118	
4.10:	Utility Routines for Birthday	119-120	
4.11:	Initialization for Birthday	120	
4.12:	Display for Calendar	121	
4.13:	Calendar	122-123	
4.14:	Frame for Calendar	124	
4.15:	Display Day Numbers	125	
4.16:	Cursor Positioning for Calendar	126	
4.17:	Initialization for Calendar	127	

CHAPTER 5

5.1:	Sample Screens for Taxman	132-133
5.2:	Taxman	134
5.3:	Display the Remaining Pie	135
5.4:	Display Current Totals	136
5.5:	Tax Computation	137
5.6:	The Taxman's Share	138
5.7:	Display Final Tax Statistics	139
5.8:	Prepare a New Pie	140
5.9:	Conditional Reverse Video	141
5.10:	Single Character Input	141
5.11:	Screen Clearing	142
5.12:	Initialization	143

CHAPTER 6

6.1:	Verbal Algorithm Description for Addition Drill	146	
6.2:	Pseudocode for Addition Drill	147	
6.3:	A Simple Free BASIC Program	148	
6.4:	A Program Using a Case Statement	154	
6.5:	BASIC for the "Document It" Game	155	
6.6:	Correspondence Between Free BASIC and BASIC	156

CHAPTER 7

7.1:	Responses to the "::" Prompt	163
7.2:	Initialization Dialog	164-165
7.3:	Group Name Editing Dialog	166
7.4:	Responses to the "*" Prompt	167

7.5:	Question Editing	168
7.6:	Adding a Question	169
7.7:	Inserting a Question	170
7.8:	Deleting a Question	171
7.9:	Responses to the "[ ]" Prompt	172
7.10:	JOHN's Answers	174
7.11:	JOHN's Preferences	175
7.12:	SUSAN's Answers	176
7.13:	SUSAN's Preferences	177
7.14:	Pairing	178
7.15:	Player Entry Editing	179
7.16:	Match-up	182
7.17:	Editing a Player Entry	183
7.18:	Creating a Player Entry	184
7.19:	Clearing the Workspace	185
7.20:	Loading the Workspace	186
7.21:	Updating a Player Entry from the Workspace	187
7.22:	A Flawed Program to Store a Player Entry	189
7.23:	Storing Workspace Contents	190
7.24:	Asking for Changes to Player Entry	191
7.25:	Asking for More	192
7.26:	Creation and Documentation of Match-up	193
7.27:	Editing the Player's Name	194
7.28:	String Input	195
7.29:	Changing Player Choices	196-197
7.30:	Asking Player for Preferences	198-199
7.31:	Editing a Player's Preferences	200
7.32:	Packing and Unpacking Preferences	201
7.33:	Pairing Players	202
7.34:	Scoring Answers vs. Preferences	203
7.35:	Scoring One Answer	203
7.36:	Clearing Top Scores	204
7.37:	Displaying Top Scoring Players	204
7.38:	Deletion of Item k	206
7.39:	Insertion Ahead of Item k	207
7.40:	Keeping Sorted Scores	209-210
7.41:	Displaying a Question	211
7.42:	Which Kind Are You?	211
7.43:	Which Kind You Aren't	212
7.44:	What's Your Number?	214-215

7.45:	Selecting the Setup Command	215
7.46:	Setting Up a New Game	216
7.47:	Getting a Question	218-219
7.48:	Editing Group Names	219
7.49:	Selecting the Question-Editing Function	221
7.50:	Adding a Question	222
7.51:	Deleting a Question	222
7.52:	Mechanics of Deleting a Question	223
7.53:	Inserting a Question	224
7.54:	Editing a Question	225
7.55:	Editing the Choices	226-227
7.56:	Getting a Question Number	227
7.57:	Single Character Input and Screen Clearing	228
7.58:	External Storage Facility	229
7.59:	Initialization	230

CHAPTER 8

8.1:	JOHN is First	234
8.2:	MARY Follows JOHN	235
8.3:	JOHN's Bet	236
8.4:	JOHN Throws the Dice	237
8.5:	MARY Makes Her Point	238
8.6:	Commands Available at Initial Display	239
8.7:	Dropping SUSAN	240
8.8:	Craps	243
8.9:	Decoding the First Input Character	244
8.10:	One Set of Throws	246-247
8.11:	One Throw	247
8.12:	Housekeeping After a Win	248
8.13:	Housekeeping After a Loss	249
8.14:	Routing the Bet Command	250-251
8.15:	Decoding the Bet Command	251
8.16:	Accepting Numeric Bet Input	252
8.17:	Better Numeric Bet Input	253
8.18:	Setting the Bet	253
8.19:	Preparing to Play	254
8.20:	Displaying Point and Throws	254
8.21:	Displaying the Throw Result	255
8.22:	Displaying the Player Statistics	256
8.23:	Preparing for Bet Input	257

8.24:	Blanking the Result Line	257
8.25:	Blanking the Statistics Area	258
8.26:	Clearing the Screen	258
8.27:	Positioning the Cursor	259
8.28:	Doing Nothing for a While	260
8.29:	Adding a Player	260
8.30:	Passing the Dice	261
8.31:	Giving the Players Numbers	261
8.32:	Quitting	262-263
8.33:	Side Betting	263
8.34:	Statistics File Handling	264
8.35:	Initializing	265
8.36:	BASIC for Initializing	266-267
8.37:	Single-Character Input	268
8.38:	Displaying the Dice	269
8.39:	Raking in the Dice	270

CHAPTER 9

9.1:	A Simple Picture Transmission	276
9.2:	Other Interpretations of the Simple Picture	276
9.3:	A Solid Interpretation of the Simple Picture	277
9.4:	First Message from an Alien	278
9.5:	Misinterpretation of the Alien Message	279
9.6:	Starting the Alien Encounter	280
9.7:	Alien Encounter Commands	282
9.8:	Starting to Draw the Alien	283
9.9:	Removing a Dot	284
9.10:	Encoding the Picture	285
9.11:	Neighboring Positions	287
9.12:	Game of Life Rules	288
9.13:	Game of Life Examples	289
9.14:	Alien Life Transformations	290
9.15:	Alien Life	291
9.16:	Setting the Frame	292
9.17:	Getting the Picture Dimensions	294
9.18:	Asking for Confirmation of Non-Prime Input	295
9.19:	Parsing the Dimension Input	296
9.20:	Checking for a Separator	297
9.21:	Testing the Dimensions for "Primeness"	298

9.22:	Checking PP's "Primeness"	299
9.23:	External Storage Stubs for Alien Life	300
9.24:	Clearing the Dot Array	301
9.25:	Centering the Picture	302
9.26:	Framing the Picture	303
9.27:	Drawing the Picture	304
9.28:	Decoding a Command	304
9.29:	Primary Command Processing	305
9.30:	Taking Miscellaneous Action	306
9.31:	Telling the Cursor Where to Go	306
9.32:	Displaying Dots and Cursors	307
9.33:	Moving the Picture Cursor	308
9.34:	Positioning the CRT Cursor	309
9.35:	Familiar Utility Routines	310
9.36:	Flashing the Picture Cursor During Input	311-312
9.37:	Showing and Erasing Dots	312
9.38:	Initializing Alien Life	313
9.39:	Building the Commands	314
9.40:	Initializing Alien Life in BASIC	315
9.41:	Counting Neighbors	316
9.42:	BASIC for Neighbors	317
9.43:	Deciding Where the Dots Shall Be	318
9.44:	Applying Life Rules to the Picture	319

Preface

The principal themes of this book spring from three sources. The themes are:

— Education through games
— Systematic programming using Free BASIC
— Playing BASIC computer games.

The first theme was a direct result of Rodnay Zaks' suggestion that I write a book on BASIC computer games. "Not just a collection of games, but an educational book that will help the reader learn to design BASIC programs," was the specification. Any reader who studies *Inside BASIC Games* will indeed be able to design BASIC programs for games and other applications.

The second theme comes from my experience in 1975 with the development of several medium-scale applications on a DTC Microfile—an early system functionally similar to a TRS-80 disk system. At that time I developed many of the ideas found in this book, including a precursor to Free BASIC.

The sources of the games are varied. I wrote all of the programs myself, from scratch. The games I invented for this book are Arithmetic Drill, Clock, Card Memory, Ten-Key Flicker, Timer, Birthday, Match-Up, and Alien Encounter. The rest are well-known computer games, but many of their sources are obscure.

Bringing a book into being is a lengthy and difficult, but rewarding, process. One of the genuine pleasures afforded the author is the opportunity to acknowledge formally some of the many contributions of others to the final work.

First, I am grateful for the high standards maintained at Sybex by the Editor-in-Chief, Rudolph Langer, and the Production Manager, Roger Gottlieb.

More specifically, I wish to thank the Sybex editors—principally Julie Sickert, then Salley Oberlin and Doug Hergert—for smoothing rough edges, resolving ambiguities, purging inconsistencies, excising irrelevancies (except when I was adamant), and, in short, turning a rough manuscript into a finished text. Some of their improvements were truly inspired.

The "mechanical" part of the job passed through many hands. Barbara Ellis managed the typing of the original manuscript. Janet Rampa and Natalie Levitt created a flawless word-processed version. Mati Sikk carried out the manual component of the transmission to the typesetting equipment. Jim Compton performed a careful proofreading of the galleys.

J. Trujillo Smith was responsible for the illustrations, layout and pasteup. This book has profited greatly from his fine combination of artistic ability and attentiveness to detail. John was ably assisted by Jeanne Tennant.

Leslie Bouffard and Chris Chambers are responsible for the high quality of packaging and promotion and for many helpful suggestions of ways to make my book responsive to the needs of the reader.

The final acknowledgement must go to my wife, Virginia Ruth Mason. I could not have written this book without her patience, understanding and sacrifices.

Berkeley, California
February, 1981

Introduction

This book is written for everyone who wants to understand computer games and to learn how to write interactive programs in BASIC. You will learn:

— How interactive programs are built, and how the principles of systems development are applied to small computers.

— How the features of specific small computer systems have been supported in BASIC.

— How Free BASIC can help you to take a systematic approach to the design of BASIC programs.

Games are used as programming examples for several reasons:

— Games are a primary interest of many small computer users and programmers.

— Games require no special expertise to understand.

— Games exhibit most of the programming situations important in other interactive computer applications.

Most readers of this book will fall into one of the following categories:

— Those who have programmed exclusively on small computers.

— Those who have been working as programmers on "mainframes" or on minicomputers, and know at least one high level language (and perhaps an assembly language).

If you never wrote a program before you acquired a home computer, then you probably read the manual that came with your computer, and spent many hours going through the examples in it. After that, you probably acquired programs from software vendors and from books of BASIC games and "practical" programs. Perhaps you even read a book on BASIC programming or on programming style.

This book will provide you with a greater understanding of computers, BASIC, and the specific features of your home computer. You will be given:

— Detailed guided tours through real BASIC programs.

— Descriptions of program structuring, information encoding, and commonly used programming algorithms.

— A detailed explanation of the "Free BASIC" program design technique to help you overcome many of the drawbacks of BASIC.

If you are an experienced programmer who has acquired a computer, then you already know how computers work and what programming languages are used for. This book will show you the strengths and weaknesses of your new programming environment. You will be shown:

— The features of both BASIC and your computer.

— A powerful program design tool called Free BASIC.

— Examples of the kinds of interactive programming that are characteristic of small systems.

A consistent pattern will be followed in presenting games throughout this book. First each game will be introduced by a description of the rules, including a sample of the dialog between the player and the game program. Then a discussion of the program used to implement the game will be presented. This discussion will focus on the principles and techniques of BASIC game programming.

Contents

— ***Arithmetic Games*** presents a very simple educational game, *Addition Drill*. Generalization, one of the most important concepts in programming, is then used to develop a more complex game, *Arithmetic Drill*. Even though the game programs presented in this chapter are simple, most readers will find the discussion interesting.

— ***Guessing Games*** presents a family of number-guessing games called One, Two, Three, Four, ..., Nine, and two word-guessing games, *Word* and *Hangman*. Generalization is again used to design a single program for both the number-guessing games and the *Word* game. An easy process of "cannibalization" transforms this program into *Hangman*. The discussion of these games also includes an example of the strategy used by a good *Four* player. This chapter should be read carefully by most readers.

— **Time Games** discusses the various techniques used for handling the time element in games and other programs. The four programs presented in this chapter are called *Clock, Card Memory, Ten- Key Flicker* and *Timer*. Program structuring and the use of "stubs" in top-down development are described. Programming techniques discussed include the synchronization of events with the "ticking" of the clock, the use of "Julian" time, and the generation of delays of fixed duration. Readers will find the *Time Games* chapter rewarding reading.

— **Date Games** contains the games *Birthday* and *Calendar*. *Birthday* demonstrates how to determine the day of the week of any given date. *Calendar* produces a calendar picture for any month and year. The importance of cursor positioning in building the calendar picture is discussed. The *Date Games* chapter should be of interest to most readers.

— **Taxman** presents an intriguing game that challenges you to discover the rules. (Once you've discovered the rules, the game is still challenging.) The program that implements *Taxman* adjusts automatically to the different idiosyncracies of number string representation on the various BASIC systems. Readers who like mathematical games will find this chapter interesting.

— **Programming in Free BASIC** presents a technique that is helpful in designing BASIC programs. Free BASIC is a "structured" BASIC that is translated manually into the actual BASIC instructions to be entered into the computer. Free BASIC is not a language; it is a program description medium (like flowcharts). A Free BASIC program description has no line numbers, and it uses symbolic names for subroutines. GOTO instructions are avoided (none appear in the Free BASIC descriptions in this book), and control structures like *if ... then ... else, repeat ... until* and *while* are used instead.

All of the programs in the first five chapters are explained in terms of their BASIC instructions, and no knowledge of Free BASIC is required to understand them. The experienced programmer should read the *Programming in Free BASIC* chapter first, in order to use the Free BASIC descriptions that appear in the first five chapters.

The programs in the final three chapters are discussed entirely in terms of their Free BASIC descriptions. Nonetheless, the actual BASIC instructions are still provided. They can be used for entering these programs by hand into your computer. In the case of the programs in Chapters 7 and 9, the BASIC appears in exactly the same compressed (no blanks) format that the author was forced to use to make them run on his 8K Pet.

— **The Match-up Game** is discussed in terms of its Free BASIC description, with no reference to the actual BASIC instructions. *Match-up* is based on pairing members of two groups. Players are provided with the means of tailoring the game to their own needs and interests. The program used to implement the *Match-up Game* is large, containing more than forty subroutines and over 500 lines of Free BASIC description. If you master all the subtleties of the material presented in this chapter, then *you are ready to design and implement interactive systems in BASIC.*

— **Craps** is simpler than *Match-up,* but it contains different elements: control of the screen display, randomization, context switching and the assurance of smooth, fast game action. If you study this chapter and apply the principles described, then the games you design will be fun to play.

— **Alien Life** introduces the *Alien Encounter* game, which includes the well-known *Game of Life.* The programs in the *Alien Life* chapter will give you insight into the techniques required to use graphic displays on your home computer.

Which Computers Does This Book Apply To?

The programs in this book are designed to be used on any computer system. The program sections that must change when the program is moved to another system are small and have been carefully identified. Each program was developed to run on some version of one or more of the following "personal" computers:

— The Radio Shack TRS-80 system, using Level II BASIC.

— The Commodore Business Machines Pet.

— The Apple Computer Model II, using Applesoft BASIC.

CHAPTER 1
Arithmetic Games

This chapter describes two simple arithmetic games, Addition Drill and Arithmetic Drill. The first game is an easy one, so that you can become accustomed to the conventions used to present the BASIC games in this book. The second game differs only slightly from the first game, but is much more interesting and challenging. The steps taken to develop the second program from the first program illustrate the principles of generalization.

Addition Drill

The Addition Drill game is played as follows. The program begins with a question like:

WHAT IS 1 + 8 ?

You reply by typing an answer and then pressing the RETURN (or ENTER) key. Since the sum of 1 and 8 is 9, you type a 9 and press the RETURN key. The program will reply

THAT'S RIGHT — NOW TRY ANOTHER ONE
WHAT IS 3 + 4 ?

Each time you type a correct answer, the program will ask another question. But let's suppose that your finger slips, and instead of typing a 7 followed by a RETURN, you answer with an 8. The program will reply

I'M SORRY, THAT'S WRONG — TRY AGAIN
WHAT IS 3 + 4 ?

That's the entire game. Each time you give a correct answer, the program will ask you a new question. Each time you give an incorrect answer, the program will repeat the question. It will continue to ask the question until you answer correctly.

The Addition Drill Program

Figure 1.1 shows a flowchart of the action of this game. To read this flowchart, begin at the circle labeled ''START'' and follow the arrows. When you reach the diamond-shaped box that has two arrows leaving

2   ARITHMETIC GAMES

it, you will either follow the arrow labeled "NO" if the answer to the question in that box is "no" or follow the "YES" arrow if the answer is "yes." Both of these paths will lead you back to a part of the diagram that you have already passed through. A flowchart segment through which you are forced to pass repeatedly is called a *loop*.

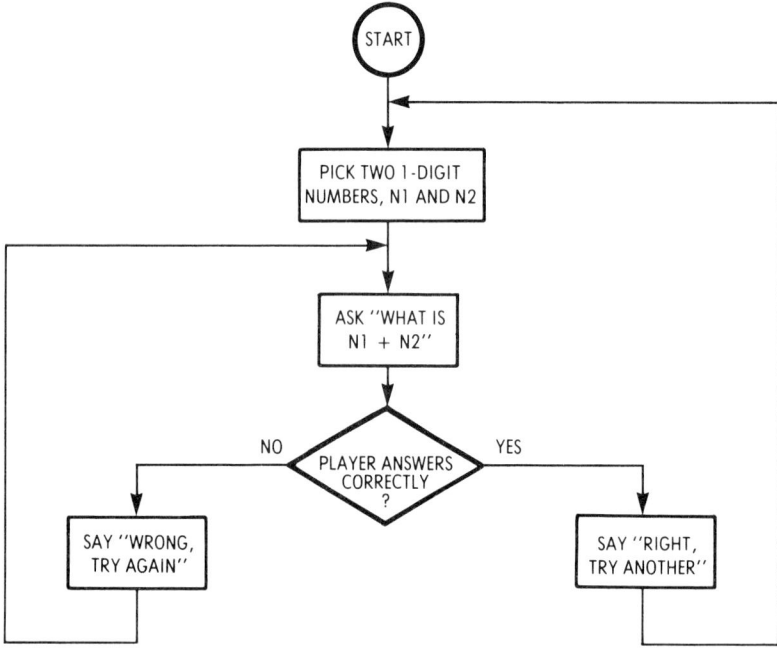

Figure 1.1:   Flowchart of Addition Drill

Figure 1.2 shows a set of actual BASIC instructions that will make the computer behave as we have just described. These instructions correspond closely to the boxes in the flowchart. The lines numbered 100 and 110 correspond to the first box. The lines numbered 120 and 130 correspond to the second box, while the line numbered 140 provides output formatting that is not shown in the flowchart. Line 150 corresponds to the diamond-shaped box and the "no" branch, while the lines numbered 160 and 170 correspond to the "yes" branch.

Let's examine each step of the program shown in Figure 1.2. The action begins at the line labeled 100, because BASIC always begins with the lowest numbered line, and continues through the lines of the program in numerical order, unless a specific instruction (like GOTO 120) changes that order.

INSIDE BASIC GAMES 3

```
100*    N1=INT(RND(1)*10)
110*    N2=INT(RND(1)*10)
120     PRINT "WHAT IS"; N1; "+"; N2;
130     INPUT A
140     PRINT
150     IF A<>N1+N2 THEN PRINT "I'M SORRY, THAT'S WRONG—TRY AGAIN":GOTO 120
160     PRINT "THAT'S RIGHT — NOW TRY ANOTHER ONE"
170     GOTO 100
```

This program randomly chooses single-digit arithmetic problems and presents them to the player. If a correct answer is given, the program goes on to another problem. If an incorrect answer is given, the program repeats the question until the player gives the correct answer.

The program shown above works for the Pet and TRS-80 BASIC systems. On an Apple, line number 120 must be changed to appear as follows:

120 PRINT "WHAT IS "; N1; " + "; N2; " ";

The difference arises from the fact that Apple BASIC neither precedes nor follows numbers with spaces, while Pet and TRS-80 precede every positive number with a space and follow every number with a space.

---

*In the TRS-80 version the term RND(0) appears in place of the RND(1). On the TRS-80, any argument other than zero has a special meaning. On the Pet and Apple, zero has a special meaning.

Figure 1.2:   Addition Drill

Generating the Random Numbers

In this program, line 100 contains one instruction:

N1 = INT(RND(1)*10)

Generally, the instruction on the line labeled 100 has the following meaning: "Choose a whole number from zero to nine and store it in the 'pigeonhole' labeled N1." This chosen number is the first of the two numbers that the program will ask you to add as the game progresses.

The line labeled 110 contains the instruction

N2 = INT(RND(1)*10)

This instruction is identical to the first instruction, except that the N1 on the left side of the equal sign has been changed to N2. Line 110 says: "Choose another whole number from zero to nine and store it in the pigeonhole labeled N2."

Even though the expression on the right side of the equal sign is the same for both of these instructions, the numbers stored in the pigeonholes N1 and N2 are not necessarily the same. This variation is due to the presence of the term RND(1) in the instruction. RND stands for the "random number" function. When the instruction is executed, the term RND(1) is replaced by an "unpredictable" number between zero and one (but never equal to zero or one).

For this game, we need whole numbers greater than or equal to zero, and less than ten. The range of numbers provided by RND can be expanded from the original zero-to-one range to a zero-to-N range, simply by multiplying by N. Thus, the term

RND(1)*10

in both of these instructions will be replaced by an unpredictable number between zero and ten (but never equal to zero or ten) when the instruction is executed.

The INT function simply "throws away" the decimal part of a number. For example, INT(5.63279) is 5, INT(0.22116) is 0, INT(9.99999) is 9. INT stands for *integer*, which is what mathematicians call whole numbers. INT takes any number and makes it into a whole number.

Thus, the term

INT(RND(1)*10)

is always replaced by a random whole number chosen from the following set of numbers: 0,1,2,3,4,5,6,7,8,9. (Do you see why this is true? Why can't the random whole number be ten? Why can it be zero?)

The PRINT Instruction

The third instruction, on the line labeled 120, is:

PRINT "WHAT IS"; N1; "+"; N2;

This instruction causes the computer to print the question "WHAT IS ____ + ____ ?" (except for the question mark, which is printed by the system when the INPUT instruction is executed). Unfortunately, what actually appears on the screen differs among the various home computer systems. For example, on a Pet computer, the output generated by this instruction will look like:

WHAT IS 9 + 7

On an Apple, it will look like:

WHAT IS9+7

This difference in spacing arises from the fact that the Pet follows each number that it prints with a space and precedes each positive number with a space. Apple inserts no spaces before or after the number. (The TRS-80 behaves like the Pet in this regard.)

There are other differences among home computer PRINT instructions, and we shall discuss them as we go along. First, let's find out how the simplest form of the PRINT instruction works. (This instruction works more or less the same for all BASIC systems.) The simplest form of this instruction consists of the word PRINT, followed by a list of *variables* and *constants*.

By definition, a constant is a number, like 5 or 9.763, or a fragment of text (like "WHAT IS") enclosed in quotes. A variable is a pigeon-hole label, that is, the name of a place in the computer's memory at which we have previously stored a constant. In more precise discussions a variable is also called a *symbolic address*, since it is the name of a place in memory. In the instruction

PRINT "WHAT IS"; N1; "+"; N2;

the list following the word PRINT consists of the constant "WHAT IS", the variable N1, the constant "+" and the variable N2. The semicolons are used to separate the items in the list, and to specify how the output will be arranged on the screen.

BASIC executes a PRINT instruction by stepping through the list of constants and variables and displaying their *values*. In other words, each time a constant is encountered in the list, BASIC displays that number or text fragment; when a variable is encountered, BASIC displays the number or text fragment that we have stored in the corresponding pigeonhole. Thus, if our first two instructions had resulted in storing a 5 at N1 and a 7 at N2, then the instruction

PRINT "WHAT IS"; N1; "+" N2;

would cause the following display:

WHAT IS 5 + 7

We have explained how BASIC determines what values to display as it executes a PRINT instruction. We shall now discuss the way in which BASIC decides upon the screen positions at which to display these values. In our example, the semicolons tell BASIC to display the values "end-to-end" without inserting additional spaces between them or starting new lines. The alternative to using semicolons is to use commas to separate the items in the list. After BASIC displays a value corresponding to a list item that is followed by a comma, it moves the cursor to the next *tab position*.

The tab positions are like the tab stops that can be set on a typewriter. With BASIC, however, they are pre-set, and cannot be controlled by the programmer. Your home computer will have either three or four tab positions per forty-character line. Thus, if we had written

   PRINT "WHAT IS", N1, "+", N2

the output would have looked like:

   WHAT IS   5   +   7

or even:

   WHAT IS   5   +
   7

In the preceding displays, we have not been able to see the effect of the final semicolon or comma. To see this we need to look at the next item displayed by our program. The instruction

   INPUT A

on the line labeled 130 is used to obtain the player's input. Use of the INPUT instructions always causes a "?" to be printed. Thus, the effect of the two lines

   PRINT "WHAT IS"; N1; "+"; N2;
   INPUT A

is to display:

   WHAT IS 7 + 5 ?

If we had omitted the final semicolon from the PRINT instruction, the display would have looked like:

   WHAT IS 7 + 5
   ?

These examples show that whenever a PRINT instruction does not end with a semicolon or a comma, BASIC finishes its execution of the instruction by moving the cursor to the beginning of the next line.

We have now covered all aspects of the simplest form of the PRINT instruction. It is a very important part of most BASIC programs, so you must be sure that you understand it. Now, try to predict the display that will result from the instructions

    PRINT "WHAT IS", N1, "+", N2
    INPUT A

and then try them on your home computer.

The INPUT Instruction

Returning to our game program, the instruction

    INPUT A

causes BASIC to print "? " (or "?" without the space on an Apple) and then wait for you to type a number and press the RETURN key (or the ENTER key on a TRS-80). After that, BASIC checks to see whether or not an actual number has been typed; if not (for example, if you type an X), BASIC will complain by displaying

    ?REENTER
    ?

or a similar message. Then it will wait for you to try again.

The IF Statement

After you have entered a number, BASIC will store it in the pigeonhole labeled A and go on to execute the line labeled 150, which contains

    IF A <> N1 + N2 THEN PRINT "I'M SORRY, THAT'S WRONG — TRY AGAIN": GOTO 120

This is a new kind of instruction — the IF instruction. In fact, since the construction beginning with "IF" can contain more than one instruction, it is called an IF *statement*. The first part of an IF statement (in this case, A <> N1 + N2) is called the *condition*. (The "<>" means "not equal to.") If the number stored in the pigeonhole labeled A is not equal to the sum of the numbers in the pigeonholes labeled N1 and N2 (that is, if you typed the wrong answer), then BASIC will execute the portion of the statement (called the *action*) following the THEN. An action consists of an instruction, or several instructions, separated by colons. In this case, the instructions are

    PRINT "I'M SORRY, THAT'S WRONG — TRY AGAIN"

and

    GOTO 120

It is very important that you understand how an IF statement works: if the condition is true, *all* of the instructions on the rest of the line are executed; if the condition is false, then *none* of the instructions on the rest of the line are executed. In either case, unless there is a GOTO among the instructions following the THEN, BASIC goes on to the next line after the line containing the IF statement. Thus, in Figure 1.2, if the condition is true (that is, if you typed the wrong answer), then BASIC will first execute the instruction

    PRINT "I'M SORRY, THAT'S WRONG — TRY AGAIN"

and then it will execute

    GOTO 120

We already know what the PRINT instruction will do, so let's look at the GOTO instruction, which causes BASIC to alter its sequence of instruction execution. Instead of going on to the line numbered 160, BASIC must go back to the line numbered 120 and begin executing from that point again. Thus, if you type the wrong answer, the sequence of printing the question and then asking for the answer will be repeated, producing a display that might look like:

    WHAT IS 5 + 7 ? **13**

    I'M SORRY, THAT'S WRONG — TRY AGAIN
    WHAT IS 5 + 7 ?

(Your answer — 13 — is shown in boldface type.)

If the condition in the IF statement on the line numbered 150 is false (that is, if you typed the correct answer), then the rest of line 150 is skipped, and BASIC proceeds to the line numbered 160. There it executes the instruction

    PRINT "THAT'S RIGHT — NOW TRY ANOTHER ONE"

and goes on to line 170 to execute

    GOTO 100

The display for this case, in which you have given the correct answer, might look like:

    WHAT IS 7 + 5 ? **12**

    THAT'S RIGHT — NOW TRY ANOTHER ONE
    WHAT IS 9 + 2 ?

Free BASIC

As we noted above, the instructions in Figure 1.2 follow the flowchart of Figure 1.1 quite closely. Figure 1.3 shows the same program in Free BASIC. (Free BASIC is described in Chapter 6 of this book.) Look at Figure 1.3 to see whether or not you can understand it. Notice that the Free BASIC Program is very similar to the program in Figure 1.2; however, it has no line numbers. Also, certain words (like "repeat" and "else") appear in lowercase letters in Figure 1.3; the words in lowercase letters do not appear at all in Figure 1.2.

Free BASIC is the tool used by the author to plan the BASIC programs that appear in this book. Chapters 1 through 5 include the Free BASIC form of each program, but no reference is made to Free BASIC in the discussion. The Free BASIC descriptions are included so that you will become accustomed to seeing Free BASIC, and so that anyone rereading these chapters after reading Chapter 6 can better understand the construction of the programs. Chapters 7, 8 and 9 refer only to the Free BASIC descriptions, not to the actual BASIC instructions.

Arithmetic Drill

The Addition Drill Program was simply a warm-up exercise for the Arithmetic Drill Program. The Addition Drill Program gave us a

```
repeat {
    N1 = INT(RND(1)*10)
    N2 = INT(RND(1)*10)
    repeat {
        PRINT "WHAT IS"; N1; "+"; N2
        INPUT A
        IF A = N1 + N2 THEN
            PRINT "THAT'S RIGHT etc"
        else
            PRINT "THAT'S WRONG etc"
    } until (A = N1 + N2)
}
```

This is a Free BASIC description of the Addition Drill program. Free BASIC is explained in Chapter 6.

Figure 1.3: **Addition Drill in Free BASIC**

chance to discuss several important BASIC instructions, but as a game it is too simple and repetitive to be interesting or challenging. Arithmetic Drill is similar, but it has two additional features that make a big difference:

— Use of numbers larger than one digit (the player selects the size).

— Random selection among addition, subtraction, multiplication and division.

Otherwise, Arithmetic Drill is very similar to Addition Drill.

Here is a sample of possible dialog between the program and the player. Note that, as earlier, the player's responses are shown in boldface type.

    HOW MANY PLACES ? **2**
    WHAT IS 240 / 16 ? **15**

    THAT'S RIGHT — NOW TRY ANOTHER ONE
    WHAT IS 8 + 1 ? **9**

    THAT'S RIGHT — NOW TRY ANOTHER ONE
    WHAT IS 23 × 77 ? **1781**

    I'M SORRY, THAT'S WRONG — TRY AGAIN
    WHAT IS 23 × 77 ?

The Arithmetic Drill Program

Figure 1.4 presents a flowchart of the program for Arithmetic Drill. Figure 1.5 shows the BASIC instructions. The program in Figure 1.5 is very similar to the Addition Drill program in Figure 1.2. Several lines of Figure 1.5 correspond almost exactly to lines in Figure 1.2, but differ slightly. For example:

— The constant "+" on line 130 in Figure 1.2 has been replaced by the variable OP$ on line 150 in Figure 1.5.

— The expression N1 + N2 on line 150 in Figure 1.2 has been replaced by the variable EX on line 170 in Figure 1.5.

— The formula INT(RND(1)*10) on lines 100 and 110 of Figure 1.2 has become FNR(D) throughout Figure 1.5, where FNR(D) = INT(RND(1)*10↑D).

These changes are called *generalizations*. A generalization is the replacement of an element by a broader category, of which the original element is a special case. For example, replacing "+" by the variable OP$ is a generalization, because "+" is one of the four possible values of OP$. Similarly, replacing the expression N1 + N2 by the variable EX is a generalization, since N1 + N2 is one of four expressions used to set the value of EX. (The other expressions are N1 − N2, N1∗N2 and N1/N2.) Finally, replacing RND(1)∗10 by RND(1)∗10↑D is a generalization, because 10 (that is, 10↑1) is one of the three possible values taken by 10↑D, as D takes the values 1, 2 and 3.

Generalization is an important technique for augmenting computer programs. Wherever possible, you should design your programs in such a way that they can be generalized easily. For example, we could have written the Addition Drill program (Figure 1.2) slightly differently. We could have added two lines at the beginning:

   80 OP$ = "+"
   90 D = 1

and then added a line after line 110:

   115 EX = N1 + N2

Then lines 120 and 150 of Figure 1.2 could have been identical to lines 150 and 170 of Figure 1.5. The advantage of this approach is that it identifies those aspects of the program that can be easily generalized. Also, the use of variables instead of constants (like OP$ and D instead of "+" and 1) reduces the number of changes that must be made when the corresponding part of the program is generalized. The chance of overlooking necessary changes is thereby reduced.

Writing your programs with variables instead of constants and planning in advance for other possible generalizations are examples of good programming practice. This was not done in the case of Figure 1.2 because doing so would have added a layer of complexity to the explanation of the first and most elementary program of the book.

Now let's look at the flowchart shown in Figure 1.4. Notice that it is virtually identical to the flowchart in Figure 1.1 except for two points:

— The box containing "Let player select. . ." does not appear in Figure 1.1.

— The box in Figure 1.1 containing the words "Pick two 1-digit numbers, N1 and N2" has been replaced in Figure 1.4 by a single box (containing the words "randomly choose OP = 1,2,3,4"), followed by a set of four paths, only one of which is taken each time through the loop.

12 ARITHMETIC GAMES

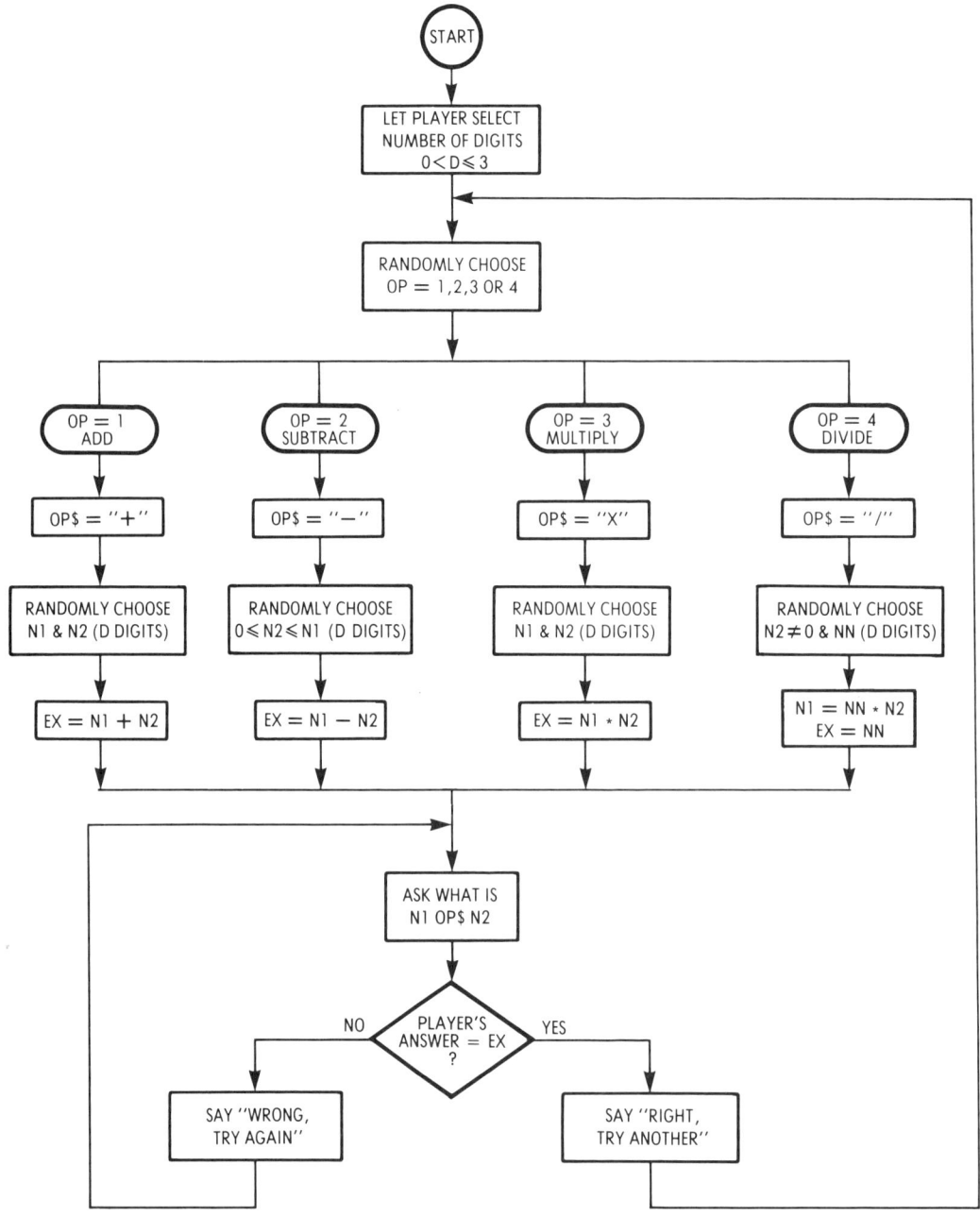

Figure 1.4: Flowchart of Arithmetic Drill

```
100   INPUT "HOW MANY PLACES";D
110   D=INT(D):IF D<=0 OR D>3 THEN 100
120*  DEF FNR(X)=INT(RND(1)*10↑X)
130*  OP=INT(RND(1)*4)+1
140   ON OP GOSUB 190,220,260,290
150   PRINT "WHAT IS";N1;OP$;N2;
160   INPUT A:PRINT
170   IF A<>EX THEN PRINT "I'M SORRY, THAT'S WRONG — TRY AGAIN":GOTO 150
180   PRINT "THAT'S RIGHT — NOW TRY ANOTHER ONE":GOTO 130
190   OP$="+"
200   N1=FNR(D):N2=FNR(D)
210   EX=N1+N2:RETURN
220   OP$="—"
230   N1=FNR(D):NN=FNR(D):IF NN<=N1 THEN N2=NN:GOTO 250
240   N2=N1:N1=NN
250   EX=N1—N2:RETURN
260   OP$="×"
270   N1=FNR(D):N2=FNR(D)
280   EX=N1*N2:RETURN
290   OP$="/"
300   N2=FNR(D):IF N2=0 THEN 300
310   NN=FNR(D):N1=NN*N2
320   EX=NN:RETURN
```

This BASIC program implements the Arithmetic Drill game. The main program consists of lines 100 through 180. The four subroutines occupy lines 190-210, 220-250, 260-280 and 290-320.

Lines 100-120 set D (the number of digits in the numbers in the drill problems) and define the FNR function that chooses the numbers used in the drill problems.

Lines 130-140 choose one of four operations (add, subtract, multiply, divide) to be used in the drill problem and call on one of the four subroutines to set:

— the numbers N1 and N2 to be included in the problem
— the symbol OP$ that is placed between them (+, —, ×, /)
— the expected answer EX.

Lines 150-180 accept an answer, evaluate it and either go on to the next problem or ask the question again.

For the same reasons given in Figure 1.2, and because Apple BASIC does not add a question mark to the string constant specified in an INPUT statement, the above program is correct for Pet and TRS-80 BASIC systems. For an Apple, lines 100 and 150 must be changed as follows:

```
100   INPUT "HOW MANY PLACES? "; D
150   PRINT "WHAT IS "; N1; OP$; N2; " ";
```

*In the TRS-80 version of this line, the term RND(0) appears instead of RND(1) (see Figure 1.2).

Figure 1.5:   **Arithmetic Drill**

14 ARITHMETIC GAMES

Each of these two points is concerned with the setting of the values of the variables D, OP$ and EX. That is, the parts of the flowchart in Figure 1.4 that handle our generalizations are expansions of the corresponding parts of the flowchart in Figure 1.1; the parts of Figure 1.4 that do not handle our generalizations are identical to the corresponding parts of Figure 1.1.

Now let's look at the program in Figure 1.5. It illustrates BASIC features that we haven't discussed yet. The first two lines are concerned with setting a value for the variable D. The first line is:

    INPUT "HOW MANY PLACES"; D

The constant "HOW MANY PLACES" tells BASIC the question to be asked; the variable D tells BASIC where to store the answer. Thus, this instruction directs BASIC to ask

    HOW MANY PLACES?

BASIC will wait for a numeric response, and will then store the response in the pigeonhole labeled D. (This is not universally true. Figure 1.5 follows the conventions of the Pet and TRS-80 computers. For an Apple computer, the same effect would be achieved by writing the instruction in the form

    INPUT "HOW MANY PLACES? "; D

On the Pet and TRS-80 systems, BASIC constructs the question by appending "? " to the end of the string constant supplied in the instruction, while on an Apple, no "? " is added.)

The next line contains two instructions separated by a colon:

    D = INT(D) : IF D <= 0 OR D > 3 THEN 100

Earlier in this chapter we learned that the action part of an IF statement can consist of several instructions separated by colons. Now we see that any line of a BASIC program can consist of several instructions separated by colons. (But remember, any instructions following an "IF" are included in the IF statement's action part.) We are familiar with INT, which makes D into a whole number by throwing away any fractional part. In the IF statement, the 100 following THEN is an abbreviation for the instruction

    GOTO 100

Line 120 of the program contains a *function definition*:

    DEF FNR(X) = INT(RND(1)*10↑X)

A function definition is simply a way of telling BASIC that we wish to abbreviate a long formula. The instruction above establishes FNR(D) as an abbreviation for the term INT(RND(1)*10↑D). This abbreviation is used on lines 200, 230, 270, 300 and 310.

The variable X that appears on both sides of the equal sign in the function definition is called a "dummy" variable. A dummy variable is simply a place holder that shows where the actual argument (D, in our case) should be plugged into the formula when the time comes to evaluate it. For example, if we wrote FNR(2), a 2 would be plugged into the formula where X appears, so that FNR(2) would be interpreted as an abbreviation for the term INT(RND(1)*10↑2).

The next line of our program, labeled 130, contains the instruction

OP = INT(RND(1)*4) + 1

This instruction causes a random choice to be made among the whole numbers 1,2,3 and 4, and the result to be stored in the pigeonhole labeled OP. (If you don't see why this is so, read the explanation given previously of why INT(RND(1)*10) takes values chosen randomly from 0,1,2,3,4,5,6,7,8 and 9.)

Incidentally, all of our uses of the INT and RND functions in combination have been for the purpose of randomly selecting one number from a set of consecutive whole numbers (0 to 9, 0 to 99, 0 to 999, 1 to 4). If BASIC were a little more flexible, we could have made a function definition that would have accommodated all of these cases:

DEF FNC(L,H) = INT(RND(1)*(H − L + 1)) + L

Then, for example,

OP = FNC(1,4)

would choose a whole number from 1 to 4, while

N1 = FNC(0,9)

would choose a whole number from 0 to 9. The expression FNR(D) would then be replaced by FNC(0,10↑D − 1). Unfortunately, such a definition is not possible, since BASIC does not allow function definitions with two dummy variables. Thus, we must use INT(RND(1)* (H − L + 1)) + L explicitly wherever we want to write FNC(L,H).

Returning to the program in Figure 1.5, the next line, labeled 140, presents another feature that we have not yet discussed, the ON statement:

ON OP GOSUB 190, 220, 260, 290

Recall that we have just set the variable OP to one of the values 1, 2, 3, or 4. There are four line numbers specified in the statement: 190, 220, 260 and 290. These lines are the initial lines of four subroutines. If OP has been set to 1, then GOSUB 190 is executed; if OP has been set to 2, then GOSUB 220 is executed; if 3, then GOSUB 260; if 4, then GOSUB 290. Exactly one of the four subroutines is called. (What happens if OP is set to a value other than 1, 2, 3, or 4 may vary among BASIC systems.)

The purpose of OP and the four subroutines is to implement the four branches shown in the flowchart in Figure 1.4. Each of the four subroutines corresponds to one of the four arithmetic operations included in the drill: addition, subtraction, multiplication and division. Each subroutine sets N1, OP$, N2 and EX; that is, the subroutine sets the first number, the symbol for the operation, the second number and the expected answer. We shall discuss the subroutines after we have finished analyzing the main program. The next line, labeled 150, contains the instruction

   PRINT "WHAT IS"; N1; OP$; N2;

This line is very similar to line 120 in Figure 1.2, but in the program in Figure 1.5, the variable OP$ has replaced the constant "+". The two instructions on the line labeled 160

   INPUT A: PRINT

are identical to the instructions on lines 130 and 140 in Figure 1.2. The INPUT instruction accepts the player's answer. The PRINT instruction spaces down one line on the screen to improve readability.

The lines labeled 170 and 180 are analogous to the lines labeled 150, 160, and 170 in Figure 1.2. The only differences are the appearance of EX instead of N1 + N2 and the different line numbers. This concludes the main program.

The four subroutines are very similar to one another. The first subroutine, occupying lines 190, 200 and 210, is the simplest:

   OP$ = "+"
   N1 = FNR(D): N2 = FNR(D)
   EX = N1 + N2: RETURN

This subroutine illustrates the general pattern exhibited by each of these four routines. First, OP$ is set. In this case, it is set to "+" to indicate addition. Next, two numbers are chosen and saved in pigeonholes N1 and N2. Finally, the expected result (the sum of the two numbers, in this case) is saved in EX.

The second subroutine, occupying lines 220, 230, 240 and 250, is only slightly more complicated:

    OP$ = "−"
    N1 = FNR(D): NN = FNR(D):
    IF NN <= N1 THEN N2 = NN: GOTO 250
    N2 = N1: N1 = NN
 250 EX = N1 − N2: RETURN

The reason for the complication is that we want the first number to be greater than or equal to the second number, so that there will be no negative answers. First N1 is chosen and a second number is saved in NN. If the second number is smaller than or equal to the first number, it is simply moved to N2. But if the number at N1 is smaller than the number at NN, the two numbers must be switched. The one at N1 must be saved at N2, and the one at NN is then saved at N1. When N1 and N2 have been chosen properly, the expected value is the difference of the two.

The third subroutine, occupying lines 260, 270 and 280, is almost identical to the first subroutine:

    OP$ = "X"
    N1 = FNR(D): N2 = FNR(D)
    EX = N1∗N2: RETURN

The only difference is the replacement of the addition symbols by multiplication symbols.

The fourth subroutine, occupying lines 290, 300, 310 and 320, is more complicated than the others because it must satisfy two requirements:

— We don't want a divisor of zero.

— We want the answer to be a whole number.

These two requirements are dealt with quite simply:

    OP$ = "/"
 300 N2 = FNR(D): IF N2 = 0 THEN 300
    NN = FNR(D): N1 = NN∗N2
    EX = NN: RETURN

Line 300 allows us to continue choosing the divisor until we get one that is non-zero. Then, instead of choosing the first number, we choose the answer instead. The first number is then computed to be the divisor multiplied by the answer.

18 ARITHMETIC GAMES

```
#Arithmetic Drill
        DEF fnc(L,H) = INT(RND(1)*(H − L + 1)) + L    #choice function
        repeat
            INPUT "HOW MANY PLACES? ";D
            until (D = 1 OR D = 2 OR D = 3)
        L = 0: U = 10↑D − 1                           #set limits for numbers
        repeat {                                      #main loop
            OP = fnc(1,4)                             #pick a number from 1 to 4
            ON OP GOSUB add, sub, mul, div
            repeat {                                  #ask until right answer
                PRINT "WHAT IS"; N1; OP$; N2;
                INPUT A: PRINT
                IF A <> EX THEN PRINT "THAT'S WRONG etc"
                } until (A = EX)
            PRINT "THAT'S RIGHT etc"
            }

add     OP$ = "+"
        N1 = fnc(L,U): N2 = fnc(L,U)                  #pick two numbers
        EX = N1 + N2: RETURN                          #expect the sum

sub     OP$ = "−"
        N1 = fnc(L,U): N2 = fnc(L,U)                  #pick two numbers
        IF N1 < N2 THEN swap N1, N2                   #put larger one first
        EX = N1 − N2: RETURN                          #expect the difference

mul     OP$ = "×"
        N1 = fnc(L,U): N2 = fnc(L,U)                  #pick two numbers
        EX = N1*N2: RETURN                            #expect the product

div     OP$ = "/"
        repeat N2 = fnc(L,U) until (N2 <> 0)          #choose divisor
        EX = fnc(L,U)                                 #choose answer
        N1 = EX*N2: RETURN                            #compute dividend
```

Figure 1.6: Arithmetic Drill in Free BASIC

This completes our discussion of the Arithmetic Drill program. Figure 1.6 shows a Free BASIC description of the program. As we explained earlier, you can ignore this information if you are not interested. However, if you do study Figure 1.6, you may notice several small differences between the programs of Figure 1.5 and Figure 1.6.

Possible Additions and Changes

Finally, for those readers who are interested in improving this game, we have included a few suggestions. Some are easy to implement, while others are more difficult.

- Keep statistics: for example, record the number of correct answers given since the last wrong answer, or the longest run of correct answers. Display these after each correct answer.

- Remember the problems the player answered incorrectly, and ask them again from time to time.

- Add other arithmetic operations like squares and cubes of numbers, powers of 2 and binary to decimal conversion.

- Use cursor control to make a fixed display: for example, always ask the question at a fixed place on the screen and always display the result and the scores at fixed places. You might even make the old question and result disappear before the new question appears.

- Allow the player to choose the operations for the drill.

- Add the time element (on a system with a built-in clock, like Pet) by only allowing so much time for the player to answer, or by keeping track of the average (or median) response time for each type of problem. If only a fixed time is allowed, display the seconds counting down while waiting for the player's answer. (See Chapter 3.)

- Add sound (on a system with a speaker, like Apple) to provide rewards for correct answers, and Bronx cheers for wrong ones. Be sure to make the sound optional so that the player can disable it with a simple keyboard command.

Summary

We have studied the simple Addition Drill game and the more complex Arithmetic Drill game. In discussing the Addition Drill program, we learned about random number generation using INT and

RND, the PRINT instruction, variables and constants, the INPUT instruction, and the IF statement.

The Arithmetic Drill game and program illustrated the meaning of generalization and the use of variables rather than constants, one of the principal means of planning for generalization. We also learned about the use of question string constants in INPUT statements, function definitions and dummy variables, the ON...GOSUB construction, and the use of subroutines.

The Free BASIC descriptions of the two programs were discussed briefly, and the use of Free BASIC in this book was explained. Free BASIC descriptions appear for all programs, but the first five chapters can be read without reference to these descriptions.

Finally, additions and improvements to the Arithmetic Drill game were suggested. Statistics, additional operations and player control, and better use of special features (like sound and a clock) were suggested.

CHAPTER 2
Guessing Games

The games presented in Chapter 1 were deliberately kept simple. Our discussion of the BASIC instructions used in their design focused on facts about BASIC rather than on principles of programming and game design. We did, however, discuss one important concept in Chapter 1: generalization. This concept will again play an important role in the games described in this chapter.

In this chapter we shall present several games based on the guessing of a secret word or number. Our discussion will begin by reviewing the general form of guessing games. Then we shall describe the rules of a game called Four and a sample series of guesses and replies. This sample will illustrate the strategy followed by a good player. Understanding games from the player's point of view will help you to design interesting and challenging games.

Rather than providing a game program that only works for Four, we shall first plan for generalization. To do this, we shall introduce a family of number-guessing games, One through Nine. (Four is a member of this family.) Then, as a further generalization, we shall introduce a word-guessing game called Word, which has rules similar to those of the number-guessing games.

Only then will we design a program. This program, called the Guessing Game program, will teach you many useful techniques and principles. Since one of these techniques is the "cannibalization" of a program in order to produce a similar program, we shall introduce the game Hangman at the outset, and then finish the chapter by cannibalizing the Guessing Game program to create a Hangman program.

General Form of Guessing Games

Games of the type that we are describing in this chapter tend to resemble one another. In general, there are two players. The first player thinks of a word or number. The second player makes a series of guesses. The first player's responses to the guesses are designed to narrow the range of possible words or numbers, so that sooner or later the second player can make the final guess with certainty. Furthermore, the number of guesses allowed is usually limited.

In the computer versions of these games, all or part of the role of the first player is performed by the program. As a result, the second player can rely upon the accuracy of the first player's replies to the series of guesses. When we go through a sample series of guesses for the game of Four, you will see how the second player uses all of the information that the first player's answers reveal. Inaccurate answers by the first player (the program) would change the game from a game of skill into a game of chance.

Four

Four is a simple game with many popular variations. The program begins the game by selecting a four-digit "secret" number, in which no two digits are the same. Then it is your turn. You make a series of four-digit guesses until you have guessed the secret number. That is, you must guess the four digits of the secret number, and you must guess them in the correct order. For example, if the secret number is 7915, then you must guess 7915; 1975 won't do.

After each of your guesses, the program replies with two numbers. The first number tells you how many of the digits in your guess appear in the secret number. For example, if the secret number is 7915 and your guess is 1975, then the first number of the program's reply will be 4, since all four of the digits of 1975 appear in 7915. The second number of the program's reply tells you how many of the digits in your guess appear in the same position in the secret number as they do in your guess. For example, if the secret number is 7915 and you guess 1975, then the second number of the program's reply will be 2, since the digits 9 and 5 appear in the same positions (the second and fourth positions) in both numbers. The screen dialog for this example would look like this:

```
WHAT DO YOU GUESS? 1975
RIGHT: 4   IN PLACE: 2
WHAT DO YOU GUESS?
```

This sequence of guesses and replies continues until you have guessed the secret number. If you have not guessed the correct number within a pre-programmed number of guesses, the program ends the game unilaterally.

A Sample Game

Figure 2.1 shows a sample game. Let's go through the dialog step-by-step to be sure we understand how the game works. We'll also discuss the player's guesses in order to give you an idea of the tricks you can use to guess the secret number quickly.

The player's first guess is made with no information other than the fact that the program has selected a four-digit number, and that no two of these digits are the same. The player's choice of the four-digit number 1234 is no better or worse than any other first guess. The

INSIDE BASIC GAMES

```
GUESS NUMBERS OF 4 DIGITS              program selects 7359

WHAT DO YOU GUESS? 1234                3 is in 7359, but in second
RIGHT: 1   IN PLACE: 0                 place, not third.

WHAT DO YOU GUESS? 5678                5 and 7 are in 7359, but in
RIGHT: 2   IN PLACE: 0                 third and first places, not first
                                       and third.

WHAT DO YOU GUESS? 9012                9 is in 7359, but in last place,
RIGHT: 1   IN PLACE: 0                 not first.

WHAT DO YOU GUESS? 3769                3, 7 and 9 are in 7359. 9 is in
RIGHT: 3   IN PLACE: 1                 last place, as guessed.

WHAT DO YOU GUESS? 6893                3 and 9 are in 7359, but in
RIGHT: 2   IN PLACE: 0                 second and fourth places,
                                       not fourth and third.

WHAT DO YOU GUESS? 3957                3, 9, 5 and 7 are all in 7359,
RIGHT: 4   IN PLACE: 1                 but only 5 is in place.

WHAT DO YOU GUESS? 7359
THAT'S RIGHT! YOU GOT IT IN 7 GUESSES.
```

The program selects a four-digit number. The player makes a series of guesses and is given information about the accuracy of each guess. As shown here, the player has designed each guess to obtain as much new information as possible, and to make maximum use of information already obtained from prior guesses.

An experienced player will usually do at least as well as this player has done. A bad break on the fourth guess — getting one in place — reduced the amount of information available at the sixth guess. If the fourth guess had been 3967, the player would have known the correct order as well as the correct digits at the time of the sixth guess.

Figure 2.1:   *Dialog for Four*

program's reply of "RIGHT: 1 IN PLACE: 0" tells the player that only one of the digits in the secret number appears among the digits 1, 2, 3 and 4, and that none of the digits of the player's guess appears in the same position in the secret number as in the guess. This last bit of information will be more useful later on in the game.

The player's second guess of 5678 contains a new set of four digits. Note that in the early guesses, the player wants to try each of the ten possible digits. This helps the player to identify the four digits as quickly as possible. The reply of "RIGHT: 2 IN PLACE: 0" tells the player that two digits of the secret number are to be found among the digits 5, 6, 7 and 8. The player can now infer that the fourth digit is either 9 or 0, since only three digits of the secret number have been identified, and the digits 9 and 0 are the only ones that have not yet been tried.

The player's third guess of 9012 is made to obtain placement information about 9 and 0, and to help to determine which of the digits 1, 2, 3 and 4 is the one included in the secret number. The reply of "RIGHT: 1 IN PLACE: 0" tells the player that the digits 1 and 2 are ruled out. The "RIGHT: 1" is either 9 or 0 (as we showed earlier), so the digits 1 and 2 are both wrong. The player now knows that one digit is a 3 or a 4, one is a 9 or a 0, and two are to be taken from 5, 6, 7 and 8.

The player's fourth guess of 3769 takes one digit from 3 and 4, two digits from 5, 6, 7 and 8, and one from 9 and 0. The reply of "RIGHT: 3 IN PLACE: 1" gives good information about which digits are in the secret number. Ironically, the fact that one digit is in place gives less information about placement than if none had been in place. We shall see why this is so when we plan the sixth guess.

The player's fifth guess of 6893 was planned by considering the following fact: on the fourth guess, either 3 and 9 were both right, or 6 and 7 were both right. (There could not be a wrong digit in each of these pairs, because there was only one wrong digit in the entire guess.) Since each of the digits 3 and 9 had a 50-50 chance of being right (because one digit is either 3 or 4 and one digit is either 9 or 0), the chance that both of the digits 3 and 9 are right is 1 in 4. On the other hand, since there are six ways of choosing two different digits from a set of four digits, the chance that the digits 6 and 7 are both right is only 1 in 6. Therefore, the player decided to assume that the digits 3 and 9 are in the secret number and that only one of the digits 6 and 7 is in the secret number. The reply of "RIGHT: 2 IN PLACE: 0" tells the player that 7 is definitely one of the digits in the secret number, since the only change (other than position) between the fourth and fifth guesses was to replace 7 by 8. This reply also tells the player that 8 is definitely not one of the digits of the secret number.

For the sixth guess of 3957, the player decided to stay with the assumption that 3 and 9 are both correct. On the basis of that assumption, 6 cannot be correct, since 3, 9 and 7 would have been the three correct digits in the fourth guess of 3769. With 6 and 8 ruled out, this leaves 5 as one of the digits, since two of the digits 5, 6, 7 and 8 are correct.

The order of the digits in the sixth guess was based upon examination of the placement information obtained from prior guesses. Notice that in the first five guesses, no digit appeared in the same position in two different guesses. Since four of those guesses received replies of "IN PLACE: 0", the player can rule out certain positions for each of our digits. The following possibilities remain:

3 can only be in the first or second position

5 can be in the second, third or fourth position

7 can be in the first, second or fourth position

9 can be in the second or fourth position.

From this, we see that only the digit 5 can be in the third position. Nothing else is pinned down, so the player has decided to try 9 in the second position. With this choice, 3 must be in the first position. This leaves only the fourth position for the digit 7.

The reply of "RIGHT: 4 IN PLACE: 1" means that the player has guessed the four digits correctly, but that the placement is wrong.

The final guess of 7359 is made with certainty. The player already knows that 5 is the third digit. The guess that 9 was second turned out to be wrong. This means that 9 must be fourth. The last choice is between 3759 and 7359. Since only one digit was correctly placed in the fourth guess of 3769, the choice of 3759 is ruled out. The number must be 7359. This completes our sample game.

Our discussion of the game of Four will focus on three points:

1. How to incorporate several similar games into one program.

2. How to "cannibalize" a working program in order to create a new program quickly.

3. How to make use of several useful programming tricks.

A Set of Games Like Four

To illustrate the first point, we need several games that are similar to Four. Two obvious examples are the games Three and Five. These games are played like Four, but with three-digit or five-digit numbers.

The Guessing Game program that we shall develop will allow secret numbers and guesses containing any number of digits from one to nine.

The generalization from Four to a set of games like Three and Five requires planning, but is essentially very simple. To make our example a little more complicated, the Guessing Game program will allow the player to choose to play the game Word as well. Word is played like Four, but there are three important differences:

— The player tries to guess a word rather than a number.

— A given letter can occur more than once in the word (in the number guessing games, no two digits are the same).

— The word to be guessed is entered into the computer by a second player (in the number guessing games the number is selected randomly by the program).

These differences will force us to plan our Guessing Game program carefully. When we have completed the design of this program it should be possible to add other games to the player's options by means of simple programming. Suggestions for additions to the Guessing Game program are given at the end of this chapter.

Hangman

To illustrate the second point—how to cannibalize a working program—we shall use the game called Hangman. Hangman is similar to the games that we have been describing, but as you will see from the following description, it also differs in important ways.

One player selects a word and specifies the number of wrong guesses that will be allowed. Then the other player tries to guess the word. The program begins by displaying a set of dashes to indicate the number of letters in the word. For example, if the word is SYZYGY and five wrong guesses are allowed, then the display will look like:

         _ _ _ _ _ _   WRONG GUESSES LEFT: 5
         WHAT DO YOU GUESS?

The player must then either guess the entire word, or guess one letter of the word. If the player guesses a letter that is in the word, then the dash is replaced by the letter in each place where the letter occurs. For example, if the player guesses Y, then the display will look like

         _ Y _ Y _ Y   WRONG GUESSES LEFT: 5
         WHAT DO YOU GUESS?

The game continues until the player either guesses the word, or exceeds the allowed number of wrong guesses.

There is no doubt that our Guessing Game program could be designed in such a way that Hangman would be included as one of its games. However, there are so many differences between Hangman and the other games that it is simply easier to provide a separate program for Hangman. We shall do this by modifying or replacing parts of our Guessing Game program.

Programming Tricks

Our third point refers to programming tricks. The most significant of these are concerned with encoding a hidden word or number, and then checking the player's guesses against it. The technique of encoding several pieces of information into one number by adding appropriate powers of two is similar to the assembly language technique of using different groups of bits within a word for different purposes. We shall discuss this technique when we discuss the "hit" array that is used to encode the hidden word or number.

Now that we have discussed the games to be implemented by our Guessing Game program, revealed our plans to cannibalize it for Hangman, and pointed out the techniques that this program illustrates, let's examine the program itself.

The Guessing Game Program

The Guessing Game Program is shown in Figures 2.2 through 2.12. The "main routine" of this program is shown in Figure 2.2. Note that it is shown twice. Appearing at the top of the figure is a Free BASIC description of the program; the actual BASIC instructions appear at the bottom. As we stated earlier, the Free BASIC is there simply to help you become accustomed to seeing it, or to aid you if you have already read Chapter 6. You can ignore it if you like, but if you compare the Free BASIC and BASIC forms of this program (and the other programs that follow), the Free BASIC form will begin to make sense to you (even if you haven't read Chapter 6).

The Guessing Game program in Figure 2.2 consists of three nested loops. The innermost loop, occupying lines 140 and 150, corresponds to the player's guesses for one hidden word or number. The loop is repeated until the player has guessed all of the letters or digits and has placed them properly (the condition IP = N), or until the player's allotment of guesses has been exhausted (the condition IP = $-$1). Inside this loop, the subroutines at lines 360, 410 and 480 are called to accept the player's guess, check it against the hidden word or number, and tell the player how many digits or letters were "RIGHT" and how many were "IN PLACE." The loop-controlling variable IP ("in place") is set in the subroutine at line 410.

```
#Guessing game

        GOSUB init                              #set up arrays, constants
        repeat {                                #main loop
            GOSUB setup                         #choose and set up for game
            repeat {                            #play one game
                GOSUB think                     #select secret
                PRINT: PRINT IN$: PRINT         #tell player which game
                G = 0                           #start counting guesses
                repeat {                        #guessing loop
                    GOSUB guess: G = G + 1      #player makes a guess
                    GOSUB check                 #program checks it
                    GOSUB hint                  #program gives hint
                } until (IP = N OR IP = -1)     #no more guesses
                GOSUB stats                     #tell player the score
                GOSUB next                      #await signal to start next
            } until (NX$ = "N" OR NX$ = "E")    #"N" to change games
        } until (NX$ = "E")                     #"E" to quit altogether
        END

100     GOSUB 670
110     GOSUB 590
120     GOSUB 200:PRINT:PRINT IN$:PRINT
130     G=0
140     GOSUB 360:G=G+1:GOSUB 410:GOSUB 480
150     IF IP <> N AND IP <> -1 THEN 140
160     GOSUB 520:GOSUB 710
170     IF NX$="N" THEN 110
180     IF NX$="E" THEN END
190     GOTO 120
```

This is the "main" routine of the guessing game program. A Free BASIC description appears first. Below that, the actual BASIC instructions are shown.

Figure 2.2:   Guessing Game

The next level of loop outside this innermost loop occupies the entire program with the exception of lines 100 and 110. This loop corresponds to a series of games of one specific type. At the end of each game, the program waits for a single-character input from the keyboard to tell it what to do next. The character typed by the player is accepted by the subroutine at line 710 (called from line 160) and stored in the variable NX$. The instructions of the loop are repeated (i.e., another game of the given type is played) until NX$ (the character typed by the player) contains either "N" or "E". "N" means that the player wishes to switch to a new type of game; "E" means that play is at an end.

The outermost loop includes line 110, which calls a subroutine that allows the player to select the type of game to be played. Each time the program is started, and whenever the player types "N" at the end of a game, the program comes to line 110 to allow a game selection.

The line labeled 100, which contains a call to an "initialization" subroutine at line 670, is executed once at the start of the program. It is not part of any loop, and it is never executed again. Similarly, the END instruction following the THEN on line 180 is outside the outermost loop, and is only executed once. If the player types "E" at the end of a game, the program leaves both the intermediate and the outermost loops, and terminates by executing the END instruction.

With this understanding of the overall structure of the program, let's look at the actual instructions and subroutines in more detail. The first line is a call to an initialization subroutine that occupies lines 670 through 700. Although modern programming practice (as advocated by Yourdon, *et al.*) discourages this style, many programmers find it helpful to begin a program with a call to a subroutine that declares array dimensions, sets values for constants and performs other, similar setup tasks. This subroutine is written after the remainder of the program has been written, since as you begin to write your program, you know that there will be a variety of tasks that will need to be performed first, but you won't know what these tasks will be until you have written the remainder of the program. We shall discuss this subroutine, which is shown in Figure 2.10, later.

The next line, labeled 110, contains the instruction

GOSUB 590

This is a call to the subroutine that establishes the type of game to be played, and assigns values to a certain set of variables. These variables have constant values during the playing of a series of games of a given type, but they vary in value among the different types of game. We shall look at this subroutine later (it appears in Figure 2.9); in order to understand the main program, we only need to know that this subroutine sets the values of the following variables:

- GM   (game type): set to 1 for number guessing, 2 for word guessing.
- N   (number of digits): set to the number of digits, for the number-guessing games. For Word, N is set to the length of the secret word by the subroutine that allows the second player to enter the secret word.
- MG   (maximum guesses): set to the maximum number of guesses that the player will be allowed to make. For Word, MG is set at the same time as N.
- VG   (very good): set to a number that represents the maximum number of guesses that will be considered a very good performance by the player. For Word, VG is set when N and MG are set.

The next line, labeled 120, contains four instructions:

GOSUB 200: PRINT: PRINT IN$: PRINT

The GOSUB 200 is a call to a subroutine (shown in Figure 2.3) that makes up the next problem. We shall study it in more detail later. For now, all we need to know is that this subroutine sets the string variable IN$, and encodes the problem in a way that is understood by the subroutine that checks the correctness of each of the player's guesses. The next three instructions on line 120 are:

PRINT: PRINT IN$: PRINT

These instructions cause the text string that was stored in IN$ (by the subroutine at line 200) to be displayed (with blank lines above and below it). This text string is the message to the player about the size and type of guess that is expected. For example, if the game selected is Four, the string stored in IN$ will be

"GUESS NUMBERS OF 4 DIGITS"

The instruction

G = 0

on line 130 serves to initialize the count of the number of guesses made by the player.

On line 140 we find the body of the program's innermost loop. It contains four instructions:

> GOSUB 360: G = G + 1: GOSUB 410: GOSUB 480

The instruction GOSUB 360 at the beginning of the line is a call to the subroutine that accepts the player's guess. This subroutine (shown in Figure 2.5) takes in the player's guess as a string (G$) and breaks it up into an array of characters (GC).

After the GOSUB 360 on line 140 is the instruction

> G = G + 1

Recall that before we entered the loop that starts on line 140, we initialized G to zero (line 130). Each time we call the subroutine at line 360 (accepting a guess), we add 1 to the number stored at G. The instruction G = G + 1 means "add 1 to the number stored at G." Thus, the number stored at G is the number of guesses that the player has made.

The third instruction on line 140 is a call to the subroutine at line 410. This subroutine, shown in Figure 2.6, checks the guess against the answer that the program expects, by using the encoded "hit" array. As we continue our study of the main program, all we need to know about this subroutine is that it sets the values of the variables RG and IP. RG (for "RIGHT") is the number of characters (letters or digits) in the player's guess that appear in the expected answer. IP (for "IN PLACE") is the number of characters that also appear in the correct position.

The final instruction on line 140 is also the final instruction in the body of the innermost loop. This instruction is a call to the subroutine at line 480. This subroutine, shown in Figure 2.7, takes the values of IP and RG set by the subroutine at 410, and uses them to generate a "hint" of the form:

> RIGHT: 3   IN PLACE: 2

This is also the subroutine that prints:

> THAT'S RIGHT!

or:

> TOO MANY GUESSES — YOU LOSE.

The first of these two statements is printed if IP = N (meaning that all digits are in place); the second statement corresponds to G = MG (meaning that the maximum number of guesses has been reached). In the latter case, IP is set to −1 as an instruction to the main program to terminate the loop that accepts and checks the player's guesses.

The line labeled 150 is the condition-testing portion of the loop that starts at line 140. The instruction

IF IP <> N AND IP <> −1 THEN 140

means "If the player has not yet guessed the answer (IP <> N) and if the player has not reached the maximum number of guesses allowed (IP <> −1), then go back to line 140 to make another pass through the guessing loop." (The symbol <> used in this instruction means "is not equal to.")

The line labeled 160 contains two instructions:

GOSUB 520: GOSUB 710

These are the two instructions that complete the game. The first is a call to the subroutine at line 520. This routine, shown in Figure 2.8, prints the "score" of the game. Using the guess counter G, which is initialized on line 130 and updated in the loop on line 140, this routine prints out a statement like

YOU GOT IT IN 4 TRIES.

If the value of G is less than or equal to the value of VG, then the routine also prints

VERY GOOD.

As a little added variant—variety is important in game program behavior—the routine says

YOU MUST BE PSYCHIC

if the player has guessed the answer on the first try. This also takes the place of the special programming that would be necessary to avoid the ungrammatical

YOU GOT IT IN 1 TRIES

The second instruction on line 160 is a call to the subroutine at line 710. This subroutine, shown in Figure 2.10, accepts a single-character input from the player and stores it in NX$. This input is the player's instruction to the program; it tells the program what to do next. The options, as interpreted by the instructions on lines 170, 180 and 190, are:

N	Allow player to select a new game.
E	Stop playing and return to BASIC.
any other key	Start another game of the same type.

The use of a single-character input at this point allows the game to proceed smoothly. For example, suppose that you have been playing Four. The program has just said

THAT'S RIGHT! YOU GOT IT IN 8 TRIES.

The rest of the screen is undisturbed, and you look back over your guesses to see what you could have done to guess the correct answer in fewer tries. When you have finished studying the screen, you depress the space bar. The screen clears and a new game of Four begins. The program gives you as much time as you wish to examine the previous game; on the other hand, you can proceed to the next game very quickly without having to answer a question like

DO YOU WISH TO PLAY ANOTHER GAME (Y OR N)?

When you depress the space bar as described in the sequence above, you are in control of the game. When you answer a question such as the one shown above, the program is in control. Maintaining control over the game contributes to the fun of playing it.

Now suppose that you wish to change the game from Four to Five. Instead of depressing the space bar at the end of the game, you type N. The screen clears, and the program allows you to select a new type of game. Compare this with the following dialog:

DO YOU WISH TO PLAY ANOTHER GAME (Y OR N)? **N**
DO YOU WISH TO PLAY A DIFFERENT KIND OF GAME (Y OR N)?

Naturally, this is a matter of personal preference. If you prefer the dialog shown above, you might find it an interesting exercise to modify the Guessing Game program accordingly. The only changes required are the removal of the GOSUB 710 from line 160, and the replacement of lines 170, 180 and 190 with instructions implementing the new version. You don't need to know anything about the operation of any of the subroutines to accomplish this change.

Now that we have gone through the Guessing Game program of Figure 2.2, let's look at its subroutines in more detail. We shall start with the subroutine at line 200, shown in Figure 2.3, and its two subroutines at lines 310 and 240, shown in Figures 2.3 and 2.4.

The subroutine at line 200 constructs a new problem. For example, if the game being played is Four, this subroutine selects the four-digit number that the player will try to guess. If the game is Word, then it asks the second player to supply the word that the first player will try to guess. After selecting a number or word, the routine encodes it, using the "hit" array HT, which we shall discuss shortly. The purpose

36 GUESSING GAMES

#Think up the next problem

think　　GOSUB clearscreen　　　　　　　　　　#begin game on clear screen
　　　　　　FOR ZZ = BH TO TH　　　　　　　　　　#clear hit array
　　　　　　　　HT(ZZ) = 0
　　　　　　NEXT ZZ
　　　　　　ON GM GOSUB number, askword
　　　　　　RETURN

#Think of a number of N digits, all different

number　　FOR ZZ = 1 TO N
　　　　　　　　repeat
　　　　　　　　　　HH = INT(RND(1)*10) + L　　　#ASCII for a digit
　　　　　　　　until (HT(HH) = 0)　　　　　　　#don't allow duplicates
　　　　　　　　HT(HH) = HT(HH) + 2↑(ZZ − 1)　　#set power for this position
　　　　　　NEXT ZZ
　　　　　　IN$ = "GUESS NUMBERS OF" + STR$(N) + "DIGITS."
　　　　　　GL = N　　　　　　　　　　　　　　　 #guess length
　　　　　　RETURN

```
200   GOSUB 740
210   FOR ZZ=BH TO TH:HT(ZZ)=0:NEXT ZZ
220   ON GM GOSUB 310,240
230   RETURN

310   FOR ZZ=1 TO N
320*  HH=INT(RND(1)*10)+L:IF HT(HH)<>0 THEN 320
330   HT(HH)=HT(HH)+2↑(ZZ-1):NEXT ZZ
340** IN$="GUESS NUMBERS OF"+STR$(N)+" DIGITS.":GL=N
350   RETURN
```

This is the subroutine that constructs the new problem. Also shown is one of its two subroutines—the one that constructs a number with N digits, each one different. Each subroutine is shown first in Free BASIC, then in BASIC.

*In the TRS-80 version of this line, the term RND(0) appears instead of RND(1).
**The Apple version of this line differs only in the spacing inside the string constants.

Figure 2.3: Think Subroutine for Guessing Game

of the encoding is to make it easy for the player's guesses to be checked.

An obvious way to check the player's guesses would be as follows. Each character of the guess could first be compared with the character in the corresponding position in the answer. If they were equal, IP and RG could be incremented, and the program could proceed to

---

#Ask for a word to become the next problem

askword   repeat {   #get a string
                INPUT "HIDDEN WORD"; H$
                N = LEN(H$)
            } until (0 < N AND N < = MX)
            IN$ = "GUESS WORDS OF" + STR$(N) + " LETTERS."
            GL = N   #guess length
            FOR ZZ = 1 TO N   #set "hit" array for chars of string
                HH = ASC(MID$(H$, ZZ, 1))
                HT(HH) = HT(HH) + 2↑(ZZ − 1)
            NEXT ZZ
            VG = N   #"very good" if this few guesses
            MG = maxguess   #"you lose" if > this many guesses
            GOSUB clearscreen   #remove word from screen
            RETURN

```
240*  INPUT "HIDDEN WORD";H$:N=LEN(H$)
250   IF N=0 OR N>MX THEN 240
260*  IN$="GUESS WORDS OF"+STR$(N)+" LETTERS.":GL=N
270   FOR ZZ=1 TO N:HH=ASC(MID$(H$,ZZ,1))
280   HT(HH)=HT(HH)+2↑(ZZ−1):NEXT ZZ
290   VG=N:MG=20
300   GOSUB 740:RETURN
```

This is the other subroutine of the first routine shown in Figure 2.3. It asks a player to enter a word to become the problem (for the other player) in the next game. It is shown first in Free BASIC, then in BASIC.

---
*The Apple versions of these lines differ only in the values of the string constants.

Figure 2.4:   Askword Subroutine for Guessing Game

the next character of the guess. If the given guess character didn't match the corresponding answer character, the program could then scan the answer looking for the guess character in other positions in the answer. If a match were found, RG could be incremented, and the program would proceed to the next character of the guess.

You might find it an interesting exercise to replace the subroutine in Figure 2.6 by a routine using the algorithm described above. You will probably find that your routine takes a noticeable amount of time to check some guesses against some answers. If you decide to carry out this exercise, you will have to modify the subroutine at line 310 in order to make a string out of the four-digit number that is constructed there. You will then have to save this string in the string variable H$. The subroutine at line 240 already uses H$ for storing the word supplied by the second player.

The method actually used for encoding the problem allows rapid checking. An array called HT is constructed. The size of HT is equal to the range of possible ASCII characters: 128 for Apple and TRS-80, 256 for Pet. (Pet uses an expanded ASCII character set that includes codes for the graphics characters that can be entered from the keyboard.)

Initially, the array HT is cleared to zero. Then, for each character of the number or word that the player must guess, an entry is made in the HT array at the position corresponding to the ASCII code for that character. For example, let's assume that the game being played is Four, and that the number selected by the program is 4726. For that number, entries are made in the HT array at positions 52, 55, 50 and 54, since the ASCII code for "4" is 52, for "7" is 55, for "2" is 50 and for "6" is 54. (See Appendix A for a list of ASCII codes.)

Placing entries in the HT array corresponding to the ASCII codes for the characters in the answer conveys no information about the order in which these characters appear in the answer. For example, the four-digit numbers 4726 and 2674 would have entries in the same four positions of the HT array. Since we need information about the order of the digits, the entries we make in the hit array must specify the positions of the corresponding characters in the answer. For example, for 4726 we could place a 1 in HT(52), a 2 in HT(55), a 3 in HT(50) and a 4 in HT(54). For 2674 we could place a 1 in HT(50), a 2 in HT(54), a 3 in HT(55) and a 4 in HT(52). That is, each location in the HT array would contain a number that specified the position of the corresponding character in the answer.

This method would work well for Four and the other number-guessing games, since these games do not allow any digit to appear more than once. But it would not work for Word, since in that game the same letter can appear in more than one place. For example, if the word selected is BILL, then the array entry HT(76) (corresponding to the letter L) would need to contain both 3 and 4. This is necessary since the letter L appears in both the third and fourth positions of the word BILL. We can store both 3 and 4 in HT(76) by simply setting HT(76) to 24, that is, $2^3 + 2^4$.

You have probably learned that a whole number (like 24) can be broken down into a sum of powers of two in one and only one way. Thus, given that HT(76) has the value 24, we can perform a computation to find that $24 = 2^3 + 2^4$, and this tells us that 3 and 4 are the two positions in which the letter L appears in the word BILL. The code that performs this computation appears in the subroutine that begins at line 410 (see Figure 2.6).

The subroutine at line 200 first clears (sets to zero) as many locations of the HT array as are needed for the given game. For example, for Four and the other number-guessing games, only the ten locations HT(48) through HT(57) (corresponding to the ASCII characters for the ten digits 0,1,...,9) are cleared. The variables BH and TH ("bottom hit" and "top hit"), which specify this range of array indices, are set by the subroutine presented in Figure 2.9. For Four, for example, BH = 48 and TH = 57. We clear only those locations of HT that are needed for the game, because clearing an entire array of 256 items can take a noticeable amount of time.

After the clearing of the appropriate part of the HT array, the routine executes the instruction

ON GM GOSUB 310,240

This instruction causes either the subroutine at line 310 or the subroutine at line 240 to be called, depending upon the value of the variable GM. The value of GM is set in the subroutine shown in Figure 2.9. The value of GM is 1 for any of the number guessing games (e.g., Four) and 2 for the game Word. Thus, the subroutine at 310 (shown in Figure 2.3) is called for number guessing, and the subroutine at 240 (shown in Figure 2.4) is called for Word. These two subroutines are short and easy to understand, but several points deserve comment.

First, notice the line labeled 320 in the subroutine called for number guessing:

    320 HH = INT(RND(1)*10) + L: IF HT(HH) <> 0 THEN 320

The first instruction on line 320 randomly chooses a number from among the ten whole numbers starting at the number whose value is stored in the variable L. The initialization routine sets L to 48, which is the ASCII code for zero, so the instruction chooses randomly from the ten numbers beginning with 48. Because of the arrangement of the ASCII character set, these ten numbers are the ASCII codes for 0,1,...,9.

The second instruction on line 320 checks the HT array entry corresponding to the selected ASCII code. If that entry is non-zero, the program goes back to the beginning of line 320 and makes another selection. In this way the program ensures that a number is chosen in which no two digits are the same.

The next line, labeled 330, contains the instruction that stores the appropriate power of two in the HT array entry that corresponds to the selected ASCII code:

HT(HH) = HT(HH) + 2↑(ZZ − 1)

Since the previous line made sure that HT(HH) has the value 0, this instruction is equivalent to

HT(HH) = 2↑(ZZ − 1)

The value of ZZ, the index for the FOR...NEXT loop, ranges from 1 to the number of digits to be chosen (e.g., 4 digits for Four). Thus, if 50 (the ASCII code for 2) is selected on line 320 the first time through the loop, then ZZ = 1 and HH = 50, so that the above instruction reduces to

HT(50) = 2↑(0)

If 54 (the code for 6) is chosen the second time through the loop, then the instruction becomes

HT(54) = 2↑(1)

If the game being played is Four, and the digits of the answer are 2649, then HT(50) will contain 1 (since 2↑0 = 1), HT(54) will contain 2, HT(52) will contain 4 and HT(57) will contain 8. The remaining six entries of the HT array (of the ten entries HT(48) to HT(57), which we cleared in the subroutine at line 200) will remain zero.

The final tasks performed by the subroutine at line 310 are to set the values of IN$ (a "prompt" message displayed both at the start of each game, and whenever the player makes the wrong kind of guess) and of GL (the number of characters expected in each guess).

The subroutine at line 240 (shown in Figure 2.4) is similar to the subroutine at line 310. Note that lines 270 and 280 contain the instructions

HH = ASC(MID$(H$,ZZ,1))
HT(HH) = HT(HH) + 2↑(ZZ − 1)

The first of these instructions shows how the ASCII codes that correspond to the characters of the hidden word are extracted one at a time from the hidden word. The expression

$$MID\$(H\$,ZZ,1)$$

represents a string one character long (because of the argument 1), taken from H$ (because H$ is the first argument), starting at the ZZ-th position. Since ZZ is the index variable for the FOR...NEXT loop, it takes values from 1 to N, where N is the number of characters in H$ (because of the instruction N = LEN(H$) on line 240). Thus, MID$(H$,ZZ,1) represents the ZZ-th character of H$ as a single-character string. The function ASC transforms this single-character string into a number, which is the ASCII code for the given character. This number is stored at HH, where it becomes an index to the HT array.

The instruction

$$HT(HH) = HT(HH) + 2\uparrow(ZZ - 1)$$

on line 280 is identical with the instruction used on line 330 (see Figure 2.3). In discussing line 330, we noted that HT(HH) was known to be zero, so the instruction was equivalent to

$$HT(HH) = 2\uparrow(ZZ - 1)$$

In the case of line 280, this is no longer true, since the same character (and hence, the same value of HH) may have occurred earlier in the hidden word. Thus, in this case, the appropriate power of two (2↑(ZZ − 1)) is being added to the previous contents of HT(HH). These previous contents will be zero if the given character has not appeared earlier in the hidden word. If the character has appeared earlier in the word, the contents will be a sum of terms of the form 2↑(ZZ − 1) for values of ZZ that correspond to the positions at which the character appeared.

The remainder of the subroutine in Figure 2.4 requires no explanation. Notice that it ends with a call to the subroutine at line 740 (shown in Figure 2.12) to clear the screen. Clearing the screen assures that the hidden word remains hidden, and allows the average game to be conducted without scrolling (i.e., disappearance of information from the top of the screen as all lines move up to accommodate a new bottom line).

Figure 2.5 shows the subroutine that asks for and accepts the player's guess. The first two lines, 360 and 370, contain the instructions

>   360 INPUT "WHAT DO YOU GUESS"; G$
>   IF LEN(G$) <> GL THEN PRINT IN$: GOTO 360

These instructions are repeated until the player enters a guess containing exactly the number of characters specified in GL (e.g., four for Four). The string IN$ (the original prompting message) is repeated if the player has entered a guess with the wrong number of characters. Then the player is allowed to try again. Notice that rejected guesses are not counted in the total of the player's guesses. The variable G is only incremented upon return from this subroutine (see line 140, Figure 2.2).

The next two lines, 380 and 390, contain the instructions that dissect the player's guess and store the ASCII codes for the individual

---

#Accept the player's guess

guess   repeat {
        INPUT "WHAT DO YOU GUESS";G$
        IF LEN(G$) <> GL THEN
           PRINT IN$
        } until (LEN(G$) = GL)
        FOR ZZ = 1 TO GL   #put chars of guess into GC array
           GC(ZZ) = ASC(MID$(G$,ZZ,1))
        NEXT ZZ
        RETURN

```
360*  INPUT "WHAT DO YOU GUESS"; G$
370   IF LEN(G$)<>GL THEN PRINT IN$:GOTO 360
380   FOR ZZ=1 TO GL
390   GC(ZZ)=ASC(MID$(G$,ZZ,1)):NEXT ZZ
400   RETURN
```

This is the subroutine of the guessing game that asks for and accepts the player's guess. Then it picks the guess apart into an array of characters. The subroutine is shown first in Free BASIC, then in BASIC.

---

*The Apple version of this line differs in the appearance of a question mark in the string constant.

Figure 2.5: *Guess Subroutine for Guessing Game*

characters of the guess in the array GC:

> FOR ZZ = 1 TO GL
> GC(ZZ) = ASC(MID$(G$,ZZ,1)): NEXT ZZ

These instructions are similar to the instructions that dissect the hidden word (see line 270, Figure 2.4).

The player's guess (now encoded into the array GC) is checked against the expected answer (encoded in HT) by the subroutine

---

#Check the guess against the hidden word or number

check   RG = 0   #start with none right, none in place
         IP = 0
         FOR ZZ = 1 TO GL   #check guess against "hit" array
            HH = HT(GC(ZZ))
            IF HH <> 0 THEN {   #char is somewhere in the string
                RG = RG + 1
                HH = INT(HH/2↑(ZZ − 1))   #look at the corresponding power of 2
                IF HH is odd THEN   #char is in place
                     IP = IP + 1
            }
         NEXT ZZ
         RETURN

```
410   RG=0:IP=0
420   FOR ZZ=1 TO GL:HH=HT(GC(ZZ))
430   IF HH=0 THEN 460
440   RG=RG+1:HH=INT(HH/2↑(ZZ−1))
450   IF HH<>2*INT(HH/2) THEN IP=IP+1
460   NEXT ZZ
470   RETURN
```

This subroutine checks the character array GC (built from the player's guess by the subroutine in Figure 2.5) against the "hit" array HT. HT has one entry for each possible ASCII character. If the character is not part of the answer, the entry is zero. If the character is part of the answer, the entry is a sum of powers of two—one for each position in the answer at which the given character appears.

The routine is shown first in Free BASIC, then in BASIC.

Figure 2.6:   *Check Subroutine for Guessing Game*

beginning at line 410. This subroutine appears in Figure 2.6. The routine begins by initializing the "right" and "in place" counters RG and IP to 0 (line 410). The remainder of the routine is a loop, the body of which occupies lines 420-450. This loop is a FOR...NEXT loop indexed by ZZ. In the loop, each character in the guess (stored in elements 1 through GL of the array GC) is checked against the corresponding entry in the "hit" array HT. The instruction

HH = HT(GC(ZZ))

on line 420 selects the HT array entry corresponding to the ZZ-th character of the guess. If HH has the value zero, then the ZZ-th character of the guess does not appear in the answer. If HH is not 0, then the character does appear in the answer. In that case RG is incremented by 1 (line 440). Then the program checks to see whether or not 2↑(ZZ − 1) is one of the terms making up the value of HH. If it is, the program increments IP. The code to perform this task is on lines 440 and 450:

HH = INT(HH/2↑(ZZ − 1))
IF HH <> 2*INT(HH/2) THEN IP = IP + 1

The first of these instructions "throws away" all of the powers (if any) that are smaller than 2↑(ZZ − 1). If you are familiar with the usual operations available in assembly language, you can picture this operation as a logical right shift that brings the bit in which we are interested into the rightmost (least significant) bit position. If you prefer to think of this instruction in mathematical terms, the original value of HH is of the form

$$a_0 + a_1 \times 2 + ... + a_{GL-1} \times 2^{GL-1}$$

where $a_0, a_1, ..., a_{GL-1}$ take values 0 or 1. The subscripts 0 to GL − 1 correspond to the ZZ values of 1 to GL, so the above operation replaces the original value by

$$a_{ZZ-1} + (a_{ZZ} \times 2 + ... + a_{GL-1} \times 2^{GL-ZZ})$$

when ZZ < GL, and by a_{GL-1} when ZZ = GL. The value of the parenthesized term is always even, so the value of a_{ZZ-1} can be obtained simply by determining whether the new value of HH is even (in which case $a_{ZZ-1} = 0$) or odd (in which case $a_{ZZ-1} = 1$). The condition

HH <> 2*INT(HH/2)

is equivalent to "the value of HH is odd."

If you have had difficulty following the discussion of the operation of this subroutine, don't worry. You won't need to understand the details of this operation in order to understand the material that follows.

The subroutine shown in Figure 2.7 was discussed briefly earlier. It is very simple. The main point to remember is that it sets IP to −1 if the player is not to be allowed any further guesses.

The subroutine in Figure 2.8 is also quite simple, but it serves an important function that can easily be refined and expanded. This subroutine tells you how well you played the game and gives you a pat on the back if you did especially well. It could (but doesn't) compare your performance in the current game to your performance in earlier games. A subroutine that gives this type of feedback makes a game more fun to play.

#Give a hint (or announce a win or loss)

hint   IF IP = N THEN   #all in place
            PRINT "THAT'S RIGHT! ";
       else IF G = MG THEN {   #maximum reached
            PRINT "TOO MANY GUESSES — YOU LOSE!"
            IP = −1   #flag to signal "game over"
            }
       else
            PRINT "RIGHT:"; RG, "IN PLACE:"; IP
       RETURN

```
480   IF IP=N THEN PRINT "THAT'S RIGHT! ";:GOTO 510
490   IF G=MG THEN PRINT "TOO MANY GUESSES — YOU LOSE!":IP=−1:GOTO 510
500*  PRINT "RIGHT:"; RG,"IN PLACE:"; IP
510   RETURN
```

This subroutine of the Guessing Game checks to see whether the player has guessed the answer or has run out of guesses. If the player hasn't done either, the routine prints the number right and the number in place. This is the routine that sets IP to −1 if the player has run out of guesses. The routine is shown first in Free BASIC, then in BASIC.

---
*The Apple version of this line differs only in the spacing in the string constants and the use of a TAB(11) instead of a comma after "RG".

Figure 2.7:   ***Hint Subroutine for Guessing Game***

46 GUESSING GAMES

The subroutine in Figure 2.9 is used to make the game selection. This routine is called at the start of the game and whenever the player types "N" at the end of a game. After clearing the screen with a call to the subroutine at line 740, the routine then asks the question "GAME: " and accepts your single-character reply. If you enter a digit from 1 to 9, you have selected the corresponding number-guessing game. A "W" selects Word; any other input is made to select Four. In such a case, we call Four the *default* selection.

```
#Give statistics on the game just concluded

stats   IF IP = N THEN {                              #all in place
            IF G = 1 THEN                             #got it on the first guess
                PRINT "YOU MUST BE PSYCHIC."
            else {
                PRINT "YOU GOT IT IN"; G; "GUESSES."
                IF G <= VG THEN                       #got it in very few guesses
                    {PRINT: PRINT "VERY GOOD."}
            }
        }
        else IF IP = -1 THEN                          #ran out of guesses
            PRINT "YOU COULDN'T GET IT IN"; MG; "TRIES."
        RETURN


  520   IF IP <> N THEN 570
  530   IF G=1 THEN PRINT "YOU MUST BE PSYCHIC.":GOTO 580
  540*  PRINT "YOU GOT IT IN"; G; "GUESSES."
  550   IF G <=VG THEN PRINT:PRINT "VERY GOOD."
  560   GOTO 580
  570*  IF IP= -1 THEN PRINT "YOU COULDN'T GET IT IN";MG;"TRIES."
  580   RETURN
```

This subroutine of the guessing game program is called when the game has been concluded. Either the player has finally guessed the answer or the game has been stopped because the player used all of the allotted guesses. In the latter case, this program has a glaring omission.

The routine is shown first in Free BASIC, then in BASIC.

---
*The Apple versions of these lines differ only in the spacing in the string constants.

Figure 2.8: Stats Subroutine for Guessing Game

INSIDE BASIC GAMES   **47**

#Set up for the upcoming game
setup   GOSUB clearscreen
             PRINT "GAME: ";
             GOSUB onech: N = VAL(X$)   #game type code
             IF N <> 0 THEN   #digit means number guessing
                GM = 1
             else IF X$ = "W" THEN   #"W" means word guessing
                GM = 2
             else
                {N = 4: GM = 1}   #default to guessing 4-digit numbers
             ON GM GOSUB setnum, setask
             RETURN

setnum VG = N + 2: MG = 2*VG   #set "very good" level and max
             BH = L: TH = L + 9   #range of hit array to be used
             RETURN

setask   VG = 0: MG = 99   #VG and MG will be set by askword
             BH = 0: TH = numascii   #range of hit array to be used
             RETURN
    590   GOSUB 740:PRINT "GAME: ";:GOSUB 720
    600   N=VAL(X$):IF N <> 0 THEN GM=1:GOTO 630
    610   IF X$="W" THEN GM=2:GOTO 630
    620   N=4:GM=1
    630   ON GM GOSUB 650,660
    640   RETURN
    650   VG=N+2:MG=2*VG:BH=L:TH=L+9:RETURN
    660*   VG=0:MG=99:BH=0:TH=255:RETURN

    This is the subroutine of the guessing game program that allows the player to choose among the available versions of the game: guessing a number of a specified number of digits, or guessing a word supplied by another player. Much of the routine is only meant to suggest possibilities (for example, better ways of setting VG and MG could be found), and to provide a framework to facilitate the addition of other types of game.
    The routine appears first in Free BASIC, then in BASIC.

---
*In the Apple and TRS-80 versions of this line, the 255 is replaced by 127.

Figure 2.9:   Setup Subroutine for Guessing Game

The operation of this routine is based on the fact that the function VAL returns the numeric value of any string that can be interpreted as a number, while it returns the value zero for a string that does not represent a number (e.g., "W").

The two subroutines at lines 650 and 660 set the values of MG (maximum guesses), VG (very good), and the HT array limits BH and TH. More sophisticated methods of choosing values for MG and VG could be devised. Also, other categories (in addition to VG) could be introduced and used by the program in Figure 2.8 in order to give the player more selective feedback.

The variable GM and the instruction

ON GM GOSUB 650,660

provide a framework into which setup routines for additional games can be integrated easily. In other words, this part of the Guessing Game program has been designed so that it can be easily generalized.

The subroutines shown in Figure 2.10 fall into the "miscellaneous" category usually relegated to the final positions in a program listing. The first of these subroutines (line 710) consists of the instructions

GOSUB 720: NX$ = X$: RETURN

The call to this routine could be replaced by a direct call to the program at line 720, and the use of NX$ could be replaced by the use of X$ directly. This would involve changes to lines 160, 170 and 180 (see Figure 2.2).

The subroutine at line 710 provides an example of a situation that often occurs in a well-designed program. A routine originally designated for a definite, significant purpose may end as nothing more than a single call to another subroutine. For clarity and efficiency, such "middleman" routines should be eliminated, but since they are harmless and usually involve little overhead, they are often spared extermination.

The second subroutine in Figure 2.10 is the initialization routine. This routine would normally appear at the very end of the program, but that position has been reserved for those few subroutines of the Guessing Game program whose Apple, Pet and TRS-80 versions differ in non-trivial ways. The initialization routine sets two array dimensions and the variable MX. MX is used in the subroutine shown in Figure 2.4 to assure that the hidden word entered for Word is not larger than the amount of space available in the GC array. (Recall that the GC array holds the individual characters of the player's current guess.) The initialization routine also sets the value L of the ASCII code for zero. In an earlier organization of the program it made sense to set L in the initialization routine. Now it would make more sense to perform this task in the subroutine at line 650 (see Figure 2.9).

#Get code from player for next game

next   GOSUB onech   #get single-character input
        NX$ = X$ #put it into NX$
        RETURN

#Initialize

init   DIM HT(numascii)   #"hit" array indexed by ASCII code
        MX = maxword: DIM GC(MX)   #array for characters of guess
        L = ASC("0")   #lower limit for digit selection
        RETURN

   710   GOSUB 720:NX$=X$:RETURN

  670*   DIM HT(255)
  680   MX=20:DIM GC(MX)
  690   L=ASC("0")
  700   RETURN

These are two small subroutines of the guessing game program. The first subroutine, which is called at the end of each game, returns a single-character string in NX$, which the main program interprets to mean one of three things: the player wishes to specify a new kind of game (NX$ = "N"), the player wishes to quit (NX$ = "E"), or the player wishes to continue with the same game (any other string). The main purpose of this routine is to provide the pause at the end of the game during which the player can look back over the output to see what can be learned.

The second subroutine is the initialization routine. Old-fashioned programmers always begin their programs with a call to an initialization routine. This one sets the dimensions of two arrays, and the values of two constants.

The routines are shown first in Free BASIC, then in BASIC.

---
*The Apple and TRS-80 versions of this line use an array dimension of 127.

Figure 2.10: Two Small Subroutines for Guessing Game

Figures 2.11 and 2.12 show the subroutines that have different versions for the Apple, Pet and TRS-80. All of the previous routines in this program use identical instructions for each of these systems (excepting the usual differences—i.e., the spacing around numbers and the "? " with the INPUT instruction—and the difference in the size of the HT array).

The first of the routines that have system-dependent versions is the single-character input routine in Figure 2.11, which uses different tools on each of the systems. For example, the Pet version of GET does not wait for keyboard input. This feature is used to advantage by including the GET instruction in a loop that also includes a call to the RND function. As we explained in Chapter 1, this approach achieves true randomization. The Apple version of the GET instruction waits for input, so it cannot (or, does not have to) be used in a loop. Since the GET instruction does not recognize the usual Apple convention of control-C to interrupt the program, this feature is added by an explicit test (CHR$(3) represents a single-character string consisting of control-C). The TRS-80 uses the INKEY$ function in exactly the same way that Pet uses the GET instruction.

The screen-clearing subroutine in Figure 2.12 consists of a single instruction in each of the three versions. On the Pet the screen is cleared by PRINTing the character corresponding to the CLR key. For the Apple and TRS-80, a separate BASIC instruction is used: HOME for Apple, CLS for TRS-80.

This concludes our discussion of the Guessing Game program. Next we shall see how this program can be transformed into Hangman.

The Hangman Program

Earlier in this chapter we learned to play Hangman. Now we shall show how to derive a Hangman program from our Guessing Game program by altering several of the subroutines of the Guessing Game program. The principal changes are shown in Figures 2.13 through 2.17. We shall now discuss them briefly.

First, note that Hangman uses the same main routine as Guessing Game. That is, the program in Figure 2.2, unchanged, will act as the main routine for Hangman. This involves a slight compromise, since the player's command "N", recognized on line 170 in the main routine, will have no meaning in Hangman. This problem is dealt with in the Hangman version of the setup routine at line 590 (called from line 110). That routine will have to be designed to ignore requests for a new game selection.

#Get single-character input

#Pet version

onech   repeat {
            X$ = CHR$(RND(1))
            GET X$
          } until (X$ <> '''')
          RETURN

    720   X$=CHR$(RND(1)):GET X$:IF X$='''' THEN 720
    730   RETURN

#Apple version

onech   GET X$
          IF X$ = CHR$(3) THEN   #make control-C work
            STOP
          RETURN

    720   GET X$:IF X$=CHR$(3) THEN STOP
    730   RETURN

#TRS-80 version

onech   repeat {
            X$=CHR$(RND(0))
            GET X$
          } until(X$ <> '''')
          RETURN

    720   X$=CHR$(RND(0)):X$=INKEY$:IF X$='''' THEN 720
    730   RETURN

   This is the important single-character non-echoing input operation. The Pet, Apple and TRS-80 provide three different ways to perform this operation.
   The routines are shown first in Free BASIC, then in BASIC. Note that the three BASIC versions have the same line numbers.

Figure 2.11: Single-Character Input

52 GUESSING GAMES

Following the same order as that shown in the figures comprising Guessing Game, we shall look at the Hangman versions of the code. Figure 2.13 shows the Hangman version of the Guessing Game routines shown in Figures 2.3 and 2.4. The Hangman version eliminates the subroutine that constructs numbers (lines 310 to 350 in Figure 2.3), since Hangman deals only with words entered by a player. The Hangman version is similar to the subroutine on lines 240

#Clear the screen and move cursor to "home" position

#Pet version

clearscreen   PRINT "clr";
                     RETURN

   740   PRINT CHR$(147);:RETURN

#Apple version

clearscreen   HOME
                     RETURN

   740   HOME:RETURN

#TRS-80 version

clearscreen   CLS
                     RETURN

   740   CLS:RETURN

This is the subroutine that clears the screen and moves the cursor to the upper left corner. Screen clearing is used in this game for two reasons: to have one game at a time on the screen, and to hide the secret word after it is entered.

The routine is shown in Free BASIC, then in BASIC for Pet, Apple and TRS-80.

Figure 2.12: Screen Clearing

to 300 (Figure 2.4) in the way that it obtains the hidden word from the first player and encodes it into the HT array. The Hangman version must also obtain from the first player the number of incorrect guesses allowed the second player.

The setting of the IN$ prompting string in Figure 2.13 is also slightly different from the way IN$ is set in Guessing Game, because each guess in Hangman is either a single character or an entire word. IP is initialized here, because this variable is used differently in the two games. In Guessing Game, IP pertains only to the current guess. In Hangman, once a character is guessed, that character is continually displayed in its proper position by the program. Accordingly, IP starts at zero and increases as each place is filled in. The setting of GL to 1 establishes that Hangman guesses are to be one character. (Special testing occurs later to accommodate guesses of the entire word.)

The GS array referred to on line 300 in Figure 2.13 is new. This array is used to accumulate the characters that the player has guessed correctly. In this routine, GS is initialized to contain dashes in as many positions as there are characters in the hidden word. Later, as characters are guessed correctly, they replace the dashes.

Figure 2.14 shows the Hangman version of the subroutine shown in Figure 2.5. The two subroutines are very similar. One difference is that the Hangman version begins by displaying a line of the form

    _ Y _ Y _ Y   WRONG GUESSES LEFT: 5

The only other difference is that in the Hangman version, if a guess is made that contains the same number of characters as the hidden word, it is passed on unchecked. Otherwise, the characters of the guess are stored in the GC array using the same instructions as were used in Figure 2.5. Of course, for Hangman, the instructions in the FOR...NEXT loop are executed only once, since GL always has the value 1. Thus, GC(1) is used to hold the ASCII code for the single character of the guess, and the remainder of the GC array is unused in Hangman.

Figure 2.15 shows the Hangman version of the subroutine shown in Figure 2.6. The Hangman version begins by checking for a guess containing the same number of characters as the hidden word. Recall that such a guess would have been passed unchecked by the routine in Figure 2.14. The routine in Figure 2.15 compares this guess directly with the hidden word, which is stored in H$. If they match exactly, then IP is set to N to signal that the player has guessed the word correctly. Otherwise the guess is treated like any other incorrect guess.

54 GUESSING GAMES

```
#Enter the hidden word—for Hangman
think   GOSUB clearscreen
        repeat {
            INPUT "HIDDEN WORD"; H$
            N = LEN(H$)
        } until (0 < N <= MX)
        INPUT "HOW MANY WRONG GUESSES ARE ALLOWED"; WG
        IN$ = "GUESS ONE LETTER OR THE WHOLE WORD"
        GL = 1: IP = 0                      #"in place" starts at 0, never declines
        MG = 99: VG = 5                     #arbitrary values
        GOSUB clearscreen                   #hide the hidden word
        FOR ZZ = 0 TO numascii              #clear hit array
            HT(ZZ) = 0
        NEXT ZZ
        FOR ZZ = 1 TO N
            HH = ASC(MID$(H$, ZZ, 1))       #set up hit array
            HT(HH) = HT(HH) + 2↑(ZZ - 1)
            GS(ZZ) = ASC("-")               #"already guessed" = all dashes
        NEXT ZZ
        RETURN
200     GOSUB 740
210*    INPUT "HIDDEN WORD";H$:N=LEN(H$)
220     IF N=0 OR N > MX THEN 210
230*    INPUT "HOW MANY WRONG GUESSES ARE ALLOWED";WG
240     IN$="GUESS ONE LETTER OR THE WHOLE WORD"
250     GL=1:IP=0:MG=99:VG=5:GOSUB 740
280**   FOR ZZ=0 TO 255:HT(ZZ)=0:NEXT ZZ
290     FOR ZZ=1 TO N:HH=ASC(MID$(H$,ZZ,1))
300     GS(ZZ)=ASC("-")
310     HT(HH)=HT(HH)+2↑(ZZ-1):NEXT ZZ
320     RETURN
```

This is the Hangman version of the subroutines shown in Figures 2.3 and 2.4. Encoding of the hit array is unchanged. A new array called GS is initialized to all dashes. Correctly guessed letters replace the corresponding dashes.

---
*Apple versions of these lines differ only in the values of the string constants.
**In Apple and TRS-80 versions, 127 replaces 255.

Figure 2.13: Think Subroutine for Hangman

INSIDE BASIC GAMES

#Accept the player's Guess

guess   PRINT   #new line
            FOR ZZ = 1 TO N   #show partially guessed word
                PRINT CHR$(GS(ZZ));
            NEXT ZZ
            PRINT " WRONG GUESSES LEFT:"; WG   #and wrong guesses left
            repeat {
                INPUT "WHAT DO YOU GUESS"; G$
                IF LEN(G$) = N THEN
                    RETURN
                else IF LEN(G$) <> GL THEN
                    PRINT IN$
            } until (LEN(G$) = GL)
            FOR ZZ = 1 TO GL   #put chars of guess into GC
                GC(ZZ) = ASC(MID$(G$, ZZ, 1))
            NEXT ZZ
            RETURN

```
360   PRINT:FOR ZZ=1 TO N:PRINT CHR$(GS(ZZ));:NEXT ZZ
365*  PRINT "  WRONG GUESSES LEFT:";WG
370*  INPUT "WHAT DO YOU GUESS";G$
375   IF LEN(G$)=N THEN RETURN
380   IF LEN(G$) <> GL THEN PRINT IN$:GOTO 370
390   FOR ZZ=1 TO GL
395   GC(ZZ)=ASC(MID$(G$,ZZ,1)):NEXT ZZ
400   RETURN
```

This is the Hangman version of the subroutine shown in Figure 2.5. The only differences from the version shown in Figure 2.5 are: (1) the initial display of the partially guessed word and the number of wrong guesses left, and (2) the immediate RETURN if LEN(G$) = N.

---
*Apple versions of these lines differ only in the values of the string constants.

Figure 2.14: Guess Subroutine for Hangman

If the guess does not contain the same number of characters as the hidden word, then the single character saved in GC(1) is checked against the corresponding HT array element. This is done in the FOR...NEXT loop indexed by ZZ. (The instructions of the loop are only executed once, since GL has the value of 1.) The contents, HH, of the HT array element are checked (just as in Figure 2.6) to determine which powers of two are present. For each position for which the corresponding power of two is present, the guessed character is filled into the "partial guess" array GS, replacing the dashes to which that array was initialized. This is accomplished by the instruction

GS(YY) = GC(ZZ)

on line 450.

An interesting error occurred in an earlier version of the subroutine shown in Figure 2.15. The test on line 410 was for LEN(G$) = GL instead of LEN(G$) <> N. This would appear to make no difference, since the only two values that LEN(G$) can have on entry into this routine are N and GL, so that the two tests should have identical results. However, ambiguity could arise if N = GL. In this case, however, it should not matter which branch is taken, since a comparison of the entire string should yield the same result as a character-by-character comparison of the GL characters in it. In fact, however, when GL = N the character-by-character comparison fails, because the subroutine shown in Figure 2.14 does not set up the GC array when LEN(G$) = N. Thus, the test at the beginning of the subroutine in Figure 2.15 has to be coordinated with the test in Figure 2.14 (or the routine in Figure 2.14 must be revised to make it foolproof—by eliminating the test for LEN(G$) = N when GL = N).

Two subroutines that need to be coordinated in this way are said to be *coupled*. The less coupling there is in your programs, the more successful you will be if you try to modify them, and the less likely you will be to discover peculiar bugs long after formal "checkout" has been completed.

Figure 2.16 is the Hangman version of the subroutine shown in Figure 2.7. The two subroutines differ at only two points. For Hangman, the new variable WG ("wrong guesses") needs to be tested, and no hint of the form

RIGHT: 3   IN PLACE: 1

needs to be generated.

The Hangman version of the subroutine shown in Figure 2.8 is not presented. The two versions are identical, except that wherever the routine in Figure 2.8 uses the word "IT" to refer to the answer, the

INSIDE BASIC GAMES

#Check the guess against the hidden word or number
check   IF LEN(G$) = N THEN
        IF G$ = H$ THEN IP = N   #guessed the whole word
        else WG = WG − 1
    else {
        IX = 0   #count number of times char appears
        FOR ZZ = 1 TO GL
            HH = HT(GC(ZZ))   #hit entry for guessed char
            IF HH <> 0 THEN
                FOR YY = 1 TO N
                    IF HH is odd THEN {
                        IP = IP + 1   #one more in place
                        IX = IX + 1   #one more for this char
                        GS(YY) = GC(ZZ) #replace dash by char
                    }
                  HH = INT(HH/2)
                NEXT YY
        NEXT ZZ
        IF IX = 0 THEN
            WG = WG − 1
    }
    RETURN

```
410    IF LEN(G$) <> N THEN 440
420    IF G$=H$ THEN IP=N:RETURN
430    WG=WG−1:RETURN
440    IX=0:FOR ZZ=1 TO GL:HH=HT(GC(ZZ))
445    IF HH=0 THEN 460
450    FOR YY=1 TO N:IF HH <>2*INT(HH/2) THEN IP=IP+1:IX=IX+1:GS(YY)=GC(ZZ)
455    HH=INT(HH/2):NEXT YY
460    NEXT ZZ
465    IF IX=0 THEN WG=WG−1
470    RETURN
```

This is the Hangman version of the subroutine shown in Figure 2.6. The routines have similar structures, but there are many small differences. The variable RG no longer appears, but IP still represents the number "in place." A correct guess of the entire word is signaled by setting IP = N.

Figure 2.15: Check Subroutine for Hangman

58 GUESSING GAMES

```
       #Announce win or loss
hint   IF IP = N THEN
           PRINT "THAT'S RIGHT! ";
       else IF G = MG OR WG < 0 THEN {
           PRINT "TOO MANY GUESSES — YOU LOSE!"
           IP = -1
           }
       RETURN


480    IF IP=N THEN PRINT "THAT'S RIGHT! ";:GOTO 500
490    IF G=MG OR WG<0 THEN PRINT "TOO MANY GUESSES — YOU LOSE!":IP=-1
500    RETURN
```

This is the Hangman version of the subroutine shown in Figure 2.7. The difference is that this subroutine checks for WG < 0 as well as for G = MG in terminating the game, and doesn't give any hints. The display of how well the player is doing in Hangman occurs in the subroutine shown in Figure 2.14.

Figure 2.16: Hint Subroutine for Hangman

Hangman version uses the string H$ that contains the answer. For example, line 540 of the Hangman version is

   PRINT "YOU GOT "; H$; " IN"; G; "GUESSES."

A similar change is made to line 570, thus eliminating the glaring omission in the program shown in Figure 2.8: in that program, the player is never told the answer.

Figure 2.17 shows the Hangman version of the routine shown in Figure 2.9. This routine has been reduced to a single call to the subroutine that clears the screen. All of the complexity disappears, since there are no other versions of the game for the player to choose from.

There is only one more change needed to convert the Guessing Game program into Hangman: the addition of GS(MX) to the dimension statement on line 680 (shown in Figure 2.10). The remainder of the Guessing Game program is carried over unmodified into the Hangman program.

You should study the creation of Hangman from Guessing Game

#Set up for the upcoming game

setup   GOSUB clearscreen
              RETURN

   590   GOSUB 740:RETURN

This is the Hangman version of the subroutine shown in Figure 2.9. It really has nothing much to do.

Figure 2.17: Setup Subroutine for Hangman

carefully, since it illustrates a useful technique: cannibalization of a working program in order to produce a similar program quickly. While the presentation of these changes has been long and detailed, the actual conversion from Guessing Game to Hangman was accomplished by the author in less than an hour. As you evaluate this conversion, pay special attention to the "tradeoff" between simplicity of conversion and clarity of the converted code. Decide for yourself about the wisdom of retaining the variable GL (which can only take the value 1) or the vestigial variable L, which is set in the initialization routine, and is never used.

Possible Additions and Changes

This chapter has presented the ten-games-in-one Guessing Game and its derivative, Hangman. Many improvements can be made to these programs. Several possibilities are listed below. Some can be implemented easily, while others may require many carefully planned changes.

- Record statistics. For example, you might wish to keep track of the smallest number of guesses, the median or average number of guesses, or the number of guesses for each of the last ten games.

- Provide for a review during the course of play. For example, print a summary of all previous guesses and replies. Use a concise format guaranteed to fit on the screen, even for an unreasonably large number of guesses. For Hangman, display an alphabet that replaces the letters that have already been ruled out with asterisks.

- Revise the display of questions and replies so that cursor control is used to assure that as much information as possible remains on the screen.

- Provide the program with minimal intelligence. For example, if the game being played is Four and the player tries to guess a digit in a position that was ruled out by a previous guess, or if the game being played is Hangman and the player guesses a previously rejected letter, the program could gently and politely chide the player. (For example, it might say YOU ALREADY TRIED J, ARE YOU A BIT ABSENT-MINDED?) This sort of program behavior should be designed carefully, since any manifestation of superior intelligence by the program risks infuriating the player.

- If sound is available on your computer, use it to chide or reward the player.

- Devise an effective method for setting the VG ("very good") and MG ("maximum number of guesses") levels. Allow them to vary with the skill of the player.

Summary

This chapter began with a discussion of guessing games. The rules of Four were explained, and eight number-guessing games and the word-guessing game, Word, were derived from Four. A sample of the play of Four was presented to illustrate the strategy of the game.

A program called Guessing Game was developed to implement Four and the nine games derived from it. The discussion of this program focused on the program's structure and the techniques used to achieve the generality necessary to accommodate all ten games into one program.

After the Guessing Game program was discussed thoroughly, the Hangman program was derived from the Guessing Game program by "cannibalization"—the replacement or alteration of a few key subroutines. The purpose of that derivation was to illustrate the techniques used to convert a working program into a similar program, since this is a good way to obtain new programs quickly.

Finally, several suggestions for improvements to the programs of this chapter were presented. The purpose of these suggestions was to illustrate the kind of critical thought that can be applied to the behavior of computer programs and to provide interesting, challenging and rewarding exercises.

CHAPTER 3
Time Games

Time is an important element in games. Fast and measured action is often essential to our enjoyment of them. As we shall see, the techniques used to control the time element are simple—once you understand them. The programs that appear in this chapter have been designed to illustrate the use of the time element.

The Pet Clock

The games in this chapter are all written for the Pet computer, since it has an easy-to-use built-in clock. First, let's review how the clock works.

Inside the machine there is a crystal-controlled clock that "ticks" sixty times per second. The Pet translates these ticks into a six-digit string stored in the variable TIME$. The six digits give the number of hours, minutes and seconds that have elapsed since midnight. For example, the string "042715" corresponds to the time 4:27 and 15 seconds a.m. The string "120000" corresponds to noon. The string "235959" corresponds to one second before midnight.

Unfortunately, the Pet has no battery or other source of power to keep the clock running when the main power is off, so every time you turn your Pet on, the clock is set to 000000, that is, to midnight. Fortunately, however, you can tell it the right time by setting the variable TIME$. For example, if you have just turned your Pet on and the correct time is 10:27 a.m., you can type

    TIME$ = "102700"

to set the clock. To set it precisely, you can type

    TIME$ = "102800"

and then wait until it is 10:28 before pressing RETURN. Once you have set the time in your Pet, you can always find out the time by typing (or inserting into your program) the instruction

    PRINT TIME$

The Pet will then print a six-digit string consisting of hours, minutes and seconds in the format described above.

That's all there is to using the Pet clock. (There is another variable, TI, in which the Pet maintains the time in another format, but we won't use it here.) Clocks are available for Apple and TRS-80 computers, and if your computer has one, you will be able to use the programs presented in this chapter. The portions of the programs that refer specifically to the Pet clock have been collected into clearly

64 TIME GAMES

defined sections of code that can easily be replaced by equivalent Apple or TRS-80 code.

Clock

Let's begin with a game that isn't actually a game: we shall make the Pet into a (very expensive) digital clock. In this "game" the computer continually displays the time in the lower right corner of the screen. This display uses the ordinary twelve-hour format of most clocks. For example, if the Pet clock says that the time is 152537, our screen display will be 3:25:37 p.m. While our clock is ticking off the seconds in the corner of the screen, it will also be waiting for commands from the keyboard. (Unfortunately, it is not possible to wait for keyboard input while simultaneously performing other tasks on an Apple, since the Apple version of GET waits for a reply, unlike the Pet version of GET or the TRS-80 INKEY$ function. On the Apple, however, a signal from a "game paddle" can be used to tell the program that the user wishes to enter a keyboard command.)

```
#Clock program
            GOSUB init
            GOSUB clearscreen
            repeat {
                GOSUB events           #check for alarm, etc.
                GOSUB positime         #place cursor for time display
                GOSUB showtime         #display the time
                GOSUB command          #process command, if any
            }
    100   GOSUB 1110:GOSUB 920
    110   GOSUB 340:GOSUB 240:GOSUB 250:GOSUB 120:GOTO 110
```

This is the main routine for the clock program. Time is kept automatically by the Pet. The program continually displays the time in a fixed screen position. Once each clock tick, the program checks on certain events that need to happen periodically or at prespecified times: display of an alarm message and adjustment of the time to compensate for an inaccurate clock circuit. The program is also continually looking for single-character commands entered from the keyboard. These can interrupt the time display while further dialog occurs, but the time is still kept automatically while the dialog proceeds.

Figure 3.1: Pet Clock

Each command consists of a single character. Their meanings are listed below:

- F   Make the clock run faster. The program will ask for a number, N. One second will be added to the time every N seconds.
- S   Make the clock run slower. The program will ask for a number, N. One second will be subtracted from the time every N seconds.
- T   Set the time. The program will ask for a six-digit string containing hours, minutes and seconds in the format described earlier. This string will be used to set the time.
- A   Set the alarm. The program will ask for a six-digit string in the same format as that used for T. When the clock reaches the specified time, an alarm will go off. (Since the Pet is silent, this will be a visual alarm.)
- Q   Turn off the alarm (if it has already gone off) and clear the alarm time, so that it won't go off again tomorrow.
- R   Turn off the alarm, but don't clear the time (so that it will go off again at the same time tomorrow).
- Z   Change time zone (or go on or off daylight savings time). The program asks how many hours to change the time by. The answer will be a positive whole number, N, to set the time forward N hours, or a negative whole number, −N, to set the time back N hours.
- C   Change the time by a few seconds. The program asks how many seconds to change the time by. The answer will be a positive whole number N, to set the time ahead N seconds, or a negative whole number, −N, to set the time back N seconds.

Of course, these commands represent only a small subset of the features that could be made available in a computer-controlled clock.

The Clock Program

Figure 3.1 shows the main routine of the Clock program. The programs for Clock will be shown both in Free BASIC and in BASIC, but no reference will be made in the text to the Free BASIC version. If you have looked at the Free BASIC versions of the previous programs and have tried to understand them (or if you have read the description of Free BASIC in Chapter 6), then you are probably already finding the Free BASIC easier to follow than the BASIC.

The Clock program consists of calls to the initialization and screen-clearing routines, followed by an infinite loop. In the loop, the time is displayed repeatedly, keyboard commands are processed, and the due times of scheduled events are checked against the current time. The actual updating of the time goes on automatically in the Pet, so that no time is lost while the program is responding to a keyboard command. If an event (e.g., the "sounding" of the alarm) has a due time that passes while a command is being processed, the event will occur immediately after the command processing is complete. We shall see how this works later, when we discuss Figure 3.4.

Figure 3.2 shows the command processing routine that is called from the main loop in Figure 3.1. The operation of this routine is very easy to understand. If no key has been pressed, the routine simply returns (that is, it executes the RETURN instruction). If a key has been pressed, the routine clears the screen, calls the appropriate command processing routine, clears the screen again and returns.

Figure 3.3 shows the two subroutines involved in the time display. The first routine simply positions the cursor at the appropriate place on the screen. The second routine has two distinct functions: (1) it converts the hours, minutes and seconds string format time into hours, minutes and seconds in the twelve hour a.m./p.m. format, and (2) it displays this newly formatted time.

The inclusion of these two distinct functions in one routine is not ideal from the standpoint of program design. Separation of the conversion and display functions would make subsequent program development easier. For example, if one routine converted from the TIME$ format to the a.m./p.m. format while a second routine did the actual display, then the conversion routine could be called by the time-setting routine (Figure 3.6) to show the user the exact time to be set. Other uses for this conversion routine can also be imagined. (As an exercise, carry out this separation.)

Figure 3.4 shows the event-checking routine. To understand this routine, you must understand the "Julian" time used for event scheduling. The Julian time assigned to any instant is the number of seconds from the most recent occurrence of midnight until the given instant. For example, if it is now 10 p.m. and you wish to set the alarm to go off at 1 a.m., then the Julian time assigned to the alarm is 90,000; since that is the number of seconds in the 25 hours from the most recent midnight (22 hours ago) to the desired alarm time (3 hours from now). When midnight arrives, two hours after the alarm has been set, the base for the Julian time changes. At that time the alarm setting must be changed to 3600, since that is the number of seconds in the one hour between the new most recent midnight and the desired alarm time.

The event-checking routine carries out this change in the time base by subtracting 86,400 (the number of seconds in a day) from each

---

#Process commands

    command   GET CM$
                    IF CM$ <> "" THEN {
                      GOSUB clearscreen
                      on case CM$ GOSUB
                            "F"   faster
                            "S"   slower
                            "T"   settime
                            "A"   setalarm
                            "Q"   killalarm
                            "R"   resetalarm
                            "Z"   changehour
                            "C"   changesec
                    GOSUB clearscreen
                  }
                  RETURN

```
120*  GET CM$:IF CM$="" THEN 230
130   GOSUB 920
140   IF CM$="F" THEN GOSUB 450:GOTO 220
150   IF CM$="S" THEN GOSUB 480:GOTO 220
160   IF CM$="T" THEN GOSUB 520:GOTO 220
170   IF CM$="A" THEN GOSUB 550:GOTO 220
180   IF CM$="Q" THEN GOSUB 590:GOTO 220
190   IF CM$="R" THEN GOSUB 600:GOTO 220
200   IF CM$="Z" THEN GOSUB 690:GOTO 220
210   IF CM$="C" THEN GOSUB 720:GOTO 220
220   GOSUB 920
230   RETURN
```

This subroutine, which is called continually in the main loop in Figure 3.1, looks for single-character commands from the keyboard. If one is entered, the appropriate processing subroutine is called.

---

*In the TRS-80 version, GET CM$ is replaced by CM$ = INKEY$.

Figure 3.2:   *Command Processing for Clock*

68 TIME GAMES

#Position cursor for time display

positime   LL = TL: CC = TC   #time line and column
               GOSUB cursor
               RETURN

#Display the time

showtime   GOSUB hms   #get hours, minutes, seconds
              HR = VAL(HR$)
              IF HR > 11 THEN
                  AP$ = "PM"
              else
                  AP$ = "AM"
              IF HR = 0 THEN
                  HR$ = "12"
              else IF HR > 12 THEN
                  HR$ = STR$(HR − 12)
              else
                  HR$ = STR$(HR)
              HR$ = RIGHT$(" " + HR$,2)
              PRINT HR$; ":"; MN$; ":"; SC$; " "; AP$;
              RETURN

```
240   LL=TL:CC=TC:GOSUB 930:RETURN
250   GOSUB 1050:HR=VAL(HR$)
260   IF HR>11 THEN AP$="PM":GOTO 280
270   AP$="AM"
280   IF HR=0 THEN HR$="12":GOTO 310
290   IF HR>12 THEN HR$=STR$(HR−12):GOTO 310
300   HR$=STR$(HR)
310   HR$=RIGHT$(" "+HR$,2)
320   PRINT HR$;":";MN$;":";SC$;" ";AP$;
330   RETURN
```

These routines display the time in the format "HH:MM:SS AP" at a fixed screen position. Time is right-adjusted with a leading space for hours 1 through 9.

Figure 3.3:   Time Display for Clock

scheduled event time whenever the clock time passes midnight.

The use of this based Julian time allows the program to distinguish between times that are 24 hours apart. For example, suppose that the alarm is set to go off at 9 a.m. and at 8:59 a.m. you decide to use one of the commands. Perhaps you noticed that your clock was running a little slow, and you used the F command to speed it up. But what if you took a little while to decide upon the proper frequency to enter, and by the time you were finished it was already 9:01 a.m. The event-checking routine is then called, and it sees that the current time (32,460 for 9:01 a.m.) is later than the alarm time (32,400 for 9:00 a.m.), so it sets off the alarm. The routine has no trouble knowing that the alarm time is 9:00 a.m. today, not 9:00 a.m. tomorrow, since 9:00 a.m. tomorrow would have Julian time 118,800.

To make this scheme work, the event-checking routine must be able to recognize that midnight has already occurred. This is the primary use for the variable OE, which keeps the time of the last previous event check. If the event-check routine finds, on entry, that the current Julian time is less than the time of the last previous check, it knows that midnight has just passed.

The other reason for keeping the time of the last check is to assure that event-checking is done only once per tick. This protection could be helpful in the implementation of additional features, but it is not needed in the current version, since the event actions are performed in such a way that there would be no harm in checking more than once per tick.

The event-checking routine checks for three events: (1) whether midnight has just passed, (2) whether the alarm is due to be "sounded," and (3) whether an adjustment is due. If the alarm is due, it is "sounded," and the alarm time variable AL is increased by the number of seconds in a day. That is, the alarm is automatically set to go off again tomorrow at the same time. If a time adjustment is due, it is made, and the time of the next adjustment is set. Notice that the time of the next adjustment is set by adding the adjustment frequency to the due time of this adjustment—not to the actual time of the adjustment. That is, if the adjustment is made a minute late because the program was processing a command, then the next adjustment will be scheduled for the same time as it would have been if the current adjustment had been made on time. Furthermore, this means that if several adjustments are missed (either because the program is processing a command for a very long time, or because adjustments are very frequent), they will all be made (at the rate of one per clock tick) as soon as the program is no longer tied up.

#Check on alarm, adjustment and other events
```
events    GOSUB julian                        #get "Julian" time
          ET = JT
          IF ET < OE THEN {                   #just passed midnight
              IF AJ <> -1 THEN
                  AJ = AJ - dayseconds
              IF AL <> -1 THEN
                  AL = AL - dayseconds
              }
          IF OE <> ET THEN {                  #check events once per tick
              OE = ET
              IF AL <> -1 AND AL <= ET THEN {
                  GOSUB setoff                #set off alarm
                  AL = AL + dayseconds        #reset to the same time tomorrow
                  }
              IF AJ <> -1 AND AJ <= ET THEN {
                  DT = AI                     #change time by AI sec
                  GOSUB bumptime
                  AJ = AJ + AF                #set time of next adjustment
                  }
              }
          RETURN
340  GOSUB 830:ET=JT
350  IF ET>=OE THEN 380
360  IF AJ<>-1 THEN AJ=AJ-86400
370  IF AL<>-1 THEN AL=AL-86400
380  IF OE=ET THEN 440
390  OE=ET
400  IF AL=-1 OR AL>ET THEN 420
410  GOSUB 610:AL=AL+86400
420  IF AJ=-1 OR AJ>ET THEN 440
430  DT=AI:GOSUB 810:AJ=AJ+AF
440  RETURN
```

This routine keeps track of whether a speed adjustment or an alarm display is due. These due times are kept in a "Julian" time that can (theoretically) be days ahead of the current time.

Figure 3.4: Event Checking for Clock

INSIDE BASIC GAMES

Figure 3.5 shows the routines that implement the adjustment commands F and S. These routines determine the adjustment frequency (AF) and the increment (AI) to be used by the event-checking routine; they then set the time (AJ) of the first adjustment.

```
#Set the clock running faster
faster      repeat
                INPUT "ADJUSTMENT INTERVAL (SEC)"; FT
                until (FT > 0 and FT is a whole number)
            IN = 1
            GOSUB setadjust
            RETURN
#Slow down the clock
slower      repeat
                INPUT "ADJUSTMENT INTERVAL (SEC)"; FT
                until (FT > 0 and FT is a whole number)
            IN = -1
            GOSUB setadjust
            RETURN
#Set up for adjustment
setadjust   AI = IN                         #set adjustment increment
            AF = FT                         #and frequency
            GOSUB julian
            AJ = JT + AF
            RETURN
 450  INPUT "ADJUSTMENT INTERVAL (SEC)";FT
 460  IF FT<=0 OR FT<> INT(FT) THEN 450
 470  IN=1:GOSUB 510:RETURN
 480  INPUT "ADJUSTMENT INTERVAL (SEC)";FT
 490  IF FT<=0 OR FT<>INT(FT) THEN 480
 500  IN=-1:GOSUB 510:RETURN
 510  AI=IN:AF=FT:GOSUB 830:AJ=JT+AF:RETURN
```
These routines implement the clock speed adjustment commands F (for "faster") and S (for "slower"). The user can request that one second be added to or subtracted from the time at a fixed frequency (e.g., once every 10,000 seconds). This can compensate for a precise but inaccurate clock circuit.

Figure 3.5: Speed Adjustment Commands for Clock

72 TIME GAMES

Figure 3.6 shows the routine that implements the time-setting command T. Notice that the operator is asked to press RETURN to start the clock. You can understand the reason for this if you have ever tried to set the Pet clock with a statement of the form

    TIME$ = 120430

After waiting until precisely 4 minutes and 30 seconds past noon, you triumphantly press RETURN, only to have the Pet announce

    TYPE MISMATCH ERROR?

because you neglected to place the string in quotes. The approach used in Figure 3.6 requires that the user press RETURN an extra time, but the second RETURN is guaranteed to start the clock—there will be no surprises.

Figure 3.7 shows the routines that implement the alarm-related commands A, Q, and R. These routines are quite primitive, and as a result, the alarm capabilities of this clock program are minimal. Several improvements are suggested at the end of this section.

Figure 3.8 shows the routines concerned with the alarm display. As you can see, there is much room for improvement.

#Set the time

settime   PRINT "SET TIME"
            GOSUB asktime
            PRINT "PRESS RETURN TO START CLOCK"
            GOSUB onech
            GOSUB newtime   #TM$ becomes new time
            RETURN

  520   PRINT "SET TIME"
  530   GOSUB 640:PRINT "PRESS RETURN TO START CLOCK"
  540   GOSUB 1030:GOSUB 1080:RETURN

This is the routine that implements the T (for "set time") command. The program requests a second RETURN to start the clock rather than using the RETURN that terminates the time input string. This approach protects the user from the frustration of discovering syntax errors at the wrong time.

Figure 3.6:   *Time Setting for Clock*

INSIDE BASIC GAMES 73

```
#Set the alarm


setalarm      PRINT "SET ALARM"
              GOSUB asktime
              AL = TM
              GOSUB julian
              IF AL < JT THEN
                  AL = AL + dayseconds
              RETURN


#Kill the alarm

killalarm     AL = -1
              GOSUB turnoff
              RETURN


#Reset the alarm

resetalarm    GOSUB turnoff
              RETURN


    550   PRINT "SET ALARM"
    560   GOSUB 640:AL=TM:GOSUB 830
    570   IF AL<JT THEN AL=AL+86400
    580   RETURN
    590   AL=-1:GOSUB 630:RETURN
    600   GOSUB 630:RETURN
```

These three routines implement the alarm-related commands A (for "set alarm"), Q (for "turn off the alarm and don't reset it"), and R (for "turn off the alarm and reset it"). The routine for setting the alarm accepts a 24-hour time, then sets the alarm to go off at the next occurrence of that time. The Q command causes AL to be set to −1, signaling the program that there is no alarm setting. The R command does not affect AL. In the routine in Figure 3.4, AL is automatically set to go off again 24 hours later. The AL mechanism could support alarm settings more than 24 hours in advance, but no commands have been provided for this purpose.

Figure 3.7:   Alarm Commands for Clock

Figure 3.9 shows the routine that asks the user for time input (for setting the clock or alarm). This routine requires input in exactly the same format as that required by the Pet for setting TIME$, except that this routine is smart enough to accept the time string with or without quotes.

Figures 3.7, 3.8 and 3.9 provide examples of stubs, a concept that arises out of the "top-down" development technique for programs. The top-down approach requires that the main structure of the program be laid out first (e.g., as shown in Figure 3.1). Then, each of the subroutines that provide the building blocks for the main structure is similarly laid out and built from other subroutines. (For example, Figures 3.2, 3.3 and 3.4 show the principal subroutines called by the main routine shown in Figure 3.1.) If at any stage in this process one of the necessary subroutines is not to be written immediately, a stub is provided. A stub is a routine that will do something plausible when it is called, that is, either it will implement the necessary action, but

---

#Set off the alarm

setoff   LL = AR: CC = AC   #alarm line and column
              GOSUB cursor
              PRINT AL$;   #print alarm string
              RETURN

#Turn off alarm

turnoff   GOSUB clearscreen
              RETURN

    610   LL=AR:CC=AC:GOSUB 930
    620   PRINT AL$;:RETURN
    630   GOSUB 920:RETURN

These routines provide the alarm display. The first displays a predefined string AL$ at a predefined position on the screen. The second clears the screen. A changing display (e.g., blinking, or, if sound is available, beeping) would require an entirely different approach, and would involve the event-checking routine shown in Figure 3.4.

Figure 3.8: Alarm Display for Clock

with a minimum of features and options, or it will perform a similar (but simpler) action. For example, the time-displaying routine at line 250 (in Figure 3.3) might first be provided by the following stub:

>PRINT "10:25:04 AM";
>RETURN

Or the command-processing routine of Figure 3.2 might first be provided in a stub version that only handles the T command.

The word "stub" was introduced to describe such routines, because the structure of a program developed using top-down structuring techniques is like that of the inverted trees used to describe the organizational charts of many business firms. A stub is often at a high level in the program's structure, so that its section of the chart stops short, as though a main trunk of the inverted tree had been cut off, leaving a stub.

The use of stubs does have one drawback. Sometimes a routine that begins as a stub is later treated as the whole branch. That is, the

---

#Ask for time

asktime repeat {
           INPUT "TIME (6 DIGITS: HHMMSS)"; TM$
           AH = VAL(LEFT$(TM$,2))
           AM = VAL(MID$(TM$,3,2))
           AS = VAL(RIGHT$(TM$,2))
           } until (0 <= AH < 24 AND 0 <= AM < 60 AND 0 <= AS < 60)
           TM = 3600*AH + 60*AM + AS
           RETURN

```
640  INPUT "TIME (HHMMSS)";TM$
650  AH=VAL(LEFT$(TM$,2)):IF AH<0 OR AH>=24 THEN 640
660  AM=VAL(MID$(TM$,3,2)):IF AM<0 OR AM>=60 THEN 640
670  AS=VAL(RIGHT$(TM$,2)):IF AS<0 OR AS>=60 THEN 640
680  TM=3600*AH+60*AM+AS:RETURN
```

This routine allows the user to set a time for the clock or for the alarm. It accepts the same format required for setting TIME$, which indicates that the routine was written with the programmer's, not the user's, convenience as a goal.

Figure 3.9:   Inelegant Time Input for Clock

76   TIME GAMES

additional features and functions that were to be added later are actually never provided.

Figure 3.10 shows the routines that implement the C and Z commands, which allow the time to be changed by a given number of seconds or hours without stopping the clock. (A routine to change the time by a given number of minutes could also have been provided, but it seemed unnecessary.)

#Change the hour

changehour   repeat
                      INPUT "CHANGE AMOUNT (HOURS)"; CH
                  until (CH is a whole number AND −24 < CH < 24)
                  CS = CH*3600   #number of seconds to change by
                  GOSUB changeclock
                  RETURN

#Change the seconds

changesec   repeat
                     INPUT "CHANGE AMOUNT (SECONDS)"; CS
                 until (CS is a whole number AND −60 < CS < 60)
                 GOSUB changeclock
                 RETURN

```
690   INPUT "CHANGE AMOUNT (HOURS)";CH
700   IF CH<>INT(CH) OR CH<=−24 OR CH>=24 THEN 690
710   CS=CH*3600:GOSUB 750:RETURN
720   INPUT "CHANGE AMOUNT (SECONDS)";CS
730   IF CS<>INT(CS) OR CS<=−60 OR CS>=60 THEN 720
740   GOSUB 750:RETURN
```

These routines implement the commands C (for "change time by a specified number of seconds") and Z (for "change time by a specified number of hours"). Each of these routines computes the total number of seconds by which the clock is to be changed, and calls the clock changing routine shown in Figure 3.11.

Figure 3.10: Time Adjustment Routines for Clock

Figure 3.11 shows the routine that is called by both of the routines in Figure 3.10. The purpose of this routine is to carry out the actual changing of the clock and to adjust the AL and AJ variables if the time change carries the time forward or backward past midnight. The actual change to the Pet clock involves two further subroutine levels.

```
#Change the clock setting
changeclock    GOSUB julian                        #remember time before change
               CT = JT
               DT = CS                             #amount to change by
               GOSUB bumptime                      #change it
               GOSUB julian                        #check for passing midnight
               IF CS < 0 AND JT > CT THEN {        #backed past midnight
                   IF AJ <> -1 THEN
                       AJ = AJ + dayseconds
                   IF AL <> -1 THEN
                       AL = AL + dayseconds
               }
               IF CS > 0 AND JT < CT THEN {        #advanced past midnight
                   IF AJ <> -1 THEN
                       AJ = AJ - dayseconds
                   IF AL <> -1 THEN
                       AL = AL - dayseconds
               }
               RETURN
750  GOSUB 830:CT=JT
760  DT=CS:GOSUB 810:GOSUB 830
770  IF CS>=0 OR JT<=CT THEN 785
775  IF AJ<>-1 THEN AJ=AJ+86400
780  IF AL<>-1 THEN AL=AL+86400
785  IF CS<=0 OR JT>=CT THEN 800
790  IF AJ<>-1 THEN AJ=AJ-86400
795  IF AL<>-1 THEN AL=AL-86400
800  RETURN
```
This routine changes the clock setting and assures that AL and AJ are adjusted properly if the time change crosses midnight.

Figure 3.11: Housekeeping for Clock Change

The first level, shown in Figure 3.12, is concerned with the algorithm for making time changes "on the fly." The level below that, shown in Figure 3.15, contains routines that refer specifically to the Pet's TIME$ mechanism. This separation of function makes the Clock program easily adaptable to other systems that have clocks.

The program shown in Figure 3.12 changes the time by a fixed (positive or negative) increment specified in the variable DT. The problem is that the clock ticks once per second. If the clock ticks between the moment at which the program obtains the time from TIME$ and the moment at which the program updates TIME$, then one second will be lost on the clock. (Do you see why? This is an example of a significant problem in the design of processes that proceed concurrently.) The technique used in Figure 3.12 is to get the current time from TIME$, then to wait for (and include in the increment computation) the next tick before updating TIME$. This method requires that all of the computations involved occupy a total of less than one second, so that the update of TIME$ is guaranteed to occur before the second tick following the moment at which the current time was originally obtained.

Figure 3.13 shows the routines that pass back and forth between the numerical Julian time format and the string format used with TIME$. These routines are simple; however, special care is required

---

#Increment the time by DT seconds

bumptime   GOSUB julian   #compute JT from old time (TT$)
                JT = JT + DT + 1   #increment, plus one for next tick
                GOSUB unjulian   #compute new time TM$ from JT
                GOSUB waitnew   #wait for next tick after TT$
                GOSUB newtime   #set new time from TM$
                RETURN

    810   GOSUB 830:JT=JT+DT+1:GOSUB 850
    820   GOSUB 1090:GOSUB 1080:RETURN

This routine increments or decrements the clock setting by a given number of seconds, DT. To avoid a subtle error, the routine always waits for the next clock tick before making the change.

Figure 3.12: Mechanics of Clock Change

#Compute "Julian" time

julian　　GOSUB hms
　　　　　　JT = VAL(SC$) + 60*VAL(MN$) + 3600*VAL(HR$)
　　　　　　RETURN

#Go from Julian time (JT) to string format (TM$)

unjulian　repeat　　　　　　　　　　　　　　　　　#adjust JT so 0<= JT< dayseconds
　　　　　　　　IF JT < 0 THEN
　　　　　　　　　　JT = JT + dayseconds
　　　　　　　until (JT >= 0)
　　　　　　repeat
　　　　　　　　IF JT >= dayseconds THEN
　　　　　　　　　　JT = JT - dayseconds
　　　　　　　until (JT < dayseconds)
　　　　　　SC = JT mod 60　　　　　　　　　　　　　　#extract seconds,
　　　　　　MN = (JT - SC)/60 mod 60　　　　　　　　　#minutes,
　　　　　　HR = (JT - SC - 60*MN)/3600　　　　　　　#hours
　　　　　　TX = SC + 100*MN + 10000*HR + 1000000　#put together with leading 1
　　　　　　TM$ = RIGHT$(STR$(TX),6)　　　　　　　　#to assure leading zeroes
　　　　　　RETURN

```
830   GOSUB 1050
840   JT=VAL(SC$)+60*VAL(MN$)+3600*VAL(HR$):RETURN
850   IF JT<0 THEN JT=JT+86400:GOTO 850
860   IF JT>=86400 THEN JT=JT-86400:GOTO 860
870   SC=JT-60*INT(JT/60)
880   MN=(JT-SC)/60:MN=MN-60*INT(MN/60)
890   HR=(JT-SC-60*MN)/3600
900   TX=SC+100*MN+10000*HR+1000000
910   TM$=RIGHT$(STR$(TX),6):RETURN
```

　　These routines translate between the string format used by TIME$ and the "Julian" format used in the clock's event checking. The trick of adding 1,000,000 to a six-digit number (line 900) in order to assure the presence of leading zeros in the string version is worth remembering.

Figure 3.13: Julian Time Routines for Clock

80 TIME GAMES

```
#clear the screen
clearscreen    PRINT "clr"; : RETURN

#Position cursor to line (0 <= LL < 24) and column (0 <= CC < 40)
cursor         LL = LL mod 24: CC = CC mod 40
               PRINT "home";
               IF LL <> 0 THEN
                   FOR XX = 1 TO LL
                       PRINT "down cursor";
                   NEXT XX
               IF CC <> 0 THEN
                   FOR XX = 1 TO CC
                       PRINT "right cursor";
                   NEXT XX
               RETURN

#Input a single character into X$
onech          repeat GET X$ until (X$ <> ""): RETURN

    920* PRINT CHR$(147);:RETURN
    930* IF LL< 0 THEN LL=LL+24:GOTO 930
    940* IF LL>23 THEN LL=LL−24:GOTO 940
    950* IF CC< 0 THEN CC=CC+40:GOTO 950
    960* IF CC>39 THEN CC=CC−40:GOTO 960
    970* PRINT CHR$(19);
    980* IF LL=0 THEN 1000
    990* FOR XX=1 TO LL:PRINT CHR$(17);:NEXT XX
   1000  IF CC=0 THEN 1020
   1010* FOR XX=1 TO CC:PRINT CHR$(29);:NEXT XX
   1020  RETURN
   1030* GET X$:IF X$="" THEN 1030
   1040  RETURN
```

These routines use the Pet facilities for screen clearing, cursor positioning and single-character input. Other systems have different facilities for these functions.

*In the TRS-80 version of these routines, screen clearing is accomplished using the CLS command (see Figure 2.11); the numbers of lines and columns are 16 and 64 rather than 24 and 40; the *home*, *down cursor* and *right cursor* characters are represented by CHR$(28), CHR$(25) and CHR$(26); the GET X$ is replaced by X$=INKEY$ (see Figures 2.11 and 2.12).

Figure 3.14: Utility Routines for Pet Clock

to assure that times in the string format have leading zeroes when needed. This is achieved by adding 1,000,000 to the number that needs to be six digits long, then keeping only the last six digits.

Note that although the string format used with TIME$ is used throughout the program, TIME$ itself is only used in the routines in Figure 3.15. There would be no need to change the string format used throughout the program, even if it were to be adapted to a system

---

#Fetch the time

hms TT$ = TIME$ #freeze it
   HR$ = LEFT$(TT$,2)
   MN$ = MID$(TT$,3,2)
   SC$ = RIGHT$(TT$,2)
   RETURN

#Set a new time from TM$

newtime TIME$ = TM$
   RETURN

#Wait for the tick after TT$

waitnew repeat { } until (TT$ <> TIME$)
   RETURN

```
1050    TT$=TIME$
1060*   HR$=LEFT$(TT$,2):MN$=MID$(TT$,3,2):SC$=RIGHT$(TT$,2)
1070    RETURN
1080**  TIME$=TM$:RETURN
1090    IF TT$=TIME$ THEN 1090
1100    RETURN
```

These routines are the only ones in the Clock program that explicitly use the TIME$ function of the Pet. These routines will need to be adapted if another system with a clock is used. The first routine reads the clock and returns hours, minutes and seconds in HR$, MN$ and SC$. The second routine sets TIME$ from the string TM$. The third routine waits until TIME$ ≠ TT$. TT$ is the string into which TIME$ is read on each call to the first routine shown above.

---

*On a TRS-80 system, the HR$ and MN$ strings are derived from MID$(TT$,10,2) and MID$(TT$,13,2).
**On a TRS-80 system, it is not possible to set TIME$ in this way.

Figure 3.15: TIME$ Routines for Clock

82 TIME GAMES

whose clock used an entirely different format. The conversion between this program's format and that system's clock format could be performed by revised versions of the routines shown in Figure 3.15.

Figure 3.16 shows the intialization routine for Clock. If no built-in speed adjustment is desired, the first constant in the data statement of line 1150 should be replaced by −1.

#Initialization routine

init OE = −1   #no previous event check
         AL = −1   #no alarm setting
         AL$ = "rvs ALARM off"   #"ALARM" in reverse video
         READ TL, TC
         DATA timeline, timecolumn
         READ AR, AC
         DATA alarmrow, alarmcolumn
         READ FT, IN   #built-in speed adjustment
         DATA frequency, increment   #FT = frequency. IN = 1 for F, −1 for S.
         GOSUB setadjust
         RETURN

1110   OE = −1:AL= −1
1120* AL$=CHR$(18)+"ALARM"+CHR$(146)
1130** READ TL,TC:DATA 22,28
1140   READ AR,AC:DATA 11,17
1150   READ FT,IN:DATA 2900,−1:GOSUB 510
1160   RETURN

This is the initialization routine for Clock. The alarm string and the display positions for the time and alarm displays are set. Then a built-in speed adjustment is set up. The idea is that after experimenting with the F and S commands, the user would find the correct setting and build it into the program. The author's Pet, for example, needs to run slower by about one second every 2900 seconds (line 1150).

---
*On a TRS-80, the *rvs* and *off* characters have no counterparts.
**On a TRS-80, the corresponding screen position for the clock display would be achieved using TL and TC settings of 14 and 52.

Figure 3.16: Initialization for Clock

This concludes our discussion of the Clock program. There are many possible improvements. Here are some of them:

- Incorporate the date. Once the program has been told the date, it can keep track of it forever.
- Use the event-checking mechanism to implement an alarm that consists of a sequence of scheduled actions—either of fixed duration, or continuing until turned off by the user (e.g., blinking on and off at one-second intervals).
- Implement multiple alarms. Allow an identifying text string to be associated with each alarm (e.g., "TIME TO WATCH STAR TREK"). Make appropriate changes to the definition and implementation of the Q and R commands.
- Allow alarms to be set more than 24 hours in advance.
- Allow the frequency of occurrence of an alarm to be specified. (Currently it is automatically reset to occur again 24 hours later, unless cancelled by the Q command.)
- Add other time input formats to the program in Figure 3.9.
- Allow simultaneous display of several different times at once, each with an identifying string like "PARIS," "MOM'S TIME" or "GMT."
- Revise the Clock program to use the Pet's TI feature rather than the TIME$ feature.

Card Memory

Now we shall look at another game in which the time element is important. Card Memory is a simple game that uses the Pet TIME$ feature to provide an interval timer. Here's how it works. The program is going to display a series of playing card names (e.g., CLUB QUEEN, DIAMOND 4). It begins by asking how many cards you wish to see and what the interval between them is to be. For example,

    NUMBER OF CARDS? **5**
    INTERVAL (SEC)? **3**

will cause the program to display a sequence of 5 distinct cards, randomly selected. Each card will be displayed for (approximately) 3 seconds, then the screen will be cleared and the next card displayed.

After all 5 cards have been displayed, the program asks you to name the cards you saw, in the order in which you saw them. Each time the program asks "CARD?", you must enter a single character identifying the suit (C, D, H, S) of the card, then a space, and after that a letter or number for the face value (A, K, Q, J, 10, 9, 8, 7, 6, 5, 4, 3, 2) of the card. When you press RETURN after entering a card in the above format, the program displays the correct card, then it asks for the next card that you saw. At any point, you can enter "Q" instead of a card code to tell the program that you give up. It will then display the remainder of the cards.

After this guessing sequence is completed, the program tells you how many cards you identified correctly and displays statistics on how well you have done in the past. For example, it might say

    YOU DIDN'T GET ANY!

    BEST FOR THIS SET: 2 OUT OF 5
    BEST LENGTH: 3 (INTERVAL: 2 SEC)

This means that since the time you asked the program to give you sequences of 5 cards for intervals of 3 seconds each, the best turn you have had has been 2 correct guesses out of 5 cards. At some time in the past you asked the program to display sequences of 3 cards for intervals of 2 seconds each, and you remembered all three at least once.

At this point the program is waiting for a single-character input from you. If you type "R", the program will list the entire sequence for your review. If you type "N", the program will allow you to enter new values for the number of cards and the interval. If you press the space bar, the program will display a new sequence.

The Card Memory Program

Figure 3.17 shows the main routine for Card Memory. The structure of this routine reflects the game structure described above.

One of the routines called from the main routine is shown in Figure 3.18. This routine simulates the shuffling of a deck of cards, but only the first NC cards are actually placed into the shuffled order in the array DK.

The shuffling is accomplished in a straightforward way. First the 52 cards are placed into a "standard" order in the array DD. Then a whole number, ZZ, between 1 and 52, is chosen randomly. The card DD(ZZ) becomes the first element of the array DK. Then all elements of DD with indices greater than ZZ are moved up one place, so that now the first 51 elements of DD contain the remaining 51 cards.

The process is then repeated, choosing ZZ between 1 and 51 to

INSIDE BASIC GAMES

#Card memory test

    GOSUB init   #initialize
    repeat {   #loop for new game types
        GOSUB clearscreen
        GOSUB gamevalues   #new interval and sequence length
        repeat {   #loop for one game type
            GOSUB clearscreen
            GOSUB startgame   #initialize for this game
            GOSUB shuffle   #shuffle the deck
            GOSUB showcards   #display the sequence
            GOSUB askplayer   #ask player to repeat it
            GOSUB stats   #tell how many right
            repeat {
                GOSUB next   #end game and proceed
                IF NX$ = "R" THEN
                    GOSUB showall
            } until (NX$ <> "R")
        } until (NX$ = "N" or "E")   #N = set new parameters
    } until (NX$ = "E")   #E = return to BASIC
    END

```
100   GOSUB 960
110   GOSUB 860:GOSUB 920
120   GOSUB 860:GOSUB 910:GOSUB 180:GOSUB 240
130   GOSUB 390:GOSUB 660
140   GOSUB 830:IF NX$="R" THEN GOSUB 270:GOTO 140
150   IF NX$="N" THEN 110
160   IF NX$="E" THEN END
170   GOTO 120
```

  The player is allowed to choose a sequence length NC between 1 and 52 and an intercard interval DC. The interval is a whole number of seconds greater than or equal to zero. Then the player is presented with sequences of cards. Each card is erased from the screen after the interval DC. Finally, the player is asked to repeat the sequence. After each card named by the player, the program displays the correct card.

Figure 3.17: Card Memory Test

obtain a card DD(ZZ) to become DK(2). Elements with indices greater than ZZ are moved up, the DD array shrinks to 50 cards, and so on.

The routines that show the sequence of cards to the player appear in Figure 3.19. These routines are simple, and require no further explanation.

The actual translation between the numeric form of card encoding (a number between 1 and 52) and the display form (a string like "SPADE 6") is done in the routines in Figure 3.20. The encoding assigns numbers 1 through 13 to cards in the club suit, 14 through 26

```
#Shuffle the deck

shuffle    FOR XX = 1 TO 52                    #first put cards in order
           DD(XX) = XX
           NEXT XX
           FOR XX = 1 TO NC                    #then pick randomly
               YY = 53 - XX                    #number of cards left
               ZZ = INT(RND(1)*YY) + 1         #1 <= ZZ <= YY
               DK(XX) = DD(ZZ)                 #next card of deck
               IF XX <> 52 AND ZZ < YY THEN
                   FOR WW = ZZ TO YY - 1       #fill hole left by
                       DD(WW) = DD(WW + 1)     #selected card
                   NEXT WW
           NEXT XX
           RETURN

180   FOR XX=1 TO 52:DD(XX)=XX:NEXT XX
190   FOR XX=1 TO NC:YY=53-XX
200*  ZZ=INT(RND(1)*YY)+1:DK(XX)=DD(ZZ)
210   IF XX=52 OR YY<=ZZ THEN 230
220   FOR WW=ZZ TO YY-1:DD(WW)=DD(WW+1):NEXT WW
230   NEXT XX:RETURN
```

This subroutine sets the first NC elements of the DK array to distinct whole numbers randomly chosen from the range 1 to 52.

---
*In the TRS-80 version of this line, the term RND(0) appears instead of RND(1).

Figure 3.18: Shuffle Subroutine for Card Memory

#Display the first NC cards at intervals of DC

showcards   FOR I = 1 TO NC
                CD = DK(I): GOSUB displaycard
                DL = DC: GOSUB delay
                GOSUB clearscreen
                NEXT I
            RETURN

#Display all of the cards

showall   FOR I = 1 TO NC
                CD = DK(I): GOSUB displaycard
                NEXT I
            RETURN

#Display one card CD

displaycard GOSUB decode   #convert CD to a string
                PRINT CD$
                RETURN

```
240   FOR I=1 TO NC
250   CD=DK(I):GOSUB 290:DL=DC:GOSUB 870:GOSUB 860:NEXT I
260   RETURN
270   GOSUB 860:FOR I=1 TO NC:CD=DK(I):GOSUB 290
280   NEXT I:RETURN
290   GOSUB 310:PRINT CD$
300   RETURN
```

    The first two routines shown here are used to display the entire set of cards for the current game. The first displays them one at a time, clearing the screen each time after the specified interval elapses. The second leaves them all on the screen for the player to review. The third routine, called by both of the first two, translates a number between 1 and 52 into a card (e.g., SPADE 6) and displays it.

Figure 3.19: Card Display Routines for Card Memory

to diamonds, 27 through 39 to hearts and 40 through 52 to spades. Within each block of 13 numbers, the face values are assigned in the order ace, deuce, trey, four, five, six, seven, eight, nine, ten, jack, queen, king.

Figure 3.21 shows the routine that asks the player to repeat the displayed sequence. The player's guess is accepted and its syntax is checked. If the player enters a "Q", then all of the cards are displayed. Otherwise, the guess is compared with the answer, and the program then notes whether it is right or wrong. After that it

---

#Decode/encode routine

decode   FV = mod(CD,13)   #face value
              SU = (CD − FV)/13 + 1   #suit
              IF FV = 0 THEN
                    {FV = 13: SU = SU − 1}
              FV$ = FV$(FV): SU$ = SU$(SU)   #make into strings
              CD$ = SU$ + " " + FV$ #combine into card
              RETURN

encode   IF NOT (1 <= SU <= 4 AND 1 <= FV <= 13) THEN
                  CD = −1   #invalid suit or value
              else
                  CD = 13∗(SU − 1) + FV
              RETURN

```
310  FV=CD−INT(CD/13)∗13
320  SU=(CD−FV)/13+1
330  IF FV=0 THEN FV=13:SU=SU−1
340  FV$=FV$(FV):SU$=SU$(SU)
350  CD$=SU$+" "+FV$:RETURN
360  IF SU<1 OR SU>4 OR FV<1 OR FV>13 THEN CD=−1:RETURN
370  CD=13∗(SU−1)+FV
380  RETURN
```

The first of the two routines shown here translates a whole number between 1 and 52 into a suit (club, diamond, heart, spade) and a face value (Ace, King, Queen, Jack, 10 through 2). The suits are encoded in the variable SU as follows: 1 = clubs, 2 = diamonds, 3 = hearts, 4 = spades. The face values are encoded in FV as follows: 1 = Ace, 2 through 10 encode themselves, 11 = Jack, 12 = Queen, 13 = King.

The second of the two routines goes from SU and FV to a number CD between 1 and 52.

Figure 3.20: Decode/Encode Routines for Card Memory

```
#Ask player to repeat the displayed sequence
askplayer   PRINT "REPEAT CARDS IN ORDER"
            FOR I = 1 TO NC
                repeat {                            #get player's guess
                    GOSUB askcard                   #CD$ is player's answer
                    GOSUB cardnumber                #convert to CD
                } until (CD <> -1)                  #-1 = bad input
                IF CD = -2 THEN                     #player has quit (Q)
                    {GOSUB showall: RETURN}         #display all cards
                else IF CD = DK(I) THEN
                    GOSUB right                     #player guessed right
                else {
                    GOSUB wrong                     #player guessed wrong
                    CD = DK(I)                      #display right card
                }
                GOSUB displaycard                   #show the card
                DL = SC: GOSUB delay                #(briefly)
                GOSUB clearscreen
            NEXT I
            RETURN

right       RG = RG + 1: RETURN

wrong       RETURN

    390  PRINT "REPEAT CARDS IN ORDER"
    400  FOR I=1 TO NC
    410  GOSUB 490:GOSUB 500:IF CD=-1 THEN 410
    420  IF CD=-2 THEN GOSUB 270:RETURN
    430  IF CD=DK(I) THEN GOSUB 470:GOTO 450
    440  GOSUB 480:CD=DK(I)
    450  GOSUB 290:DL=SC:GOSUB 870:GOSUB 860:NEXT I
    460  RETURN
    470  RG=RG+1:RETURN
    480  RETURN
```

This routine accepts the player's guesses. After each guess (whether right or wrong), the right card is displayed. If the player types "Q", all of the cards are displayed, and the process is terminated.

Figure 3.21: Player Input for Card Memory

displays the correct answer, clears the screen and asks for the next card. The routine in Figure 3.21 calls the routines in Figures 3.22 and 3.23 for the actual card input and syntax checking. These programs are largely self-explanatory.

#Ask for card from player

#Expected format: suit (1 character), space, face value (1 or 2 characters).
Suit is C, D, H or S.
Face value is A, 2, 3, 4, 5, 6, 7, 8, 9, 10, J, Q, K.

askcard   INPUT "CARD"; CD$
                RETURN

   490* INPUT "CARD";CD$:RETURN

This routine asks for and accepts the player's card guess. Checking is done in Figure 3.23.

---
*The Apple version of this line differs only in the value of the string constant.

Figure 3.22: Ask-for-Card Routine for Card Memory

```
500   IF LEN(CD$)>=3 THEN 530
510   IF CD$="Q" THEN CD=-2:RETURN
520   CD=-1:RETURN
530   L$=LEFT$(CD$,1):R$=MID$(CD$,3):VR=VAL(R$)
540   IF L$="C" THEN SU=1:GOTO 590
550   IF L$="D" THEN SU=2:GOTO 590
560   IF L$="H" THEN SU=3:GOTO 590
570   IF L$="S" THEN SU=4:GOTO 590
580   SU=-1
590   IF VR>0 THEN FV=VR:GOTO 650
600   IF R$="A" THEN FV=1:GOTO 650
610   IF R$="K" THEN FV=13:GOTO 650
620   IF R$="Q" THEN FV=12:GOTO 650
630   IF R$="J" THEN FV=11:GOTO 650
640   FV=-1
650   GOSUB 360:RETURN
```

Figure 3.23a: Actual BASIC for Parsing Player's Answer

#Translate player guess CD$ into a number CD

cardnumber IF LEN(CD$) < 3 THEN   #guesses are 3 or 4 chars
                 IF CD$ = "Q" THEN
                     CD = −2   #player quits
                 else
                     CD = −1   #invalid guess
             else {
                 L$ = LEFT$(CD$,1)   #suit string
                 R$ = MID$(CD$,3)   #face value string
                 VR = VAL(R$)
                 IF case
                     L$ = "C" THEN SU = 1
                     L$ = "D" THEN SU = 2
                     L$ = "H" THEN SU = 3
                     L$ = "S" THEN SU = 4
                 else
                     SU = −1   #invalid suit
                 IF VR > 0 THEN   #R$ is numeric
                     FV = VR
                 else IF case
                     R$ = "A" THEN FV = 1
                     R$ = "K" THEN FV = 13
                     R$ = "Q" THEN FV = 12
                     R$ = "J" THEN FV = 11
                 else
                     FV = −1   #invalid face value
                 GOSUB encode   #SU,FV are combined into CD
             }
             RETURN

The routine "parses" CD$ into SU and FV fields, then calls one of the subroutines shown in Figure 3.20 to compute CD.

Figure 3.23b: Free BASIC for Parsing Player's Answer

Figure 3.24 shows the routine that announces the result (that is, how good your guesses were). As with other such programs in this book, this routine has been designed to provide variety and add interest to the game. Figure 3.25 shows the subroutine called from the program in Figure 3.24 in order to display the best scores from previous turns.

```
#Display game results and compare with "best"
stats   PRINT                                    #space down one line
        IF RG = NC THEN                          #player got them all
           IF NC = 1 THEN
              PRINT "YOU GOT IT!"
           else IF NC = 2 THEN
              PRINT "YOU GOT BOTH OF THEM!"
           else
              PRINT "YOU GOT ALL"; NC; "OF THEM!"
        else IF RG = 0 THEN                      #player didn't get any
           IF NC = 1 THEN
              PRINT "YOU MISSED IT!"
           else
              PRINT "YOU DIDN'T GET ANY!"
        else
           PRINT "YOU GOT"; RG; "OUT OF"; NC
        PRINT                                    #space down one line
        IF RG = NC AND NC > BC THEN {
           PRINT "A NEW RECORD LENGTH!"
           BC = NC: BD = DC                      #new "best" length & interval
           }
        IF RG > BR THEN
           BR = RG                               #new best for this set
        PRINT
        GOSUB saybest                            #state the current records
        RETURN
```
This routine tells the player the score for the sequence just finished. Then the best scores to date (longest string correct and most right in the current set) are stated.

Figure 3.24: Give the Score for Card Memory

INSIDE BASIC GAMES

```
660   PRINT:IF RG <> NC THEN 700
670   IF NC=1 THEN PRINT "YOU GOT IT!":GOTO 740
680   IF NC=2 THEN PRINT "YOU GOT BOTH OF THEM!":GOTO 740
690*  PRINT "YOU GOT ALL";NC;"OF THEM!":GOTO 740
700   IF RG <> 0 THEN 730
710   IF NC=1 THEN PRINT "YOU MISSED IT!":GOTO 740
720   PRINT "YOU DIDN'T GET ANY!":GOTO 740
730*  PRINT "YOU GOT";RG;"OUT OF";NC
740   PRINT
750   IF RG <> NC OR NC<=BC THEN 780
760   PRINT "A NEW RECORD LENGTH!"
770   BC=NC:BD=DC
780   IF RG> BR THEN BR=RG
790   PRINT:GOSUB 800:RETURN
```

*The Apple versions of these lines differ only in the spacing inside the string constants.

Figure 3.24a: Actual BASIC for Card Memory Score

#Display best scores

saybest   PRINT "BEST FOR THIS SET:"; BR; "OUT OF"; NC
        IF BC > 0 THEN
            PRINT "BEST LENGTH:"; BC; " (DELAY:"; BD; "SEC)"
        RETURN

```
800*  PRINT "BEST FOR THIS SET:";BR;"OUT OF";NC
810*  IF BC>0 THEN PRINT:PRINT "BEST LENGTH:";BC;" (DELAY:";BD;"SEC)"
820   RETURN
```

This routine displays the best scores achieved thus far.

*The Apple versions of these lines differ only in the spacing inside the string constants.

Figure 3.25: Recall Record Performances for Card Memory

TIME GAMES

Figure 3.26 shows several routines that we have used in other programs in this book. The first two routines implement the waiting routine that is called after the score of the game has been displayed. The third routine is called to clear the screen.

Figure 3.27 shows a routine that waits until the time given in TIME$ has changed by DL seconds. An alternative to this routine appears in Figure 3.34. Incidentally, the routine shown in Figure 3.27 is the only place in the Card Memory game at which the TIME$ feature is used. In a computer system without a clock, this routine's function could be accomplished by using a loop of the form:

FOR XX = 1 TO N*DL: NEXT XX

The value of N can be determined by trial and error.

```
#Some familiar routines

next       GOSUB onech                    #end game, specify next action
           NX$ = X$
           RETURN


onech      repeat {                       #single character input
              X$ = CHR$(RND(1))           #randomize while waiting
              GET X$
           } until (X$ <> '''')
           RETURN


clearscreen  PRINT "clr";                 #clear screen
             RETURN


   830    GOSUB 840:NX$=X$:RETURN
   840    X$=CHR$(RND(1)):GET X$:IF X$="" THEN 840
   850    RETURN
   860    PRINT CHR$(147):RETURN
```

These are the subroutines for accepting a single-character command at the end of the game, and for clearing the screen. The versions shown are for the Pet. Apple and TRS-80 versions are like those shown in Figures 2.11 and 2.12.

Figure 3.26: Old Favorite Routines for Card Memory

Finally, Figure 3.28 shows three initialization routines for Card Memory. The first routine (line 960) is called once when the program is started. The second routine (line 920) is called each time new game parameters are to be set, i.e., initially, and whenever the player types "N". The third routine (line 910) is called at the start of each new sequence display.

#Wait until time changes by DL seconds — Pet version

delay   DX$ = TIME$
               GOSUB convertsec: DZ = DX
               repeat {
                  DX$ = TIME$
                  GOSUB convertsec
               } until (DX − DZ >= DL)
               RETURN

convertsec   DX = 3600*VAL(LEFT$(DX$, 2)) +
                      60*VAL(MID$(DX$, 3, 2)) +
                      VAL(RIGHT$(DX$, 2))
               RETURN

```
870  DX$=TIME$:GOSUB 890:DZ=DX
880  DX$=TIME$:GOSUB 890:IF DX−DZ<DL THEN 880
885  RETURN
890  DX=3600*VAL(LEFT$(DX$,2))+60*VAL(MID$(DX$,3,2))+VAL(RIGHT$(DX$,2))
900  RETURN
```

This routine checks the time on entry, then waits until it has changed by DL seconds before returning.

The Apple and TRS-80 versions must use a "delay loop." The actual BASIC instructions for those systems are:

```
870  FOR DX = 1 TO DL * SS: NEXT DX: RETURN
```

The value of the variable SS is determined empirically and set in the initialization routine (Figure 3.28).

Figure 3.27: Delay Routine for Card Memory

As with other games presented in this book, there are many ways in which Card Memory can be improved. Here are several possibilities:
- Display pictures of the cards instead of their names. On the Pet, this change could be made in Figure 3.19 by replacing the

    GOSUB 310: PRINT CD$

  on line 290 with

    PRINT CP$(CD)

```
#Initialization routines

startgame   RG = 0
            RETURN

gamevalues  repeat
                INPUT "NUMBER OF CARDS"; NC
                until (1 <= NC <= 52)
            INPUT "INTERVAL (SEC)"; DC    #length of delay for test
            SC = 1                        #length of delay for showing answer
            BR = 0                        #new "best for this length"
            RETURN

init        DIM SU$(4), FV$(13), DK(52), DD(52)
            FOR I = 1 TO 4
                READ SU$(I)
                DATA "CLUB", "DIAMOND", "HEART", "SPADE"
                NEXT I
            FOR I = 1 TO 13
                READ FV$(I)
                DATA "ACE", "2", "3", "4", "5", "6", "7",
                     "8", "9", "10", "JACK", "QUEEN", "KING"
                NEXT I
            BC = 0: BD = 0                #initialize "best"
            RETURN
```

Figure 3.28: Initialization for Card Memory

where CP$ is an array of card picture strings (using the Pet's cursor-moving and graphic characters). CP$ would be initialized in the initialization routine at line 960 (in Figure 3.28). Of course, you would have to arrange that the Q and R inputs still resulted in lists of card names, not pictures. On an Apple, you could use the color graphics capabilities. This would require substantial reorganization and new coding.

- Improve the player input routines. For example, you might print a number indicating which card (e.g., first, second) is to be guessed next. You might allow a code that allows the player to say "I've forgotten this one. Show it to me, then let me go on to try the next one."

- Improve the program's ability to keep statistics. Remember the best lengths for each interval. Remember the last five scores. Compute averages.

```
 910  RG=0:RETURN
 920* INPUT "NUMBER OF CARDS";NC:IF NC<1 OR NC>52 THEN 920
 930* INPUT "INTERVAL (SEC)";DC
 940  SC=1
 950  BR=0:RETURN
 960  DIM SU$(4),FV$(13),DK(52),DD(52)
 970  FOR I=1 TO 4:READ SU$(I)
 980  DATA "CLUB","DIAMOND","HEART","SPADE"
 990  NEXT I
1000  FOR I=1 TO 13:READ FV$(I)
1010  DATA "ACE","2","3","4","5","6","7","8","9","10","JACK","QUEEN","KING"
1020  NEXT I
1030  BC=0:BD=0:RETURN
```

*The Apple versions differ only in the values of the string constants. Also, the Apple and TRS-80 versions have an additional line to set the value of SS (see Figure 3.27.) The Apple version is:

965   READ SS: DATA 700

The TRS-80 version is:

965   READ SS: DATA 200

Figure 3.28a: Actual BASIC for Card Memory Initialization

- Synchronize the display of the first card. As the program is now written, the instructions

    DL = DC: GOSUB 870

 on line 250 (in Figure 3.19) can cause a delay of almost a full second less than desired. To alleviate this problem, the instruction

    FOR I = 1 TO NC

 on line 240 can be preceded by a call to a routine that waits for the clock to change.

Ten-Key Flicker

Ten-Key Flicker is one of those frustrating games in which the program gives you a problem and a fixed length of time to give an answer. If you don't give an answer in the allotted time, the program will make a snide remark and cut you off. Here's how it works.

Figure 3.29 shows a portion of the Pet keyboard. The digits 1 through 9 are arranged in a square, providing a convenient way for the player to "point to" a position in a square of nine digits displayed on the screen. The program begins by asking you:

    DELAY INTERVAL?

You respond with a whole number, indicating the number of seconds (from zero on up) you wish to be given in which to answer. (If you can play this game with a delay interval of zero, your reflexes are truly superhuman.) After you have entered a delay interval, the program will display a pattern of nine digits in a square on the screen. Either all nine of the digits will be the same, in which case you must press the "0" key before the delay interval elapses, or one of the nine digits will be different from the other eight, in which case you must "point to" the odd digit by pressing the digit key whose position in the square pattern on the keyboard is the same as the position on the screen of the one odd digit. (See Figure 3.30.)

If you give an answer before the time is up, the program will say THAT'S RIGHT or THAT'S WRONG, as appropriate. If you don't give an answer, the program will say YOU'VE GOT TO BE FASTER. If your answer was incorrect, the program will also tell you the correct answer, and will display the pattern again. This time you can study the pattern for as long as you like. The program will wait for your

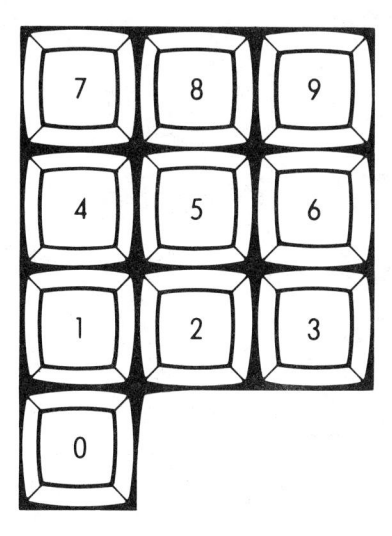

> This is a portion of the "number pad" area of the Pet keyboard. The Ten-Key Flicker game uses the square arrangement of the digits 1 through 9 in this number pad to allow the player to "point to" the one digit that differs from the others in a square arrangement of 9 digits displayed briefly on the screen.

Figure 3.29: Pet Digit Arrangement

single-character input—either "N" to set a new delay time, or any other key to display the next pattern.

Incidentally, if you did try to answer by pressing a key but were just a little late, the program will interpret your answer as a signal to start the next game. A simple program change could correct this bug, but the behavior it causes is in keeping with this program's personality. (You can change this behavior by inserting GET X$ after the GOSUB 200 on line 170 in Figure 3.31.)

Figure 3.31 shows the main routine of Ten-Key Flicker. This routine follows the same general outline seen in many other programs in this book.

100 TIME GAMES

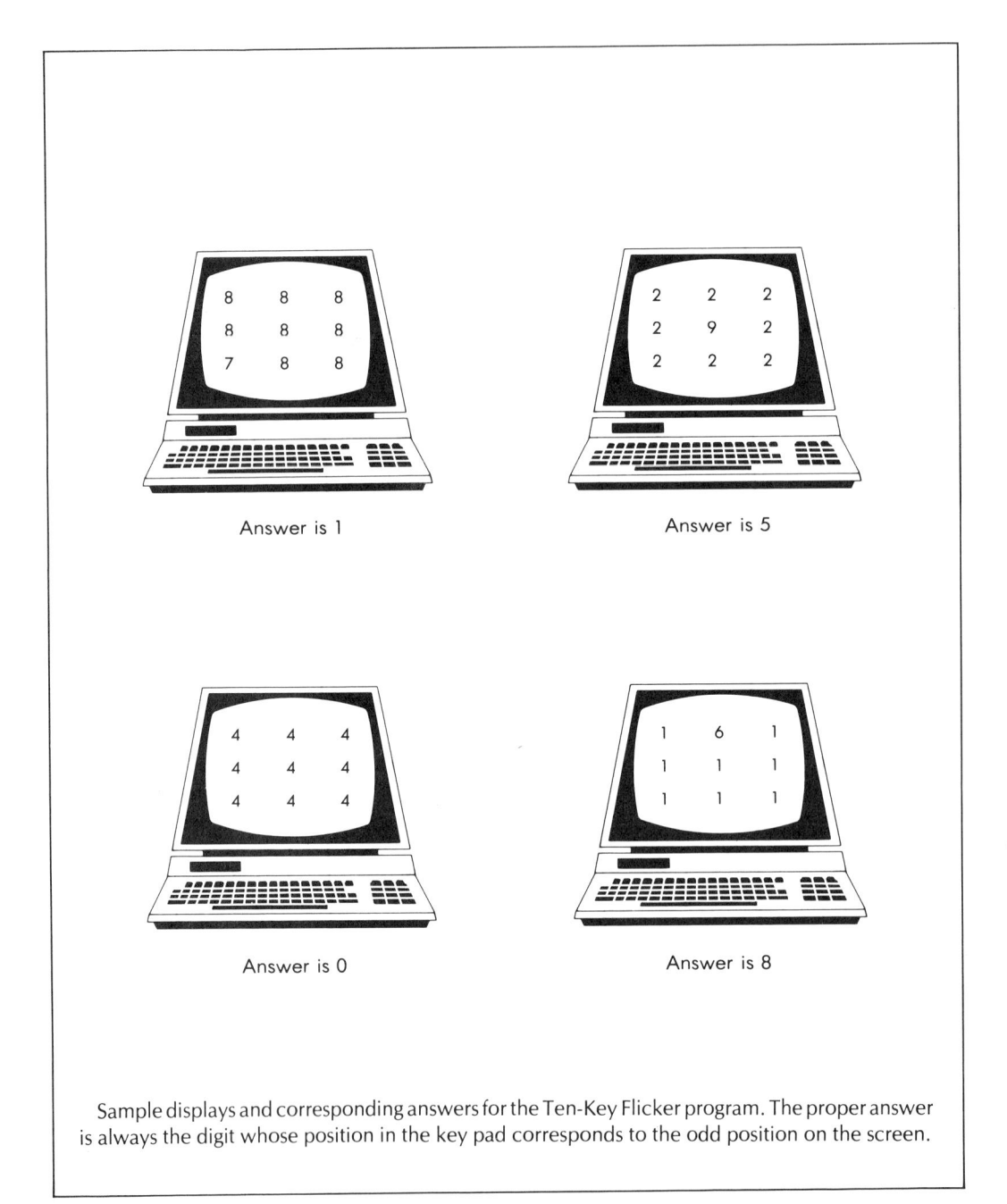

```
    8   8   8              2   2   2
    8   8   8              2   9   2
    7   8   8              2   2   2
```
Answer is 1   Answer is 5

```
    4   4   4              1   6   1
    4   4   4              1   1   1
    4   4   4              1   1   1
```
Answer is 0   Answer is 8

Sample displays and corresponding answers for the Ten-Key Flicker program. The proper answer is always the digit whose position in the key pad corresponds to the odd position on the screen.

Figure 3.30: Samples of Ten-Key Flicker Displays

```
#Ten-Key Flicker

        GOSUB init
        repeat {
            GOSUB setdelay
            repeat {
                GOSUB pattern
                GOSUB show
                GOSUB keyin
                IF XX = 1 THEN
                    PRINT "THAT'S RIGHT!"
                else IF XX = 0 THEN
                    PRINT "THAT'S WRONG! etc."; EX$;
                else                                    #XX = -1
                    PRINT "BE FASTER etc."; EX$;
                DL = 2: GOSUB delay         #wait 2 seconds
                GOSUB show                  #then show pattern again
                GOSUB next                  #wait for player input NX$
            } until (NX$ = "N" or "E")
        } until (NX$ = "E")
        END
```

```
100   GOSUB 430
110   GOSUB 420
120   GOSUB 340:GOSUB 200
130   GOSUB 230:IF XX=1 THEN PRINT "THAT'S RIGHT!":GOTO 160
140   IF XX=-1 THEN PRINT "YOU'VE GOT TO BE FASTER! — IT WAS ";EX$;:GOTO 160
150   PRINT "THAT'S WRONG — THE ANSWER IS ";EX$;
160   DL=2:GOSUB 300
170   GOSUB 200:GOSUB 390:IF NX$="E" THEN END
180   IF NX$="N" THEN 110
190   GOTO 120
```

This is the main routine for the "Ten-Key Flicker" game. It is similar to the main routines of other games in this book.

Figure 3.31: Ten-Key Flicker

102 TIME GAMES

Figure 3.32 shows the pattern display routine, which uses the strings L$ and B$. These strings contain Pet cursor-moving characters. On an Apple, the pattern display could be accomplished by using the HTAB and VTAB instructions.

Figure 3.33 shows the routine that accepts the player's input while watching the clock. This routine uses the variable XX to encode the three possible outcomes:

- 1: Right answer
- 0: Wrong answer
- −1: No answer in time

Figure 3.34 shows a delay routine, which performs the same function as the routine in Figure 3.27. This routine counts ticks; the routine in Figure 3.27 compares times.

#Show the pattern

```
show    PRINT "clr"
        PRINT A(7); L$; A(8); L$; A(9); B$;
              A(4); L$; A(5); L$; A(6); B$;
              A(1); L$; A(2); L$; A(3);
        RETURN

  200*  PRINT CHR$(147)
  210   PRINT A(7);L$;A(8);L$;A(9);B$;A(4);L$;A(5);L$;A(6);B$;A(1);L$;A(2);L$;A(3);
  220   RETURN
```

This routine displays the square pattern of digits on the screen.

*The TRS-80 version uses CLS (see Figure 2.12).

Figure 3.32: Pattern Display for Ten-Key Flicker

```
#Accept the player's input
keyin   DL = DC
        TX$ = TIME$
        repeat {
            GET X$
            IF X$ <> "" THEN {
                IF X$ = EX$ THEN
                    XX = 1
                else
                    XX = 0
                break
            }
            else if DL <= 0 then {
                XX = -1
                break
            }
            else if TX$ <> TIME$ THEN {
                TX$ = TIME$
                DL = DL - 1
            }
        }
        PRINT "clr"
        RETURN
```

```
230   DL=DC:TX$=TIME$
240*  GET X$:IF X$<>"" THEN 280
250   IF DL<=0 THEN XX=-1:GOTO 290
260   IF TX$=TIME$ THEN 240
270   TX$=TIME$:DL=DL-1:GOTO 240
280   XX=0:IF X$=EX$ THEN XX=1
290** PRINT CHR$(147):RETURN
```

This routine accepts a single-character input from the player, if the input occurs before the clock has ticked DC times.

---
*In the TRS-80 version of this line, GET X$ is replaced by X$ = INKEY$. An entirely different approach would be needed in an Apple version, since the Apple GET X$ command waits until X$ <> "".
**In the TRS-80 version, CLS is used (see Figure 2.12).

Figure 3.33: Time-Limited Input for Ten-Key Flicker

104 TIME GAMES

Figure 3.35 shows the routine that creates the pattern of digits that will appear on the screen. This routine selects three digits at random in the range 1 through 9:

— The "main" digit that will appear in eight places.
— The position for the odd digit.
— The odd digit.

If the odd digit is the same as the main digit, then the player's answer must be zero. If the odd digit differs from the main digit (the usual case), then the player's number must be the same as the second randomly selected digit.

Figure 3.36 shows the routines for accepting the next game code, and the two initialization routines. Included in the initialization is a function definition:

$$DEF\ FNR9(X) = INT(RND(1)*9) + 1$$

This is used by the program in Figure 3.35 to select digits in the range 1 through 9.

Delay until the clock ticks DL times

delay   IF DL <> 0 THEN
            FOR I = 1 TO DL
                TX$ = TIME$
                repeat { } until (TX$ <> TIME$)
            NEXT I
        RETURN

```
300  IF DL=0 THEN RETURN
310  FOR I=1 TO DL:TX$=TIME$
320  IF TX$=TIME$ THEN 320
330  NEXT I:RETURN
```

This routine waits for the clock to tick DL times. This is an alternative to the method used in Figure 3.27.

Figure 3.34: Delay Routine for Ten-Key Flicker

Improvements to the Ten-Key Flicker game are possible. For example:

- When displaying the correct answer, show the odd digit in reverse video.

- Add sound to make the program even ruder and more obnoxious.

- Synchronize the start of the display. (See the similar suggestion for Card Memory.)

#Create the pattern

pattern   N1 = FNR9(1)   #choose three whole numbers
        N2 = FNR9(1)   #in the range 1 to 9
        N3 = FNR9(1)
        FOR I = 1 TO 9   #set all 9 numbers to N1
          A(I) = N1
          NEXT I
        A(N2) = N3   #then change the N2-nd to N3
        IF N1 = N3 THEN   #set single-character answer
          EX$ = "0"   #expected from player
        else
          EX$ = RIGHT$(STR$(N2), 1)
        RETURN

```
340*  N1=FNR9(1):N2=FNR9(1):N3=FNR9(1)
350   FOR I=1 TO 9:A(I)=N1:NEXT I:A(N2)=N3
360   IF N1=N3 THEN EX$="0":GOTO 380
370   EX$=RIGHT$(STR$(N2),1)
380   RETURN
```

This routine creates the pattern that the player must recognize. A square of 9 digits will be displayed (3 rows of 3 digits each). If all digits are the same, the player must input a zero. Otherwise, all but one digit will be the same, and the player must input the digit that is in the same position in the keyboard number pad as the one odd digit is in the display.

---

*The argument 1 that appears in the function calls is ignored. (See Figure 3.36.)

Figure 3.35: Pattern Creation for Ten-Key Flicker

106   TIME GAMES

#Accept code for next game

next   GOSUB onech
             NX$ = X$
             RETURN

onech   repeat {
                 X$ = STR$(RND(1))
                 GET X$
             } until (X$ <> '''')

#Ask for delay interval

setdelay INPUT ''DELAY INTERVAL'';DC
             RETURN

#Initialize

init   DIM A(9)
             L$ = ''sixteen cursor right characters''
             B$ = ''one cursor left character followed by ten cursor down characters''
             DEF FNR9(X) = INT(RND(1)*9) + 1   #ignore X
             RETURN

```
390    GOSUB 400:NX$=X$:RETURN
400*   X$=STR$(RND(1)):GET X$:IF X$='''' THEN 400
410    RETURN
420    INPUT ''DELAY INTERVAL'';DC:RETURN
430    DIM A(9)
440**  L$='''':FOR LL=1 TO 16:L$=L$+CHR$(29):NEXTLL
450**  B$=CHR$(157):FOR LL=1 TO 10:B$=B$+CHR$(17):NEXTLL
460*   DEF FNR9(X)=INT(RND(1)*9)+1
470    RETURN
```

---

*In the TRS-80 version of these lines, the term RND(0) appears instead of RND(1), and GET X$ is replaced by X$=INKEY$.
**In the TRS-80 version these lines are:

```
440    L$='''':FOR LL=1 TO 27:L$=L$+CHR$(25):NEXT LL
450    B$=CHR$(24):FOR LL=1 TO 6:B$=B$+CHR$(26):NEXTLL
```

Figure 3.36: Odds and Ends for Ten-Key Flicker

Timer

This is not a timer game, it's a game timer. For example, you might use it to time chess moves.

The program (shown in Figure 3.37) displays a number in the upper left corner of the screen. Initially this number is zero. As soon as you press the space bar, the number will begin to increase at the rate of one number per second, i.e., the program displays the number of seconds that have elapsed since you hit the space bar. If you press the space bar again, the number will stop increasing. If you press it once again, the program will resume counting where it left off. At any time when the number is not increasing, you can press R to cause it to be reset to zero.

There are several ways to improve upon this program:

- Make the program display in the format HH:MM:SS.

- Give it two timers in opposite corners of the screen. Pressing the space bar will start the first timer. Thereafter, pressing the space bar will stop whichever timer is running, and start the other.

- Allow a time limit to be set, and display an alarm or signal when this limit has been reached.

- Provide a version of Timer that uses the Pet's TI feature.

- Integrate a timer facility into the Clock program developed earlier in this chapter.

```
100    TX=0:GOSUB 190
110*   GET X$:IF X$="" THEN 110
120    IF X$="R" THEN TX=0:GOSUB 190:GOTO 110
130    IF X$=" " THEN GOSUB 150:GOTO 110
140    GOTO 110
150    TX$=TIME$
160*   GET X$:IF X$<>"" THEN RETURN
170    IF TX$=TIME$ THEN 160
180    TX=TX+1:GOSUB 190:GOTO 150
190*   IF TX=0 THEN PRINT CHR$(147):GOTO 200
195*   PRINT CHR$(19)
200    PRINT TX:RETURN
```

*The TRS-80 version uses X$=INKEY$ rather than GET X$. The PRINT "clr" is replaced by CLS, and the "home" character is represented by CHR$(28). An Apple version would require an entirely different approach because the Apple GET X$ command waits until X$ <> "".

Figure 3.37a: *Actual BASIC for Timer*

108 TIME GAMES

```
#Timer
            TX = 0                              #start counting at zero
            GOSUB showcount
            repeat {
                GET X$                          #wait for command
                IF X$ = "R" THEN {              #R resets count to zero
                    TX = 0
                    GOSUB showcount
                }
                else IF X$ = "space" THEN      #space continues count
                    GOSUB count
            }
count       repeat {
                TX$ = TIME$
                repeat {
                    GET X$
                    IF X$ <> "" THEN            #anything stops the count
                        RETURN
                } until (TX$ <> TIME$)
                TX = TX + 1                     #increment once per second
                GOSUB showcount
            }
showcount   IF TX = 0 THEN                      #if starting fresh,
                PRINT "clr"                     # clear the screen
            else
                PRINT "home"
            PRINT TX
            RETURN
```

This is the program that implements Timer.

Figure 3.37b: Free BASIC for Timer

Summary

The purpose of this chapter was to provide examples of the techniques used to manage the time element in games. We began by describing the clock facility of the Pet. We then examined four games that illustrate several different ways that the time element can enter into games.

Clock illustrates the techniques used to maintain a time display simultaneously with other activities. Card Memory and Ten-Key Flicker illustrate the use of the clock facility to provide measured action at varying speeds. Timer provides a model for a "stopwatch" style of time measurement.

CHAPTER 4
Date Games

> *Days of the Week*
> *Monday's child is fair of face,*
> *Tuesday's child is full of grace,*
> *Wednesday's child is full of woe*
> *Thursday's child has far to go,*
> *Friday's child is loving and giving,*
> *Saturday's child works hard for its living.*
> *And a child that's born on the Sabbath day*
> *Is fair and wise and good and gay.*
> —Author unknown; quoted in Bray's Traditions of Devon

Every programmer should be familiar with the techniques used for the input and output of dates. The games in this chapter, Birthday and Calendar, are designed to illustrate these techniques.

Birthday

This game is designed to teach you:

— How to determine the day of the week on which a given date falls. (This ability is useful in many commercial settings as well as in games.)

— How to design a convenient input format for the date.

The program begins by asking for your date of birth (see Figure 4.1). You reply by entering a date in the form "month, day, year," where month, day and year are numbers separated by slashes (or spaces or hyphens). If the year that you enter is two digits in length, then the program will automatically add 1900 to it. In the example in Figure 4.1, the player has specified the year 1941 by typing "41". (If you had wished to specify the year 41 A.D., you would have had to enter "041".) The program replies to the specification of a birth date by saying "YOU WERE BORN ON", followed by the day of the week and date in the form shown in Figure 4.1. It then follows with a gratuitous remark that varies with the day of the week (for example, "MONDAY'S CHILD IS FAIR OF FACE").

Finally, the program waits for the player to press any key on the keyboard. This will cause the screen to be cleared and a new turn to begin.

The Birthday Program

The Birthday program is shown in Figures 4.2 to 4.11. Many of the subroutines shown will be used again in other date games.

The main routine, shown in Figure 4.2, is quite simple, and we do not need to comment on it. Its structure is similar to that of other main routines in this book.

The date input routine shown in Figure 4.3 is designed for general use. It accepts as an argument the prompting string that it will use when asking for date input.

The date string examination routine is shown in Figure 4.4. First, the routine sets up the length of the string in the variable L. This length is used by the routine called to break the string into three fields. Next the routine verifies that a valid date has been entered. If a valid date has been entered, then it sets XX to 1 and returns the month, day and year (expanded by adding 1900, if necessary) in the variables MO, DA and YR. If the date entered is invalid, the routine returns XX = 0.

The routine shown in Figure 4.5 breaks the date string into three fields separated by slashes, hyphens or spaces.

The routine that actually determines the day of the week is shown in Figure 4.6. Despite the fact that it has special code for dealing with dates before the date change in 1582, this routine actually gives incorrect results for dates before 1600. As an exercise, you might try to find and repair this "bug," which affects this routine and the routine in Figure 4.9. The routine in Figure 4.6 is simpler than it would have been if a base date of January 1, 2000 had been used instead of January 1, 2001. Try to figure out why this is so.

```
BIRTHDATE: 8/31/41
YOU WERE BORN ON
SUNDAY, AUGUST 31, 1941
HOW NICE!

YOU'RE FAIR, AND WISE, AND GOOD, AND GAY
```

This is the screen display for one turn at Birthday. The program is waiting for the player to press a key, which will clear the display and begin a new turn.

Figure 4.1: Screen Display for Birthday

INSIDE BASIC GAMES   113

```
#Birthday
        GOSUB init
        repeat {
            GOSUB clearscreen           #start on clear screen
            DP$ = "BIRTHDATE: "         #argument for datein
            GOSUB datein                #ask for and input date
            GOSUB day                   #convert date to day of week
            GOSUB daypatter             #recite poem, etc.
            GOSUB onech                 #wait for key
        }

100  GOSUB610
110  GOSUB510:DP$="BIRTHDATE: ":GOSUB120:GOSUB300:GOSUB400:GOSUB520:GOTO110
```

This is the main routine for Birthday.

Figure 4.2:   Birthday

```
#Date input — gets DA,MO,YR
datein    repeat{
              PRINT DP$;                #prompt for date
              GOSUB stringin            #get answer up to RETURN
              GOSUB parsedate           #convert to DA,MO,YR
          } until (XX = 1)
          RETURN

120  PRINT DP$;:GOSUB540:GOSUB140:IF XX<>1 THEN120
130  RETURN
```

This is the date input routine. It accepts the prompting string DP$ as an argument, so that it can be called to accept dates for other purposes. In Birthday, DP$ is set to "BIRTHDATE: " before this routine is called.

Figure 4.3:   Date Input

114 DATE GAMES

```
#Transform input string XX$ into MO,DA,YR
parsedate  L = LEN(XX$)
           IF L >= 5 THEN {                      #don't bother if too short
              GOSUB fields                       #break XX$ into MO$, DA$, YR$
              MO = INT(VAL(MO$))
              DA = INT(VAL(DA$))
              YR = INT(VAL(YR$))
              IF LEN(YR$) = 2 THEN               #2-digit year YR means 19YR
                 YR = YR + 1900
              GOSUB leapyear                     #if year is a leapyear
              IF XX = 1 THEN
                 ML(2) = 29                      #then Feb hath 29
              else
                 ML(2) = 28                      #otherwise 28
              IF 1 <= MO <= 12 AND YR >= 1 THEN  #check month and year
                 IF 1 <= DA <= ML(MO) THEN       #check day
                    IF NOT (YR = 1582 AND        #make sure Gregory didn't drop it
                    MO = 10 AND 4 < DA < 15) THEN {
                       XX = 1
                       RETURN
                    }
           }
           XX = 0
           RETURN

140   L=LEN(XX$):IF L<5 THEN210
150   GOSUB220:MO=INT(VAL(MO$)):DA=INT(VAL(DA$)):YR=INT(VAL(YR$))
160   IF LEN(YR$)=2 THEN YR=YR+1900
170   GOSUB480:ML(2)=28:IF XX=1 THEN ML(2)=29
180   IF NOT (1<=MO AND MO<=12 AND YR>=1) THEN210
190   IF NOT (1<=DA AND DA<=ML(MO)) THEN210
200   IF NOT (YR=1582 AND MO=10 AND 4<DA AND DA<15) THEN XX=1:RETURN
210   XX=0:RETURN
```

  This routine searches through the string XX$ for three fields that make up the month, day and year of a valid date after January 1, 1 A.D. This task is more difficult than it appears, because the dates October 5, 1582 through October 14, 1582 do not exist.

Figure 4.4:   Date String Examination

```
#Get MO$,DA$,YR$ fields from XX$ — MO$ = "" signals bad format

fields  XS = 0                                    #number of separators seen
        FOR XL = 2 TO L - 1                       #can't start or end with a separator
            IF MID$(XX$,XL,1) =                   #if character is a separator,
            "/" or " " or "-" THEN {
                XS = XS + 1                       # then count it
                IF XS = 1 or 2 THEN
                    SP(XS) = XL                   # and remember where it was
                else {
                    MO$ = ""
                    RETURN
                }
            }
        NEXT XL
        IF XS <> 2 THEN
            MO$ = ""
        else {                                    #use the two separators to derive
            MO$ = LEFT$(XX$, SP(1) - 1)           # three fields
            DA$ = MID$(XX$, SP(1) + 1, SP(2) - SP(1) - 1)
            YR$ = MID$(XX$, SP(2) + 1)
        }
        RETURN

220  XS=0:FOR XL=2 TO L-1:ZZ$=MID$(XX$,XL,1)
230  IF ZZ$<>"/" AND ZZ$<>" " AND ZZ$<>"-" THEN260
240  XS=XS+1:IF XS=1 OR XS=2 THEN SP(XS)=XL:GOTO260
250  MO$="":RETURN
260  NEXT XL:IF XS<>2 THEN MO$="":GOTO290
270  MO$=LEFT$(XX$,SP(1)-1):DA$=MID$(XX$,SP(1)+1,SP(2)-SP(1)-1)
280  YR$=MID$(XX$,SP(2)+1)
290  RETURN
```

This routine breaks the string XX$ into three fields, separated by slashes, hyphens or spaces. The fields are placed into three string variables: MO$,DA$,YR$. No validity checking is done, except for the existence of three fields in the given format. The routine returns MO$ = "" if this check fails.

Figure 4.5: Breaking Date Strings into Fields

DATE GAMES

```
#Convert MO,DA,YR to DW (0 = Sunday, ..., 6 = Saturday)
#Gregorian version — assumes a calendar change in 1582
day     YZ = YR mod 400              #400 years = exactly 20,871 weeks
        DW = DZ                      #Jan 1, 2001
        while (YZ >= 101) {          #add 5 days per century
            YZ = YZ - 100
            DW = DW + 5
        }
        while (YZ >= 5) {            #add 5 days per 4 years
            YZ = YZ - 4
            DW = DW + 5
        }
        IF YZ = 0 THEN
            DW = DW - 2              #2 days for A.D. 2000 (e.g.)
        else
            DW = DW + (YZ - 1)       #1 day each for 2002-2004
        GOSUB julian                 #JD = day number within year
        DW = DW + JD - 1             #add one day per day
        IF (MO,DA,YR) <= (10,4,1582) THEN
            DW = DW + 3              #add 3 for missing 11 days
        DW = DW (mod 7)
        RETURN
300   YZ=YR-400*INT(YR/400):DW=DZ
310   IF YZ<101 THEN330
320   YZ=YZ-100:DW=DW+5:GOTO310
330   IF YZ<5 THEN350
340   YZ=YZ-4:DW=DW+5:GOTO330
350   IF YZ=0 THEN DW=DW-2:GOTO356
352   DW=DW+YZ-1
356   GOSUB450:DW=DW+JD-1:IF YR>1582 THEN390
360   IF YR=1582 AND MO>10 THEN390
370   IF YR=1582 AND MO=10 AND DA>=15 THEN390
380   DW=DW+3
390   DW=DW-7*INT(DW/7):RETURN
```

This routine takes any date on or after January 1, 1 A.D. and determines the day of the week upon which it falls. This routine takes account of the calendar change decreed by Pope Gregory XIII, which caused October 4, 1582 to be followed by October 15, 1582.

Figure 4.6: Determine Day of Week

The routine in Figure 4.7 provides the game element. It is extremely simple, consisting entirely of PRINT instructions.

The routine in Figure 4.8 shows the computation of the "Julian" date. This is the number that provides a sequence number from 1 to 365 (or 366 on leap years) for each day of the year. The routine uses the array J, which is set up in the initialization routine in Figure 4.11. J contains one entry for each month. The value of that entry is the total number of days from January 1 up to the first of the given month (assuming a non-leap year). For example, J(1) = 0, since there are no days preceding January. J(2) = 31, J(3) = 59 (that is, 31 + 28), and so on.

The routine in Figure 4.9 determines whether or not a given year is a leap year, and returns XX = 1 if the year is a leap year, XX = 0 if it is not. The routine contains a "bug" that arises from "rewriting history" incorrectly. Try to find and correct this error. A corresponding change will need to be made to the routine in Figure 4.6. (Hint: these routines treat 1500 like 1900.)

#Say something nice about the birthdate

daypatter　　PRINT: PRINT "YOU WERE BORN ON"
　　　　　　　PRINT DW$(DW);", "; MO$(MO); DA; ","; YR
　　　　　　　PRINT "HOW NICE!"
　　　　　　　PRINT
　　　　　　　PRINT PM$(DS)
　　　　　　　RETURN

```
400   PRINT:PRINT "YOU WERE BORN ON"
410*  PRINT DW$(DW);", ";MO$(MO);STR$(DA);",";YR
420   PRINT "HOW NICE!"
430   PRINT
440   PRINT PM$(DW):RETURN
```

This is the routine that makes a remark about the player's date of birth. The array PM$ of remarks specific to the given day is initialized by DATA statements in the initialization routine (Figure 4.11).

---
*The Apple version of this line differs only in the spacing in the string constant preceding the variable YR.

Figure 4.7: Gratuitous Remark about Day of Birth

118 DATE GAMES

#Day number within year: convert (MO,DA,YR) into JD

julian   JD = J(MO) + DA
          IF MO > 2 THEN {
              GOSUB leapyear
              IF XX = 1 THEN
                  JD = JD + 1
          }
          RETURN

    450   JD=J(MO)+DA:IF MO<=2 THEN RETURN
    460   GOSUB480:IF XX=1 THEN JD=JD+1
    470   RETURN

This routine determines the "Julian" date within a year. The variable JD takes the value 1 through 365 to 366, starting with 1 for January 1, and ending with 365 or 366 for December 31.

Figure 4.8:   Julian Date

#Is YR a leap year? — XX = 1 if so.

leapyear IF 4 divides YR AND   #Every fourth year is
           (100 does not divide YR OR   #a leap year,
           400 divides YR) THEN   #except centenary years
              XX = 1   #that aren't divisible
           else   #by 400.
              XX = 0
           RETURN

    480   XX=0:IF YR<>4*INT(YR/4)THEN RETURN
    490   IF YR=100*INT(YR/100)AND YR<>400*INT(YR/400)THEN RETURN
    500   XX=1:RETURN

This routine determines whether or not the year specified in YR is a leap year. It returns XX = 1 if a leap year, XX = 0 if not. Leap years occur every fourth year, except centenary years that are not divisible by 400 (e.g., 1700, 1800, 1900 are not leap years.) Try to find the "bug" in this routine.

Figure 4.9:   Leap Year Checking

Figure 4.10 shows several utility routines that have appeared in similar or identical versions in other programs in this book. The string input routine shown here implements the BASIC command

LINE INPUT XX$

Few home computer systems have LINE INPUT commands. Figure 4.11 shows the initialization for the routines used in Birthday.

```
#String input
stringin      XX$ = ""                              #start with null string
              repeat {
                  GOSUB onech                       #get next character
                  IF X$ <> "delete" THEN {          #if not a rubout,
                      PRINT X$;                     #echo it
                      IF X$ <> "return" THEN        #if not a return,
                          XX$ = XX$ + X$            #add to string
                      else break                    #return terminates
                  }
                  else IF LEN(XX$) <> 0 THEN {      #if rubout,
                      PRINT "delete string";        #erase from screen
                      XX$ = LEFT$(XX$,LEN(XX$) - 1) #and pull from string
                  }
              }
              RETURN
     540  XX$=""
     550* GOSUB520:IF X$=CHR$(20) THEN580
     560  PRINT X$;:IF X$=CHR$(13) THEN600
     570  XX$=XX$+X$:GOTO550
     580  IF LEN(XX$)=0 THEN550
     590**PRINT X$;:XX$=LEFT$(XX$,LEN(XX$)−1):GOTO550
     600  RETURN
```

*In the Apple and the TRS-80 versions of this line, the delete character is represented by CHR$(8).
**In the Apple version, this line is replaced by two lines:

```
     590  PRINT X$;" ";X$;:IF LEN(XX$)=1 THEN XX$ = "":GOTO550
     595  XX$ = LEFT$(XX$,LEN(XX$)−1): GOTO550
```

Apple BASIC will not accept a zero length argument in the LEFT$ function, and the Apple "back arrow" character does not erase the previous character from the screen.

Figure 4.10a: *Utility Routines for Birthday*

120 DATE GAMES

#Clear the screen

clearscreen PRINT "clr";
RETURN

#Single character input

onech repeat
GET X$
until (X$ <> '''')
RETURN

```
510 PRINTCHR$(147);:RETURN
520 GET X$:IF X$="" THEN520
530 RETURN
```

These utility routines are similar to others in this book. (See Figures 2.11 and 2.12.)

Figure 4.10b: Utility Routines for Birthday

#Initialization for Birthday

```
610 READ DZ:DATA 1
620 DIM DW$(6),J(12),MO$(12),ML(12),PM$(6)
630 FOR XX=0 TO 6:READ DW$(XX),PM$(XX):NEXT XX
640 DATA SUNDAY,"YOU'RE FAIR, AND WISE, AND GOOD, AND GAY"
650 DATA MONDAY,"MONDAY'S CHILD IS FAIR OF FACE"
660 DATA TUESDAY,"TUESDAY'S CHILD IS FULL OF GRACE"
670 DATA WEDNESDAY,"WEDNESDAY'S CHILD IS FULL OF WOE"
680 DATA THURSDAY,"THURSDAY'S CHILD HAS FAR TO GO"
690 DATA FRIDAY,"FRIDAY'S CHILD IS LOVING AND GIVING"
700 DATA SATURDAY,"SATURDAY'S CHILD WORKS HARD FOR ITS LIVING"
710 FOR XX=1 TO 12:READ MO$(XX):NEXT XX
720 DATA JANUARY,FEBRUARY,MARCH,APRIL,MAY,JUNE,JULY,AUGUST,SEPTEMBER
730 DATA OCTOBER,NOVEMBER,DECEMBER
740 DT=0:FOR XX=1 TO 12:J(XX)=DT:READ ML(XX):DT=DT+ML(XX):NEXT XX
745 IF DT<>365 THEN STOP
750 DATA 31,28,31,30,31,30,31,31,30,31,30,31
760 RETURN
```

This is the initialization for Birthday. The BASIC and Free BASIC versions are nearly identical, so only the BASIC version is shown.

Figure 4.11: Initialization for Birthday

Calendar

This game is designed to show you how the routines developed for Birthday can be used in other programs. The Calendar game is derived from Birthday by the process of "cannibalization" that we described in Chapter 2.

The program begins by asking you to enter a month and year in the same numeric form as that used for Birthday. The program responds by displaying a calendar for the specified month (see Figure 4.12). The calendar remains displayed until any key is pressed. The screen then clears, and the program requests another month and year.

The Calendar Program

The Calendar program was derived from the Birthday program by the addition of the programs in Figures 4.13 through 4.17 and the deletion of the main program (Figure 4.2), the date input routine (Figure 4.3), and the gratuitous remark routine (Figure 4.7). The Birthday program has otherwise been left largely intact. Although almost all of the initialization code for Birthday (Figure 4.11) is irrelevant to Calendar, that code has been left intact, and a separate, additional initialization routine has been provided for Calendar.

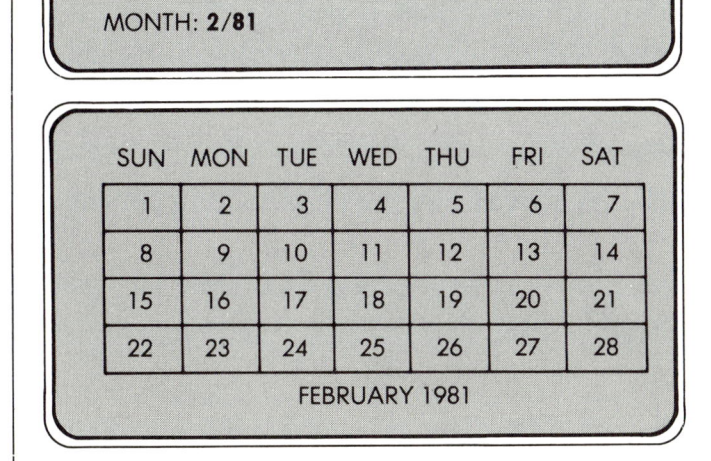

MONTH: 2/81

SUN	MON	TUE	WED	THU	FRI	SAT
1	2	3	4	5	6	7
8	9	10	11	12	13	14
15	16	17	18	19	20	21
22	23	24	25	26	27	28

FEBRUARY 1981

This is the screen display for the Calendar game. The player specifies a month and year, and the program displays a calendar for that month.

Figure 4.12: Display for Calendar

DATE GAMES

The main routine for Calendar (shown in Figure 4.13) exhibits several "contortions" that are necessary because the Birthday routines have been used for purposes slightly different from those for which they were intended. For example, after the program has obtained a month and year in the format "2/81" (meaning February

```
#Calendar

    GOSUB init                              #"Birthday" initialization
    GOSUB calinit                           #"Calendar" initialization
    repeat {
        GOSUB clearscreen                   #start on clear screen
        repeat {
            PRINT "MONTH: ";                #get month & year ("MM/YY")
            GOSUB stringin
            XX$ = "1/" + XX$                #replace by "1/MM/YY"
            L = LEN(XX$)
            GOSUB fields
            XX$ = DA$ + " " + MO$ + " " + YR$ #make into "MM 1 YY"
            GOSUB parsedate                 #check validity
        } until (XX = 1)                    #XX = 1 if valid
        GOSUB day: SZ = DW                  #find starting point
        NL = 1                              #first line
        DL = ML(MO) - (7 - SZ)              # contains 7 - SZ days
        while (DL > 0) {
            DL = DL - 7: NL = NL + 1        #each additional line takes 7,
        }                                   # until the last
        GOSUB clearscreen: GOSUB frame      #put up NL lines of boxes
        ND = ML(MO): GOSUB daydisp          #print 1 to ND
        GOSUB onech                         #leave on screen until key pressed
    }
    END
```

This is the main routine for Calendar. It makes use of several subroutines taken from Birthday.

Figure 4.13: Calendar

1981), it first converts the date to "1/2/81" (which would ordinarily mean January 2, 1981). The program then calls the routine in Figure 4.5, which breaks the date into three fields: MO$ = "1", DA$ = "2", YR$ = "81". These fields are rearranged into a new string in the correct format: "2 1 81", meaning February 1, 1981. This string is then passed to the date string examination routine in Figure 4.4, just as if the player had typed "2/1/81" to begin with.

The routines taken from Birthday can be used from this point on. The routine in Figure 4.6 is called to determine the day of the week upon which February 1, 1981 falls. This information allows the calendar display to be constructed in a simple manner, using calls to the routines in Figure 4.14 and 4.15. The first of these routines displays the calendar frame (the day names, the boxes, and the month and year). The second routine fills in the day numbers.

Cursor positioning makes possible the use of separate functions to display the frame and to fill in the numbers. Both of the routines in Figures 4.14 and 4.15 make calls to a routine that positions the cursor to a line and column specified in the variables LL and CC. (The "cursor" is the name given to the position at which the next character display will occur. Its location is sometimes indicated by a flashing solid square on the display screen. This square itself is also called the cursor.) The routine in Figure 4.15 uses cursor positioning instructions to move the cursor successively to each empty calendar box destined to receive a number. The two-character right-adjusted number string is printed, using the PRINT instruction with the semicolon terminator. This assures that nothing already on the screen is disturbed by the printing of the numbers.

```
100   GOSUB610:GOSUB900
105   GOSUB510
110   PRINT "MONTH: ";:GOSUB540:XX$="1/"+XX$:L=LEN(XX$):GOSUB220
112   XX$=DA$+" "+MO$+" "+YR$:GOSUB140:IF XX<>1 THEN110
115   GOSUB300:SZ=DW:NL=1:DL=ML(MO)+SZ-7
120   IF DL>0 THEN DL=DL-7:NL=NL+1:GOTO120
125   GOSUB510:GOSUB130:ND=ML(MO):GOSUB400
128   GOSUB520:GOTO105
```

Figure 4.13a: Actual BASIC for Calendar

124 DATE GAMES

The separate display of the box and the numbers (made possible by cursor positioning), leads to clear, simple programming. The alternative would be to combine the functions of Figures 4.14 and 4.15 into one routine that builds lines of boxes with numbers in them, then prints them a line at a time. This in turn would mean that the construction of the string VL$ (done in advance in the initialization routine) would have to be integrated with this function. That is, three simple tasks that are presented clearly in the current version of the program would be replaced by one involuted program that is difficult to understand. This example helps to illustrate the usefulness of cursor positioning in the construction of display programs.

#Frame display

frame	PRINT HD$: PRINT TL$	#day names and top line
	FOR LX = 1 TO NL	
	FOR VV = 1 TO VS	
	PRINT VL$	#box sides
	NEXT VV	
	PRINT BL$	#bottom line
	NEXT LX	
	CC = TC: LL = LL(NL) + BH	#position cursor
	GOSUB cursor	
	PRINT MO$(MO); YR	#print month and year under frame
	RETURN	

```
130    PRINT HD$:PRINT TL$:FOR LX=1 TO NL:FOR VV=1 TO VS:PRINT VL$:NEXT VV:
       PRINT BL$:NEXT LX
135*   CC=TC:LL=LL(NL)+BH:GOSUB800:PRINT MO$(MO);YR;:RETURN
```

This routine displays the empty calendar box.

---

*The Apple version of this line differs only in the appearance of the string constant " " between the MO$(MO) and YR variables.

Figure 4.14: Frame for Calendar

Figure 4.16 shows the routine used for positioning the cursor to line LL, column CC. On a Pet this is accomplished by repeated printing of "cursor right" and "cursor down" characters, after an initial printing of the "home" character. This function is implemented on an Apple using HTAB and VTAB instructions. The TRS-80 version is similar to the Pet version, but with different cursor-moving characters and screen dimensions.

```
#Display the days 1 to ND starting with the SZ-th box

daydisp    DY = 1
           FOR LX = 1 TO NL
               LL = LL(LX)                              #line for LX-th row
               IF LX = 1 THEN
                   ZC = SZ + 1                          #ZC = first position
               else
                   ZC = 1
               FOR CX = ZC TO 7
                   CC = CC(CX)                          #column for this number
                   GOSUB cursor
                   PRINT RIGHT$(" " + STR$(DY),2);      #2 characters, right adjusted
                   DY = DY + 1                          #count days printed
                   IF DY > ND THEN
                       RETURN
               NEXT CX
           NEXT LX
           RETURN

   400  DY=1:FOR LX=1 TO NL:LL=LL(LX):ZC=1:IF LX=1 THEN ZC=SZ+1
   405  FOR CX=ZC TO 7:CC=CC(CX):GOSUB800:PRINT RIGHT$(" "+STR$(DY),2);
   410  DY=DY+1:IF DY>ND THEN RETURN
   415  NEXT CX:NEXT LX:RETURN
```

This is the routine that fills in the calendar boxes with day numbers.

Figure 4.15: Display Day Numbers

126 DATE GAMES

Figure 4.17 shows the special initialization used for Calendar. This routine is called in addition to the initialization routine taken over from Birthday.

#Cursor positioning — move to Line LL, Column CC
#Pet version

cursor   PRINT "home";   #move to line 0, column 0
        CC = CC mod columns   #adjust arguments to be
        LL = LL mod lines   #within range of screen size
        IF CC > 0 THEN   #move right by CC
            FOR ZZ = 1 TO CC
              PRINT "right";
            NEXT ZZ
        IF LL > 0 THEN   #move down by LL
            FOR ZZ = 1 to LL
              PRINT "down";
            NEXT ZZ
        RETURN

```
800* PRINT CHR$(19);:CC=CC−40*INT(CC/40):LL=LL−24*INT(LL/24)
805* IF CC>0 THEN FOR ZZ=1 TO CC:PRINT CHR$(29);:NEXT ZZ
810* IF LL>0 THEN FOR ZZ=1 TO LL:PRINT CHR$(17);:NEXT ZZ
815  RETURN
```

This is the cursor-positioning routine for Calendar. The Apple version is simpler because cursor-positioning commands already exist in that system.

---

*For the Apple version, lines 805 and 810 are:

```
805  HTAB CC+1
810  VTAB LL+1
```

For the TRS-80 version, the *columns* and *lines* values are 64 and 16, not 40 and 24. The *home*, *right* and *down* characters are represented by CHR$(28), CHR$(25), and CHR$(26).

Figure 4.16: Cursor Positioning for Calendar

#Special initialization for Calendar

calinit   READ BW,BH: DATA boxwidth,boxheight #box width and height
        READ LZ,CZ: DATA boxline,boxcol   #upper left corner of box
        FOR XX = 1 TO 6
            LL(XX) = LZ + (XX − 1)*BH   #set lines for numbers
            NEXT XX
        FOR XX = 1 TO 7
            CC(XX) = CZ + (XX − 1)*BW   #set columns for numbers
            NEXT XX
        TL$ = "": BL$ = "": VL$ = ""   #top, bottom and mid lines
        FOR XX = 1 TO 7
            TL$ = TL$ + "four top line characters"
            BL$ = BL$ + "one lowerleft corner and three bottom line characters"
            VL$ = VL$ + "one vertical line character and three blanks"
            NEXT XX
        VL$ = VL$ + "one vertical line character"
        BL$ = BL$ + "one vertical line character"
        HD$ = " SUN MON TUE WED THU FRI SAT"
        VS = BH − 1   #number of VL$ lines per row
        TC = CZ + 5   #column for caption
        RETURN

```
 900*  READ BW,BH:DATA4,3
 905   READ LZ,CZ:DATA3,1
 910   FOR XX=1 TO 6:LL(XX)=LZ+(XX−1)*BH:NEXT XX
 915   FOR XX=1 TO 7:CC(XX)=CZ+(XX−1)*BW:NEXT XX
 920   TL$="":BL$="":VL$=""
 925** FOR XX=1 TO 7:TL$=TL$+"_____":VL$=VL$+"|   ":BL$=BL$+"|_____":
       NEXT XX
 930** VL$=VL$+"|  ":BL$=BL$+"| "
 935   HD$=" SUN MON TUE WED THU FRI SAT"
 940   VS=BH−1:TC=CZ+7:RETURN
```

This is the special initialization for Calendar. This routine is called in addition to the initialization routine for Birthday.

*In the TRS-80 version the two values in line 900 are 4 and 2.
**The Apple and TRS-80 versions of these lines differ in the use of hyphens and exclamation marks for the vertical and horizontal lines of the calendar boxes.

Figure 4.17: Initialization for Calendar

Summary

Two games, Birthday and Calendar, have been designed to illustrate the important techniques used for the input and output of dates.

Birthday illustrates the algorithm used to determine the day of the week upon which a given date falls. It also provides an example of a date input routine with enough flexibility to make it easy for anyone to use.

Calendar uses the routines that make up the Birthday program to display a calendar for a given month and year. The program illustrates the kinds of contortion that are sometimes necessary to adapt an existing routine to a new situation. It also illustrates the structural simplification that can result from the use of cursor positioning instructions to achieve separation of logically independent display functions.

CHAPTER 5
Taxman

> *Let me tell you how it will be,*
> *There's one for you, nineteen for me...*
> *Should five percent appear too small,*
> *Be thankful I don't take it all,*
> *'Cause I'm the Taxman*
>
> —Song by George Harrison

The game presented in this chapter has been around for a long time, but the version described here probably makes much better use of the facilities of your home computer system than the older teletype-based versions. Taxman is a different type of game from the other games presented in this book. One of the objects of this game is to figure out what the rules are.

Instructions for Taxman

The game begins with a question by the program asking you to specify a "pie" size. This is an allusion to the diagrams you may have seen in which a "silver" dollar has been sliced up like a pie—so much goes for rent, so much for food, so much for taxes, etc. You respond by entering a whole number.

The pie displayed on the screen will not be circular. It will be a rectangular array of numbers. In fact, if you answered the request for a pie size by entering the whole number N, then the pie will consist of the whole numbers 1,2,...,N. The smallest value of N that is allowed is 4. The largest value depends upon the size of your screen, and is computed from values in DATA statements in the initialization routine (see Figure 5.12).

Figure 5.1 shows a sample turn at the game. You have selected a pie size of 15; the numbers 1,2,3,...,15 appear first on the screen. All of the numbers except the number 1 are in "reverse" video. You are not told the meaning of this distinction, but most players discover the meaning quickly enough.

Below the display in Figure 5.1 appear the current totals:

    YOU: 0%   TAXMAN: 0%   LEFT: 100%

Below that appears the question:

    YOUR SLICE?

This question prompts you to enter a number between 1 and 15. (Again, you are never told this, but most players learn quickly not to enter numbers that are not on the screen.)

In Figure 5.1 you have entered the number 12. The screen clears, and new screen information appears. Several of the numbers that made up the original pie are no longer shown, and others no longer appear in reverse video. The new totals now appear below the pie display:

YOU: 10%   TAXMAN: 13.3%   LEFT: 76.7%

It is your task to figure out how these numbers are related to the slice of the pie that you have chosen, to the missing numbers, and to the numbers that are no longer in reverse video. (Hint: The sum of the numbers 1,2,...,15 is 120.)

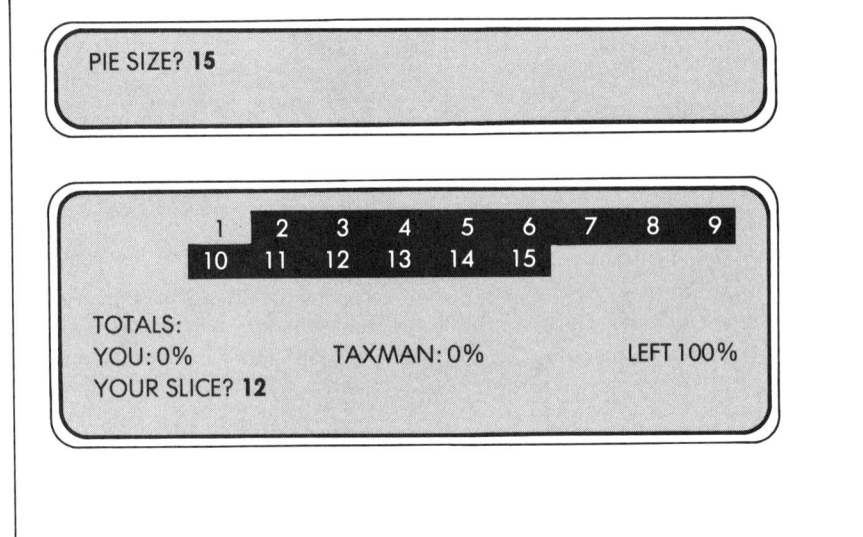

```
PIE SIZE? 15


   1  2  3  4  5  6  7  8  9
  10 11 12 13 14 15

TOTALS:
YOU: 0%        TAXMAN: 0%              LEFT 100%
YOUR SLICE? 12
```

In the first display, the player selects the pie size of 15. In the next display, the whole pie is shown, and the player selects slice 12. The third display reveals that the player's choice of 12 resulted in 10% of the pie for the player and 13.3% for the Taxman. The remaining 76.7% is represented by a changed pie; some numbers are missing, others no longer appear in reverse video. The player's choice of 15 for the next slice leads to the fourth screen. There the player makes the mistake of choosing 10, and the Taxman takes the rest of the pie.

Figure 5.1a: Sample Screens for Taxman

For your next slice, you have selected 15, and the screen shows the resulting new pie and new totals. Your next selection of 10 results in confiscation of the remainder of the pie by the Taxman. The final totals are displayed on the next to last line of the final screen display:

YOU KEPT: 22.5%   TAXMAN GOT: 77.5%

Below that appears a reminder of your best score of all previous turns:

THE RECORD IS 54% OF 30

This means that in your most successful turn, you specified a pie size of 30 and managed to keep 54%.

Once the final totals and the best score have been displayed, the program waits for a single character input from the keyboard. It then clears the screen and requests a new pie size for the next turn.

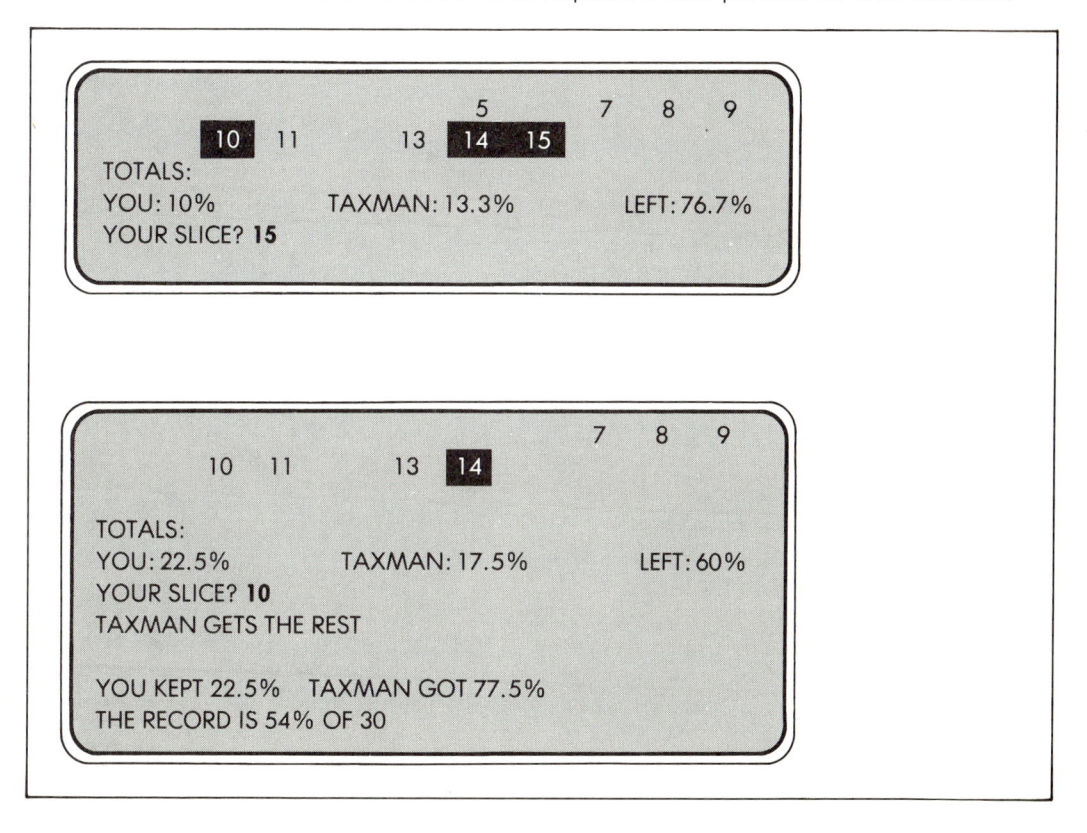

```
                       5      7   8   9
      10   11      13  14  15
TOTALS:
YOU: 10%         TAXMAN: 13.3%         LEFT: 76.7%
YOUR SLICE? 15
```

```
                              7   8   9
      10   11      13  14
TOTALS:
YOU: 22.5%       TAXMAN: 17.5%         LEFT: 60%
YOUR SLICE? 10
TAXMAN GETS THE REST

YOU KEPT 22.5%   TAXMAN GOT 77.5%
THE RECORD IS 54% OF 30
```

Figure 5.1b: Sample Screens for Taxman

The Taxman Program

The Taxman program is shown in Figures 5.2 through 5.12. Its structure is similar to that of other programs in this book. The "secret" part of the game is contained in the subroutines shown in Figures 5.6 and 5.8. These routines use the array HT to encode information about which numbers appear on the screen, which ones are in reverse video, and which ones the player may choose. The original setting of HT, the changes to HT after each choice and the method of scoring are all based on the arithmetic algorithms that appear in these two routines.

```
#Taxman

        GOSUB init                      #set array sizes and constants
        repeat {
            GOSUB turninit              #set pie size, initialize HT
            IF N = 0 THEN
                break                   #N = 0 is the quitting signal
            repeat {
                GOSUB showpie           #display what's left of the pie
                IF CH > 0 THEN
                    INPUT "YOUR SLICE"; S   #let the player pick a piece
                GOSUB taxgrab           #Taxman's turn
            } until (PL = 0)            #done when no pie left
            GOSUB stats                 #display final statistics
        }
    END
100   GOSUB530
110   GOSUB420:IF N=0 THEN END
120*  GOSUB150:IF CH>0 THEN INPUT "YOUR SLICE"; S
130   GOSUB260:IF PL>0 THEN120
140   GOSUB360:GOTO110
```

   This is the main routine of the Taxman program. To begin each turn, the player chooses a "pie" size. Then the player and the Taxman alternately take pieces of the pie until none remain. The Taxman follows certain rules, but the player is never told what they are. A clever player can guess the rules, but knowing the rules does not assure success when dealing with the Taxman.

---
*The Apple version of this line differs only in the spacing in the string constant.

Figure 5.2: Taxman

```
#Display what's left of the pie
showpie  GOSUB clearscreen                    #start on a clear screen
         CH = 0                               #count the player's choices
         FOR YY = 1 TO N STEP LS              #one screen line
            TP = min(N, YY + LS − 1)          #last line may not be full
            FOR XX = YY TO TP                 #individual numbers
               IF HT(XX) < 0 THEN
                  NM$ = L2$ + SZ$             #blank if not available
               else {
                  NM$ = STR$(XX)              #otherwise, the number
                  IF XX < 10 THEN             #add leading blanks
                     NM$ = L2$ + NM$          #   to right adjust
                  else IF XX < 100 THEN       #   in 3-digit field
                     NM$ = L1$ + NM$
                  else NM$ = LB$ + NM$
               }
               IF HT(XX) > 0 THEN             #if "hit" entry not zero
                  { RV = 1: CH = CH + 1 }     #   reverse video, count choice
               else
                  RV = 0                      #don't use reverse video
               GOSUB displaynum               #display the field
            NEXT XX
            PRINT                             #new line
         NEXT YY
         GOSUB totals: RETURN                 #display totals under the pie
```

```
150   GOSUB520:CH=0:FOR YY=1 TO N STEP LS:TP=YY+LS−1:IF TP>N THEN TP=N
160   FOR XX=YY TO TP:IF HT(XX)<0 THEN NM$=L2$+SZ$:GOTO190
170   NM$=STR$(XX):IF XX<10 THEN NM$=L2$+NM$:GOTO190
180   IF XX<100 THEN NM$=L1$+NM$:GOTO190
185   NM$=LB$+NM$
190   RV=0:IF HT(XX)>0 THEN RV=1:CH=CH+1
200   GOSUB480:NEXT XX:PRINT:NEXT YY:GOSUB210:RETURN
```

This routine displays "the pie," which is the set of numbers from 1 to N (the "size" of the pie). The numbers already taken by the player or by the Taxman do not appear. Certain numbers appear in reverse video for reasons that the player must guess.

Figure 5.3: **Display the Remaining Pie**

136 TAXMAN

The most complicated subroutine in the program appears in Figure 5.3. This is the routine that displays the new pie. One of the programmer's objectives was to make the routine independent of the actual BASIC system that it runs on. To further this end, the following techniques were used:

— The variable LS (set in the initialization routine in Figure 5.12) was introduced. The value of LS is the number of digits that will appear in each line of the pie display. In the example shown in Figure 5.1, LS has the value 9.

— The variables L1$ and L2$, both of which have been defined in terms of the variable LB$, are used to provide right justification. LB$ is defined in the initialization routine (lines 550 and

---

#Display the totals at this stage of the game

totals   PK = FNP(KP/PZ): PK$ = LB$ + STR$(PK)   #"keep" percentage
        PT = FNP(PY/PZ): PT$ = LB$ + STR$(PT)   #"pay" percentage
        PP = FNP(PL/PZ): PP$ = LB$ + STR$(PP)   #"left in pie" percentage
        PRINT: PRINT "TOTALS:"
        PRINT "YOU:"; PK$; "% ";
            "TAXMAN:"; PT$; "% ";
            "LEFT:"; PP$; "%"
        RETURN

```
210   PK=FNP(KP/PZ):PK$=LB$+STR$(PK)
220   PT=FNP(PY/PZ):PT$=LB$+STR$(PT)
230   PP=FNP(PL/PZ):PP$=LB$+STR$(PP)
240*  PRINT:PRINT "TOTALS:":PRINT "YOU:";PK$;"%  TAXMAN:";PT$;"%  LEFT:";PP$;"%"
250   RETURN
```

This routine prints the current percentages of the pie belonging to the player, the Taxman, or still remaining in the pie. The string variable LB$ is used to make sure that a leading space precedes the number, regardless of which BASIC system the program runs under. LB$ is set by the initialization routine in Figure 5.12.

*The Apple version of this line differs only in the spacing in the string constants.

*Figure 5.4: **Display Current Totals***

560 in Figure 5.12) by the instructions:

L = LEN(STR$(1))
LB$ = " ": IF L > 1 THEN LB$ = ""

This assures that LB$ is the null string on systems that provide a leading blank for positive numbers (e.g., Pet and TRS-80), while on systems that do not provide the blank, LB$ consists of a single blank.

— A call to a separate routine (GOSUB 480 on line 200 in Figure 5.3) is used to provide the reverse video display. The routine at 480 (in Figure 5.9) is system-dependent.

You should be able to understand the remainder of the Taxman program simply by reading through it.

#Tax computation

taxgrab   IF CH > 0 THEN {   #if there are any choices
                    S = INT(S)
                    IF 2 <= S <= N THEN   #and if the player's choice
                        IF HT(S) >= 1 THEN {   #   is valid,
                            GOSUB playfair   #then play by the rules
                            RETURN
                        }
                }
                PRINT "TAXMAN GETS THE REST."   #otherwise, ...
                PY = PY + PL
                PL = 0
                RETURN

260 IF CH=0 THEN300
270 S=INT(S):IF S<2 OR S>N THEN300
280 IF HT(S)<1 THEN300
290 GOSUB310:RETURN
300 PRINT "TAXMAN GETS THE REST.":PY=PY+PL:PL=0:RETURN

This routine decides whether or not to give the player a sporting chance. If the Taxman doesn't take it all, then the subroutine in Figure 5.6 is called to determine the Taxman's share.

Figure 5.5: **Tax Computation**

138 TAXMAN

```
#Compute the Taxman's share


playfair  TX = 0                                    #initialize Taxman's share
          KP = KP + S: PL = PL - S                  #player's share is S
          HT(S) = -1                                #remove S
          IF INT(N/S) >= 2 THEN                     #its multiples now
              FOR YY = 2 TO INT(N/S)                #   all have one
                  HT(YY*S) = HT(YY*S) - 1           #     fewer divisor
              NEXT YY
          FOR XX = INT(S/2) TO 1 STEP -1            #find the Taxman's share
              IF XX is a divisor of S
              AND HT(XX) >= 0 THEN {
                  HT(XX) = -1                       #another for the Taxman
                  TX = TX + XX
                  FOR YY = 2 TO INT(N/XX)           #adjust its multiples'
                      HT(YY*XX) = HT(YY*XX) - 1     #   divisors
                  NEXT YY
              }
          NEXT XX
          PY = PY + TX: PL = PL - TX                #update "pay" total
          RETURN


310  TX=0:KP=KP+S:PL=PL-S:HT(S)=-1:IF INT(N/S)<2 THEN330
320  FOR YY=2 TO INT(N/S):HT(YY*S)=HT(YY*S)-1:NEXT YY
330  FOR XX=INT(S/2) TO 1 STEP -1:IF HT(XX)<0 OR S<>XX*INT(S/XX) THEN350
340  HT(XX)=-1:TX=TX+XX:FOR YY=2 TO INT(N/XX):HT(YY*XX)=HT(YY*XX)-1:
     NEXT YY
350  NEXT XX:PY=PY+TX:PL=PL-TX:RETURN
```

This routine computes the Taxman's share of the pie. The algorithm used in this routine is a secret that the player must guess as a first step to playing the game effectively.

Figure 5.6: The Taxman's Share

#Display final statistics

```
stats   PK = FNP(KP/PZ)                    #"keep" percentage
        PT = 100 - PK                      #"tax" percentage
        IF PK > BK THEN {                  #is it a new record?
            BK = PK: BZ = N
            NR = 1                         #yes
            }
        else
            NR = 0                         #no
        PK$ = LB$ + STR$(PK)               #make percentages into strings
        PT$ = LB$ + STR$(PT)               #   with leading blanks
        PRINT: PRINT "YOU KEPT";PK$;
        "% TAXMAN GOT";PT$;"%"
        IF NR = 1 THEN                     #compare with old record
            PRINT "THAT'S A NEW RECORD!"
        else {
            BK$ = LB$ + STR$(BK)
            BZ$ = LB$ + STR$(BZ)
            PRINT "THE RECORD IS"; BK$;  "% OF";BZ$
            }
        GOSUB onech                        #leave on screen until key pressed
        RETURN
```

```
360  PK=FNP(KP/PZ):PT=100-PK:NR=0:IF PK>BK THEN BK=PK:BZ=N:NR=1
370  PK$=LB$+STR$(PK):PT$=LB$+STR$(PT)
380  PRINT:PRINT "YOU KEPT";PK$;"%    TAXMAN GOT";PT$;"%"
390  IF NR=1 THEN PRINT "THAT'S A NEW RECORD!":GOTO410
400  BK$=LB$+STR$(BK):BZ$=LB$+STR$(BZ):PRINT "THE RECORD IS";BK$"% OF";BZ$
410  GOSUB500:RETURN
```

This routine displays the final score and its relationship to scores from previous turns.

Figure 5.7: Display Final Tax Statistics

140 TAXMAN

#Set up for this turn

turninit   GOSUB clearscreen   #start on a clear screen
            repeat {
                INPUT "PIE SIZE"; N   #let player enter pie size
                IF N = 0 THEN   #N = 0 is quitting code
                     RETURN
                else
                     N = INT(N)   #need whole numbers
            } until (4 <= N <= MX)
            PL = N*(N+1)/2: PZ = PL   #sum of 1,2,...,N
            FOR XX = 1 TO N   #clear "hit" array
                HT(XX) = 0
            NEXT XX
            FOR XX = 1 TO INT(N/2)   #count divisors
                FOR YY = 2 TO INT(N/XX)
                     HT(YY*XX) = HT(YY*XX) + 1
                NEXT YY
            NEXT XX
            KP = 0: PY = 0   #initialize keep and pay amts
            RETURN

```
420   GOSUB520
430*  INPUT "PIE SIZE";N:IF N=0 THEN RETURN
440   N=INT(N):IF N<4 OR N>MX THEN430
450   PL=N*(N+1)/2:PZ=PL:FOR XX=1 TO N:HT(XX)=0:NEXT XX
460   FOR XX=1 TO INT(N/2):FOR YY=2 TO INT(N/XX):HT(YY*XX)=HT(YY*XX)+1:
      NEXT YY:NEXT XX
470   KP=0:PY=0:RETURN
```

This routine allows the player to enter the size of the "pie" to be divided between the player and the Taxman. The routine then initializes the HT array, which is used to keep track of which numbers are still available and which are not.

---
*The Apple version of this line differs only in the value of the string constant.

Figure 5.8:   *Prepare a New Pie*

#Display NM$—using reverse video if RV = 1
#Pet version

displaynum   IF RV = 1 THEN
           PRINT "rvs"; NM$; "off";
      else
           PRINT NM$;
      RETURN

```
480* IF RV=1 THEN PRINT CHR$(18);NM$;CHR$(146);:RETURN
490  PRINT NM$;:RETURN
```

This is the first of several routines that have different versions for the various BASIC systems. The Pet version uses the "rvs" and "off" character codes to bracket the string to be displayed in reverse video. The Apple version is very similar, bracketing a PRINT NM$; instruction between the INVERSE and NORMAL instructions. A different approach is needed on the TRS-80, since reverse video commands are not available in TRS-80 BASIC.

*The Apple version of this line is:

```
480 IF RV=1 THEN INVERSE: PRINT NM$;: NORMAL: RETURN
```

Figure 5.9:   Conditional Reverse Video

#Single-character input
#Pet version

onech   repeat
        GET X$
        until (X$<> "")
    RETURN

```
500 GETX$:IF X$="" THEN500
510 RETURN
```

This routine performs single-character input for the Pet. The same routine works for Apple, but there the loop is unnecessary. It is also advisable to insert code to recognize control-C on the Apple, since the Apple GET instruction does not do so. On TRS-80, a loop similar to this one must be used, but with the INKEY$ function rather than GET (GET means something entirely different on a TRS-80).
See Figure 2.11 for Apple and TRS-80 versions.

Figure 5.10: Single Character Input

Suggestions for Improvements and Additions

Here are several ways in which this game can be improved:

- Use cursor control to alter the pie display rather than clearing the screen and redisplaying the entire pie.

- Remember the best and worst percentage for each pie size.

- Keep track of how well the player performs and how much the player improves. Make encouraging or congratulatory remarks after each turn (or after some turns.)

Summary

Taxman is a game that challenges the player to discover the rules. The Taxman program illustrates one way of achieving independence from the small differences that exist between the various BASIC systems. Improvements to the program are possible, especially in the areas of cursor control and record keeping.

#Clear the screen
#Pet version

clearscreen   PRINT "clr";
                  RETURN

520   PRINT CHR$(147);:RETURN

   This routine clears the screen. The Apple version uses the HOME instruction. The TRS-80 version uses the CLS instruction.
   See Figure 2.12 for the Apple and TRS-80 versions.

Figure 5.11: Screen Clearing

#Initialization

```
init  READ LS: DATA linesize              #numbers/line in display
      READ DL: DATA screenlines           #lines/screen
      MX = LS*(DL-textunder)              #subtract lines for text below
      DIM HT(MX)                          #dimension HT accordingly
      L = LEN(STR$(1))                    #see if BASIC puts leading blank
      SZ$ = ""                            #SZ$ = a one-digit blank number
      FOR LL = 1 TO L
          SZ$ = SZ$ + " "
      NEXT LL
      IF L > 1 THEN                       #LB$ assures a leading blank
          LB$ = " "                       #LB$ is null if BASIC provides blank
      else
          LB$ = " "
      L1$ = LB$ + " "                     #extra leading blanks—2-digit value
      L2$ = L1$ + " "                     # 1-digit value
      DEF FNP(X) = INT(1000*X + .5)/10    #function for XX.X%
      RETURN


530*  READ LS:DATA 9:READ DL:DATA 24
540   MX=LS*(DL-5):DIM HT(MX)
550   L=LEN(STR$(1)):SZ$="":FOR LL=1 TO L:SZ$=SZ$+" ":NEXT LL
560   LB$=" ":IF L>1 THEN LB$=""
570   L1$=LB$+" ":L2$=L1$+" "
580   DEF FNP(X)=INT(1000*X+.5)/10
590   RETURN
```

This is the initialization routine for the Taxman program. The variables SZ$, LB$, L1$, L2$ are set with the same code on Apple, Pet or TRS-80, but they do not have the same values. This is an example of system-independent programming.

*In the TRS-80 version, the LS and DL values are 15 and 16.

Figure 5.12: Initialization

CHAPTER 6

Programming With Free BASIC

In the previous chapters of this book, you have seen many examples of Free BASIC program descriptions. Each program presented appeared in two forms: a Free BASIC description and a set of actual BASIC instructions. Our discussions thus far have referred only to the latter form. Now, the rules for using Free BASIC will be explained. You must learn these rules if you wish to read further in this book, because our program discussions will no longer make reference to the actual BASIC instructions. Hereafter, only the Free BASIC will be referred to. We shall begin by examining the program design process.

Programming in BASIC is simple—deceptively simple. It's like learning how to pound nails and saw wood; you can acquire these skills quickly, but they won't make you a carpenter. A skilled programmer, like a skilled carpenter, must do a great deal more than simply use the available tools competently. The skilled craftsman —carpenter or programmer—always relies on forethought and careful planning. Sitting at a keyboard and beginning to write a program is like starting to build a house without a blueprint.

Program Design Techniques

The three principal techniques that programmers use to plan and describe their programs are:

— Flowcharts

— Verbal algorithm descriptions

— Pseudocode.

We shall use the Addition Drill program of Chapter 1 to illustrate these techniques.

Since you are probably already familiar with flowcharting, we shall not discuss this technique in detail. Figure 1.1 shows a flowchart of the Addition Drill program presented in Figure 1.2.

A verbal algorithm description is a flowchart that does not use pictures. The boxed phrases of the flowchart become numbered "steps," and the arrows of the flowchart become numbered steps of the form

"Go back to step 39"

Figure 6.1 shows a verbal algorithm description corresponding to the flowchart of the Addition Drill program.

Figure 6.2 shows the pseudocode that a programmer might have written before writing the Addition Drill program. While the content of the pseudocode is the same as that of the flowchart and the verbal description, its form is a little different. First of all, the loop structure of the program is indicated by the word "repeat" followed by the set of actions that are to be repeated. These actions are enclosed between braces (also known as curly brackets). Note that the pseudocode version contains neither arrows nor step numbers. In the pseudocode representation there are no analogs of the BASIC GOTO instruction. Looping and branching structures that cannot be implemented as FOR...NEXT loops are provided for by means of "super instructions" like "repeat." Thus, GOTO is not needed in the pseudocode representation.

Experience with "structured" languages like Pascal and C shows that this style of program design, known as *GOTO-less programming*, leads to programs that are easier to write, easier to get running, and easier to understand than those designed with GOTO instructions.

Free BASIC

Free BASIC is a form of pseudocode. It is much more precise than the pseudocode shown in Figure 6.2, since Free BASIC uses actual BASIC instructions rather than verbal descriptions of the steps.

---

1. Pick two single-digit numbers, N1 and N2.
2. Ask "What is N1 + N2?"
3. If the player's answer is correct, skip ahead to step 6.
4. Say "Wrong, try again."
5. Go back to step 2.
6. Say "Right, try another."
7. Go back to step 1.

This is a verbal description of the algorithm for the Additional Drill program presented in Chapter 1. The sequence of operations begins at step 1 and proceeds through the steps in ascending order, except when a step (like 3, 5 or 7 above) is encountered that explicitly alters this sequence.

Figure 6.1: Verbal Algorithm Description for Addition Drill

Free BASIC was developed by the author to satisfy two main objectives:
- To free the BASIC programmer from BASIC line numbers.
- To allow the GOTO-less program design techniques to be applied to BASIC.

Figure 6.3 illustrates the way that Free BASIC has been used in this book. A Free BASIC description appears first in each figure. In Figure 6.3, this description is of a simple program in the form of an infinite loop. The program takes diameter values from a DATA list and prints the corresponding circumference values.

Below the Free Basic description appear the actual BASIC instructions derived from the Free BASIC. The translation from Free BASIC into BASIC was carried out by hand, using simple mechanical rules. In most cases, two different programmers would translate a given Free BASIC description into identical BASIC programs.

Since Free BASIC was designed to be translated by hand into BASIC, it was not necessary to include the formal mechanisms necessary to allow a computer program to perform the translation. As we examine the program shown in Figure 6.3 in more detail, we shall see examples of the "informality" of Free BASIC that makes machine translation impossible.

We shall now discuss the conventions of Free BASIC.

```
repeat {
    pick two single-digit numbers N1 and N2
    repeat {
        ask "What is N1 + N2?"
        if answer is incorrect
            say "Wrong, try again"
        else
            say "Right, try another"
    } until (answer is correct)
}
```

This is a pseudocode version of the Addition Drill program. The "steps" are the same as for the flowchart (Figure 1.1) and the verbal algorithm description (Figure 6.1), but the sequence of operations is exhibited in an entirely different way.

Figure 6.2: Pseudocode for Addition Drill

Use of Lowercase Characters

The first point to notice about Free BASIC is the use of lowercase characters to represent the non-BASIC elements of Free BASIC, and the use of uppercase characters to represent material that will be carried over, unchanged, into the BASIC program. For example, in Figure 6.3, lowercase characters are used for the words "repeat" and "pi." The word "repeat" signals one of the control structures of Free BASIC. (We shall discuss all of Free BASIC's control structures shortly.) The word "pi" is an example of Free BASIC's informality: it is a symbolic name for the constant that will appear in the BASIC program (3.14159 in Figure 6.3). However, nowhere in the Free BASIC program have we made the definition "pi := 3.14159" that would be necessary to enable a computer program to derive the BASIC program from the Free BASIC description in Figure 6.3. No mechanism for making such definitions has been provided, since such formality would not add to the usefulness of Free BASIC in this book.

```
            repeat {
                READ D
                C = pi*D
                PRINT "DIAMETER: "; D; " CIRCUMFERENCE: "; C
            }
            DATA 4,6,10
10      READ D
20      C=3.14159*D
30      PRINT "DIAMETER: ";D; " CIRCUMFERENCE: ";C
40      GOTO 10
50      DATA 4,6,10
```

This figure shows a Free BASIC description for a very simple program. Below the Free BASIC description is a set of actual BASIC instructions derived from the Free BASIC. Comparison of the Free BASIC and BASIC versions reveals several facts about Free BASIC:

— Free BASIC is "GOTO-less" and does not use line numbers.
— Indentation is used in Free BASIC to convey information about the program structure to the reader.
— Lowercase letters are used for components of the Free BASIC description that are not BASIC instructions.
— Parameters, i.e., symbolic names for constants, are used in Free BASIC, but no indication of the actual values to be assigned to these parameters appears in the Free BASIC.

Figure 6.3: A Simple Free BASIC Program

A constant that has been given a symbolic name is called a *parameter* of the program. This name is derived from the terminology of mathematical statistics. A parameter of a statistical distribution is a constant that occurs in the formula that defines the distribution. Different values of the constant give rise to different, but related distributions. Similarly, in the program described in Figure 6.3, different values of the parameter "pi" give rise to different, but related, programs. A value of 3.14 will generate a different set of circumference values from those generated with a value of 3.14159.

Indentation

Note that the Free BASIC program description that appears in Figure 6.3 uses indentation to reflect the program structure. The program consists of two instructions: the repeat instruction and the DATA instruction. Thus, the words "repeat" and "DATA" are the only words written at the leftmost position on the page. The three instructions that occur within the repeat instruction are indented from this leftmost position; if any of these instructions contained instructions within it, then those subinstructions would be indented even further. Notice that the closing right brace appears alone on a line, at the same indentation as the sequence of instructions that it is delimiting.

These conventions for indentation are intended to clarify the visual presentation of the material. In fact, the meaning of the Free BASIC description would be unchanged if nothing were indented, or if the braces were placed on the same lines as the instructions that they enclose.

Comments

Most programming languages allow the programmer to include comments in the source code of a program. A comment is an "aside" to the reader that clarifies the purpose or meaning of the instructions. Since Free BASIC is the form of the BASIC program that is designed to be read and understood, a provision for comments is essential. In Free BASIC, any text between a "#" and the end of a line is a comment. Examples of comments appear in all of the game programs in this book.

Translating from Free BASIC into BASIC

There is no program that translates the Free BASIC program description into the actual BASIC instructions. This means that you must perform the translation yourself. Let's look at the rules of Free BASIC and discuss the translation techniques that you can use.

The Repeat Instruction

The general form of the repeat instruction is:

>repeat instruction until (condition)

For example,

>repeat X = X + 1 until (X = 9)

would translate into the following BASIC code:

>10   X = X + 1
>   IF X <> 9 THEN 10

In general, the meaning of

>repeat instruction until (condition)

is: execute the instruction repeatedly, following each execution with a test of the condition. If the condition is true, then stop executing the instruction. For any instruction and any condition, the general form of the repeat instruction can be translated into the following BASIC code:

>N   instruction
>   IF NOT condition THEN N

where N is the next available line number. If your home computer does not recognize the logical operator NOT (all of the major home computers do), or if you simply want to streamline the BASIC program, you can transform "NOT condition" into an equivalent condition. In the example above, we used "X <> 9" instead of "NOT X = 9."

The general definition we have given here allows the repetition of only a single instruction. Braces allow this construct to be extended to the repetition of many instructions. Braces play the same role in Free BASIC descriptions that parentheses play in arithmetic expressions. For example, in an arithmetic expression, if we wish to multiply a term by 5, we can write

>5 * term

If the term we wish to multiply by 5 is the variable A, then we can write

>5 * A

but if the term is A + B + C, then we must write

>5 * (A + B + C)

The parentheses cause everything that appears within them to be treated as a single term in the expression. Similarly, when we write instructions between braces, as in Figure 6.3, everything within the braces is treated as a single instruction in the Free BASIC description.

The repeat instruction in Figure 6.3 does not include the "until (condition)" portion. Figure 1.3 shows a Free BASIC description of the Addition Drill program. The instruction

    150 IF A <> N1 + N2 THEN ... GOTO 120

in Figure 1.2 corresponds to the

    until (A = N1 + N2)

portion of the inner repeat instruction in Figure 1.3.

Note that the condition following the word "until" is always enclosed in parentheses. This is done so that we know exactly what terms make up the condition. In fact, in most cases ambiguity would be unlikely if the parentheses were not used, but including them helps to make the program description clearer, especially when the condition consists of several terms. The conditions following the "until" in a repeat instruction are identical to those used in IF statements, but no parentheses are needed in an IF statement, because BASIC always knows that the condition consists of everything between the IF and the THEN.

Another way that a repeat instruction can be terminated is through the use of a break instruction. The break instruction will stop repetition of the instruction of which it is a subinstruction. An example of the use of break occurs in the main routine of the Match-up program (see Figure 7.16 in Chapter 7). In many cases, the use of the break instruction can be avoided by a reorganization of the program's structure. Reorganization will often represent an improvement, since the need for a break instruction can be a sign of inappropriate program organization.

IF...THEN...else

More than 2,000 years ago, Aristotle set down the rules of logic that have become the central reasoning tool in all Western thought. One of these rules is called the *law of the excluded middle*, which states: "any assertion is either true or false—it must be one of these, and it can't be both." This rule is the model for the IF...THEN...else construction of Free BASIC. The general form of this construction is

    IF condition THEN instruction 1 else instruction 2

The "else instruction 2" portion is optional. The effect of this instruction is that if "condition" is true, then "instruction 1" is executed; if "condition" is false, then "instruction 2" is executed. The result is that one and only one of the instructions is executed. (If "condition" is false and the "else instruction 2" portion has been omitted, then no instruction is executed.) As with the repeat instruction, "instruction 1" and "instruction 2" can be single instructions or sets of instructions enclosed within braces.

Figure 1.3, which contains a Free BASIC description of the Addition Drill program, gives an example of this construction and the convention that we shall use for the indenting and grouping of its parts. In this figure the "condition" is "A = N1 + N2", "instruction 1" is 'PRINT "THAT'S RIGHT etc."' and "instruction 2" is 'PRINT "THAT'S WRONG etc."'

Note the uppercase and lowercase letter conventions used with the IF...THEN...else construction. We could have written if...then...else (all lowercase) to distinguish the Free BASIC construction from the BASIC IF...THEN construction. However, the IF...THEN of the Free BASIC IF...THEN...else is always translated directly into the BASIC IF...THEN. Our Free BASIC usage emphasizes this by writing IF...THEN in uppercase. Since most home computer BASIC systems do not allow the use of ELSE with IF...THEN (TRS-80 has ELSE, Pet and Apple do not), we have used lowercase letters for writing the else.

Cases

The case statement, a useful form of the IF statement, is based upon another of Aristotle's favorite techniques, the enumeration of possibilities. The general form of the case statement is

```
    IF case
        condition 1 THEN instruction 1
        condition 2 THEN instruction 2
                    .
                    .
                    .
        condition n THEN instruction n
    else
        instruction n + 1
```

The "else instruction n + 1" portion of this statement is optional.

The case statement is a great aid to program clarity. When you see the above statement, you know that one and only one of the instructions that appear in it will be executed. The instruction executed will be the one corresponding to the first of the listed conditions that is true. If all of the conditions are false, then the instruction executed will be instruction n + 1.

Figure 6.4 shows a Free BASIC program that uses a case statement. In this statement, there are three conditions and an else portion. What the rest of this game program does is left for you to discover. In fact, the object of the game (called "Document It If You Can") is to discover what the program does. Figure 6.5 shows a BASIC program derived from Figure 6.4. If you have worked as a programmer, or if you have received game programs from friends or obtained them through books or magazines, then you have probably had to decipher programs similar to the "Document It If You Can" program.

While

The last control structure that we shall introduce is used for implementing loops that begin with a test. The general form of a while structure is:

> while (condition) instruction

The instruction in a while structure will not be executed if the condition is false to begin with. (Figure 4.6 in Chapter 4 shows a program that uses a while construction.)

Subroutines

The principal design goal of Free BASIC was to free the BASIC programmer from having to think about line numbers. The control structures that we have introduced eliminate the need to use line numbers in IF and GOTO statements. One problem remains, however: how to eliminate line numbers from GOSUB instructions. The solution that we have chosen is to use symbolic line numbers for the initial lines of subroutines. (If you feel that you need a GOTO in your program, you can also place a symbolic line number on the line that you wish to refer to.) These symbolic line numbers are indicated by writing a name (in lowercase letters, of course) to the left of the instruction, exactly where a line number would appear.

Mechanical Translation

Throughout this book there are many programs for which both a Free BASIC description and the actual BASIC instructions have been

```
         INPUT A$, B$
         IF case
            A$ = "A" THEN
               IF B$ = "1" THEN
                  PRINT "A1"
               else
                  PRINT "AX"
            A$ = "B" THEN
               repeat
                  INPUT C$
                  until (B$ = C$)
            A$ = "C" THEN
               IF VAL(B$) <= 0 THEN
                  PRINT "???"
               else {
                  FOR J = 1 TO VAL(B$)
                     PRINT "*";
                  NEXT J
                  repeat {
                     INPUT C$
                     IF C$ = "Q" THEN
                        break
                     else
                        PRINT "XYZ"
                  }
               }
            else
               PRINT "HA HA"
         END
```

  This program implements a game called "Document It If You Can." The object of the game is to figure out what the program does. You are allowed to operate the program and read the listing. (For the advanced version, you are only allowed to read the BASIC listing, not the Free BASIC listing.) Skill in playing this game can be useful in real-life situations.

Figure 6.4: A Program Using a Case Statement

provided. These programs provide practical examples of the translation from Free BASIC into BASIC. In many cases the translation was performed more cleverly by hand than if a straightforward computer program had been used. For each of the constructions used in this chapter, there is an "obvious" mechanical translation into BASIC. Figure 6.6 shows this correspondence.

Free BASIC, Structured Programming and Pascal

Free BASIC is a "structured" BASIC. You have surely heard a great deal about "structured programming." This term derives from "Notes on Structured Programming" by Edsger W. Dijkstra (appearing in "Structured Programming" by Dahl, et al.; Academic Press, 1972). In that monograph, Dijkstra made two main points:

— GOTO-less programs are easier to understand and easier to prove correct than programs that contain GOTO instructions.

— Many benefits are to be realized from the arrangement of programs into "layers," each representing a different level of abstraction and using an "ideal" machine suitable for that level of abstraction.

The first of these points is easier to understand. As a result, many books, articles and seminars have purported to teach "structured programming," but have actually been concerned primarily with the

```
10   INPUT A$,B$
20   IF A$<>"A" THEN 40
30   IF B$ = "1" THEN PRINT "A1":GOTO 110 ELSE PRINT "AX":GOTO 110
40   IF A$<>"B" THEN 60
50   INPUT C$:IF B$<>C$ THEN 50 ELSE 110
60   IF A$<>"C" THEN 100
70   IF VAL(B$)<=0 THEN PRINT "???":GOTO 110
80   FOR J = 1 TO VAL(B$):PRINT "*";:NEXT J
90   INPUT C$:IF C$ = "Q" THEN 110 ELSE PRINT "XYZ":GOTO 90
100  PRINT "HA HA"
110  END
```

This is the "Document It If You Can" game shown in Figure 6.4. It is written in TRS-80 BASIC, which allows ELSE clauses in IF statements. The program would be more complicated on an Apple or a Pet, since these systems do not allow ELSE clauses.

Figure 6.5: BASIC for the "Document It" Game

Free BASIC	BASIC
repeat 　　instruction until (condition)	nnn　instruction 　　　IF NOT condition THEN nnn
while (condition) 　　instruction	nnn　IF NOT condition THEN mmm 　　　instruction 　　　GOTO nnn mmm　...

IF condition THEN
　　instruction 1
else
　　instruction 2
	IF NOT condition THEN nnn
	　　　instruction 1
	　　　GOTO mmm
	nnn　instruction 2
	mmm　...

IF case 　　condition 1 THEN instruction 1 　　　. 　　　. 　　condition 2 THEN instruction 2 　　　. 　　　. 　　　. 　　condition n THEN instruction n else 　　instruction n + 1	IF NOT condition 1 THEN nn2 　　　instruction 1 　　　GOTO mmm nn2　IF NOT condition 2 THEN nn3 　　　instruction 2 　　　GOTO mmm 　　　. 　　　. 　　　. nnn　IF NOT condition n THEN eee 　　　instruction n 　　　GOTO mmm eee　instruction n + 1 mmm　...

A simple correspondence between Free BASIC control structures and BASIC instructions is shown here. A straightforward computer program could perform such a translation.

Figure 6.6:　*Correspondence Between Free BASIC and BASIC*

techniques of GOTO-less programming. One of the major objectives of Free BASIC is to bring the benefits of GOTO-less programming to BASIC programmers; in that sense, Free BASIC supports structured programming techniques.

The second of these points is more difficult to grasp, and the technique suggested is difficult to learn to practice. The programs in this book have not been designed by a conscious effort at "layering," or by the definition of abstract "ideal" machines; in that sense, the programs of this book do not illustrate the techniques of structured programming. Rather, the related technique of "top-down" design has been employed. Top-down structuring techniques are discussed in several chapters of this book. (For example, see Chapter 7, the Match-up Game.)

Pascal is a programming language designed to facilitate GOTO-less programming. Pascal's control structures are essentially identical to those used in Free BASIC, so that a Free BASIC program description resembles a "hybrid" language derived by crossing BASIC with Pascal. Thus, many of the Free BASIC program descriptions in this book could form the basis for Pascal programs. There are, however, two major obstacles to such a translation:

— Not all of the features of BASIC are available in "standard" Pascal.

— Pascal programs require the "declaration" of all data and subroutines. In Pascal, you must say what an object is before you use it, and thereafter, Pascal checks carefully to ensure that you never use the object as if it were a different type of object.

Summary

There are three principal techniques used by programmers to plan and describe their programs: flowcharts, verbal algorithm descriptions and pseudocode. Free BASIC is a version of pseudocode that is so precise that it is almost a computer language. Free BASIC employs a convention that assigns different roles to lowercase and uppercase letters. Uppercase letters are used for parts of the Free BASIC description that will appear verbatim in the actual BASIC program. Lowercase letters are used for all other parts of the Free BASIC description. The principal uses of lowercase letters are for program parameters, and for the key words used in specifying control structures. Indentation is used in Free BASIC descriptions to reflect the program's structure. Use of this convention is optional, and different programmers

are free to adopt different indentation styles from the author's.
The control structures that are used in Free BASIC are

- repeat...until
- IF...THEN...else
- IF case...else
- while

The break instruction provides for "unscheduled" terminations of the loops defined by repeat and while constructions.

These features eliminate the need for GOTO instructions in Free BASIC descriptions. Furthermore, symbolic line numbers are used in GOSUB instructions for subroutine calls. This is the final step in freeing BASIC programmers from the use of line numbers—the principal design objective of Free BASIC.

Free BASIC supports the GOTO-less aspect of structured programming. Its control structures are the same as those used in Pascal, so that a Free BASIC description can be readily translated into Pascal. The main obstacles to such translations are the lack of specific BASIC features in Pascal and the Pascal requirement that subroutines and data be "declared."

CHAPTER 7
The Match-up Game

This chapter presents the Match-up Game, a full-scale game that illustrates many important points about programming interactive computer systems. A description of how to play the game will be given first, then the programs used to implement the game will be presented.

The object of the Match-up game is to "match up" compatible pairs of members from two different groups. The groups can be men and women, sellers and buyers, pets and owners, jobs and job seekers, and so forth. Compatibility is determined by designing two sets of questions, and then comparing each player's answers with the stated preferences of members of the other group.

For example, if the two groups being used are the men and the women attending a party, then the object of the game is to select compatible pairs of men and women. This is done by first assigning names to the two groups, MEN and WOMEN. Then, a set of multiple choice questions is prepared for each group, and a numeric weight is assigned to each question. The weight is used to determine the degree to which one player's answers match the stated preferences of members of the other group.

Let's run through a sample game. The game begins with the appearance of the prompt "::" in the upper left corner of the screen. Figure 7.1 shows the commands that can be entered in response to the "::" prompt. Each consists of a single letter or digit.

The Game-Building Phase

The first phase of the game is the "game-building" phase, in which group names and the multiple-choice questions are entered. Figure 7.2 shows a sample dialog that might follow the entry of an "I" command. All of the player responses are followed by RETURN (or ENTER). The player responses appear in boldface type to distinguish them from the questions and prompts output by the computer.

162 THE MATCH-UP GAME

In Figure 7.2, the player has entered "1", and the program has responded with the prompt "GROUP 1:" to indicate that the player must now enter the name of the first group. The player enters "MEN", and the program immediately responds with "Q1 FOR MEN" to request the first question for group 1. If the player had responded with "DOGPATCH" as the name for group 1, the program would have asked "Q1 FOR DOGPATCH" instead.

For the first question for MEN, the player has entered "WHOM DO YOU MOST RESPECT?". Notice that the player has been able to type this string, which contains blanks, without enclosing it in quotes. This suggests that the LINE INPUT instruction is being used. Later on, when we discuss the program, we shall see that LINE INPUT is being simulated by a subroutine.

Now the program begins to prompt with numbers, and the player responds by specifying the choices for the first question. In our example, after five choices have been entered, the program stops asking for choices and asks "WEIGHT: ". This is a signal to the player to enter the numeric weight to be assigned to this question. In Figure 7.2, the program stops asking for choices after five have been specified, because the maximum number of choices for any question has been set to 5 by a DATA statement in the initialization subroutine.

We shall discuss the setting of this parameter and others later, but note that if you have a small home computer system (e.g., with only 8K of RAM), you may have to assign very small values to the maximum number of choices per question, the number of questions and the number of players, because these parameters enter into the dimension statements for multi-dimensional arrays.

After the player has entered a weight of 4 for the first question, the program prompts "Q2 FOR MEN," and the player replies with the next question: "WHAT DO YOU WANT MOST FROM LIFE?". Again the program prompts for specification of the possible answers by displaying choice numbers, but this time, after entering three choices, the player has responded with "Q" to the fourth prompt.

Throughout the Match-up game, Q is frequently used as a "terminate" or "abort" command. In this case it means "I don't wish to enter any more choices for this question. Let's go on to the next item (the weight given to the question)." The program responds by displaying "WEIGHT: ", and the player enters the value 6. Then the program responds with the prompt "Q3 FOR MEN." The player enters "Q" again; this time the Q means "I don't wish to enter any more questions for group 1." The program obliges by proceeding to the group 2 entry section. The dialog for group 2 is similar to the dialog for group 1.

Figure 7.2 does not show an aspect of the display that is difficult to represent in print. At three points in the dialog, "Q" has been entered. In order to reduce the chance of confusion, the program actually removes the character Q from the screen if the player's input consists of the single character "Q" followed by a RETURN. Thus, the actual screen display would look like Figure 7.2, but the terminating Q's would be replaced by blanks.

Let's see how the N, 1 and 2 commands (see Figure 7.1) can be used to modify the set of questions entered in Figure 7.2. Since S and L commands are not implemented in the version of Match-up shown in this book, we shall not include them in our discussion.

Command	Action
I	*Initialize:* The game is started from scratch. The program requests names for the two groups, then accepts a different set of questions for each group.
L	*Load:* This command is not implemented in the version of the program shown in this book. A group name and a set of questions for that group are loaded from cassette or disk.
N	*Edit Names:* The group names are displayed, and replacement names are accepted.
Q	*Quit:* The "[ ]" prompt is displayed and the playing phase is begun.
S	*Save:* This command is not implemented in the version of the program shown in this book. A group name and a set of questions for that group are saved on cassette or disk.
1	*Edit Group 1:* The "*" prompt is displayed, and edit commands for group 1 questions are accepted.
2	*Edit Group 2:* The "*" prompt is displayed, and edit commands for group 2 questions are accepted.

These are the commands that can be entered in response to the "::" prompt. The commands I, L, and S lead to fixed tasks. The commands Q, 1, and 2 lead to further command selection routines.

Figure 7.1: *Responses to the "::" Prompt*

164   THE MATCH-UP GAME

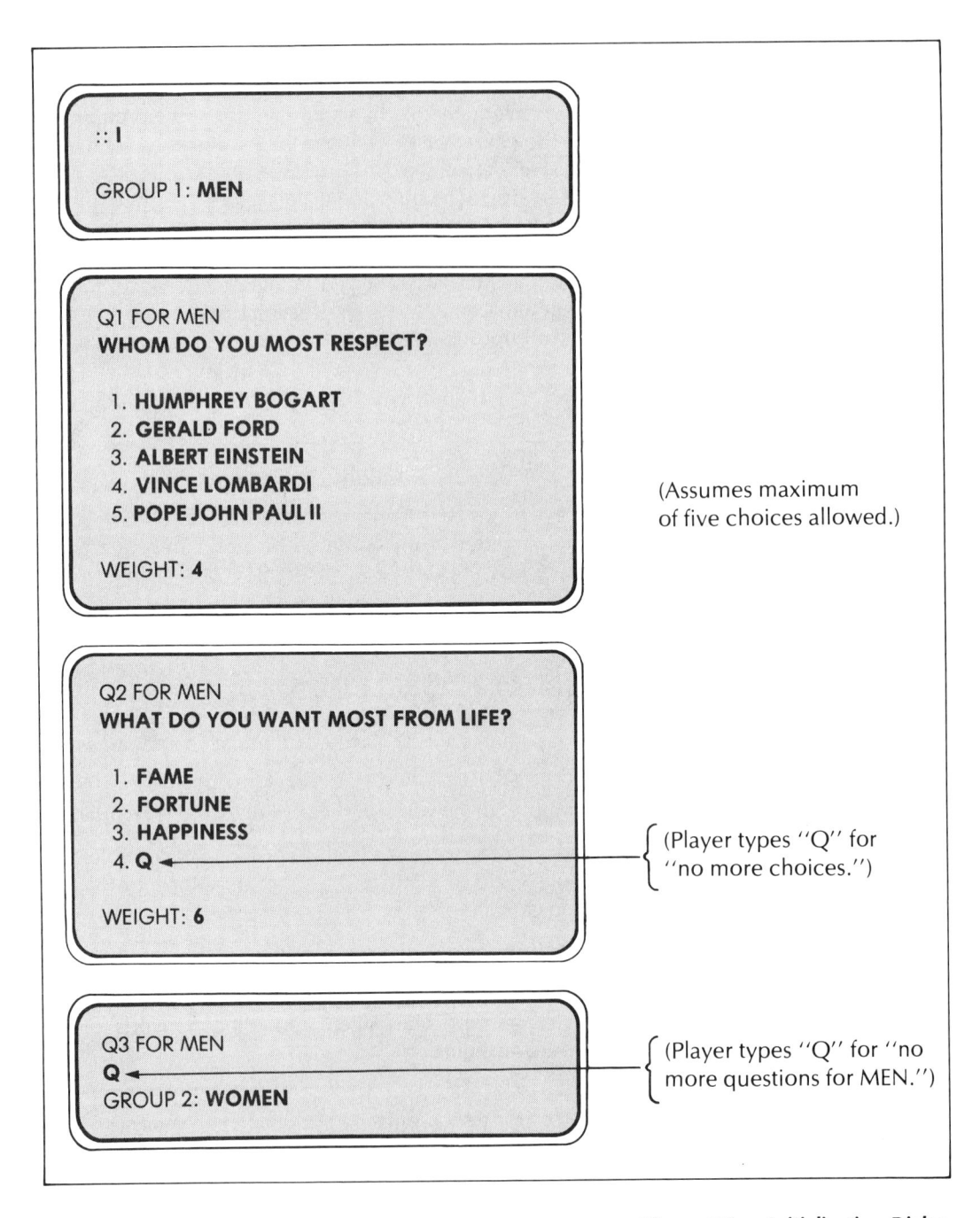

```
:: I
GROUP 1: MEN
```

```
Q1 FOR MEN
WHOM DO YOU MOST RESPECT?

 1. HUMPHREY BOGART
 2. GERALD FORD
 3. ALBERT EINSTEIN
 4. VINCE LOMBARDI
 5. POPE JOHN PAUL II

WEIGHT: 4
```
(Assumes maximum of five choices allowed.)

```
Q2 FOR MEN
WHAT DO YOU WANT MOST FROM LIFE?

 1. FAME
 2. FORTUNE
 3. HAPPINESS
 4. Q

WEIGHT: 6
```
(Player types "Q" for "no more choices.")

```
Q3 FOR MEN
Q
GROUP 2: WOMEN
```
(Player types "Q" for "no more questions for MEN.")

Figure 7.2:   Initialization Dialog

The N Command

Figure 7.3 shows sample dialog for the N command. In the example, two group names are displayed in turn. As each name is displayed, the player is allowed to enter either a RETURN, which will leave the name unchanged, or a replacement name. In this example, the group 1 name is changed from "MEN" to "MENFOLK", while the group 2 name is left unchanged.

The 1 and 2 Commands

The 1 and 2 commands are similar to each other in effect. In either case, the screen clears, the new prompt "*" appears, and the program waits for the player to enter another single-character command.

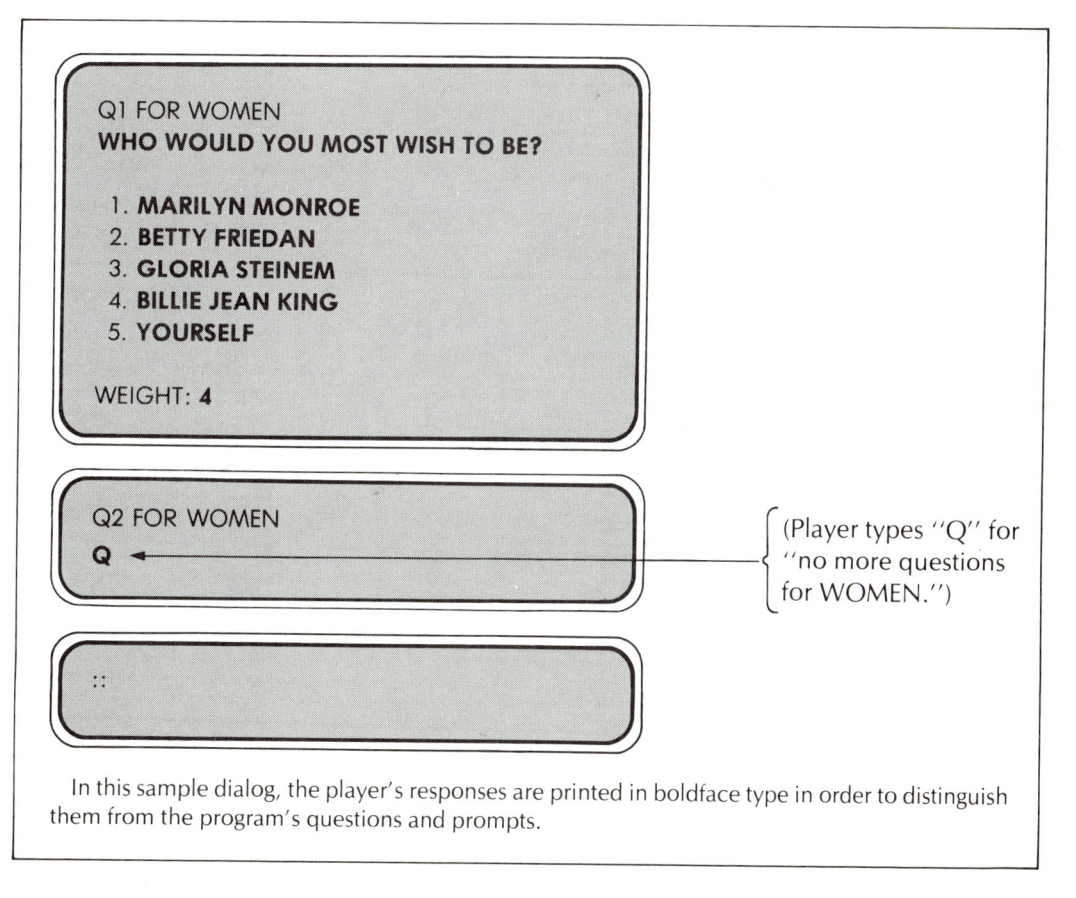

```
Q1 FOR WOMEN
WHO WOULD YOU MOST WISH TO BE?

1. MARILYN MONROE
2. BETTY FRIEDAN
3. GLORIA STEINEM
4. BILLIE JEAN KING
5. YOURSELF

WEIGHT: 4
```

```
Q2 FOR WOMEN
Q
```
(Player types "Q" for "no more questions for WOMEN.")

```
::
```

In this sample dialog, the player's responses are printed in boldface type in order to distinguish them from the program's questions and prompts.

Figure 7.2a: Initialization Dialog

Figure 7.4 shows the commands that can be entered at this point. The commands pertain to the specified group—group 1 if the "1" command was used to enter this question-file-editing mode, or group 2 if "2" was used.

The E Command

Figure 7.5 shows an example of the use of the E command. In this example, the player edits the question that was entered for group 2 in Figure 7.2. The question itself is left unaltered. The first choice is replaced by a new choice, the second choice is left unaltered, and the third choice is deleted, so that the answers that were originally choices four and five are now choices three and four. The new

```
:: N
```

```
1 MEN
MENFOLK
2 WOMEN
```
(Player changes group 1 name to "MENFOLK.")

(Player types RETURN and group 2 name remains "WOMEN.")

```
::
```
(Screen clears and prompt re-appears.)

This sample dialog for the N command shows how the screen looks as the player changes the group 1 name from "MEN" to "MENFOLK." The group 2 name is left unchanged.

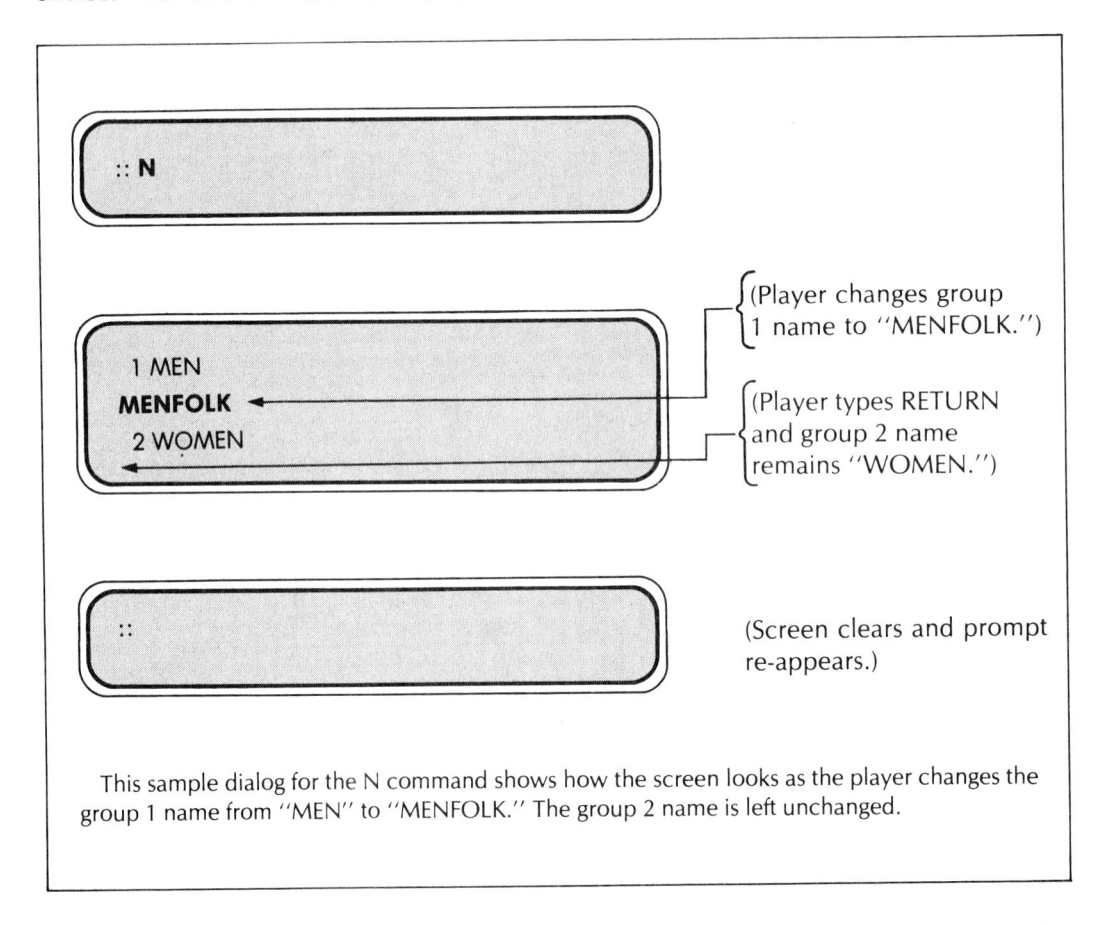

Figure 7.3:   Group Name Editing Dialog

choice three (previously choice four) is altered, and the new choice four (previously choice five) is left unaltered. Since there are now only four choices, the program prompts for a fifth choice, but the player presses RETURN so that no fifth choice is added. The program then displays the numeric weight assigned to the question, and the player enters a different value.

The A and B Commands

The A and B commands are used to incorporate additional questions into the set of questions for the given group. The A command is

Command	Action
A	*Add:* The player is allowed to add a question to the end of the set of questions for the given group. The program prompts with the question number and choice numbers, then it requests the numerical weight.
B	*Before:* The program asks the player for the number of the question ahead of which the new question is to be inserted. Then the question information is requested, just as for the add function.
D	*Delete:* The program asks for a question number and deletes the specified question, moving subsequent questions up in numbering so that there is no gap in the set of question numbers.
E	*Edit:* The program asks for a question number and displays the current statement of the specified question, allowing the player to replace it or leave it unchanged. Then the program steps through the current choices, allowing the player to replace them, delete them or let them remain. Next, the program accepts additional choices for the question (up to the maximum allowed). Finally, the current weight assigned to the question is displayed by the program; the player can change it or leave it unchanged.
Q	*Quit:* Editing of the current group is complete. The screen is cleared, and the "::" prompt is displayed.

These are the commands that can be entered in response to the "*" prompt that occurs when either the "1" or the "2" command has been given.

Figure 7.4: Responses to the "" Prompt*

168 THE MATCH-UP GAME

```
:: 2
```

```
*E
QUESTION NUMBER: 1
WHO WOULD YOU MOST WISH TO BE?
  1. MARILYN MONROE
  BO DEREK
  2. BETTY FRIEDAN
  3. GLORIA STEINEM
  D
  3. BILLIE JEAN KING
  ELIZABETH HOLTZMAN
  4. YOURSELF
  5.
WEIGHT:   4 :  7
```

(Player presses RETURN.)

(Player alters first choice.)
(Player presses RETURN.)

(Player deletes third choice.)

(Player alters new third choice.)
(Player presses RETURN.)
(Player presses RETURN.)
(Player alters numeric weight.)

```
*E
QUESTION NUMBER: 1
WHO WOULD YOU MOST WISH TO BE?
  1. BO DEREK
  2. BETTY FRIEDAN
  3. ELIZABETH HOLTZMAN
  4. YOURSELF
  5.
WEIGHT:   7:
```

(Player checks altered question by requesting the edit function again and pressing RETURN at each opportunity.)

```
*
```

This sample dialog shows how the first question entered for group 2 in Figure 7.2 might be edited. Player responses are shown in boldface type. (Except in answer to the "::" and "*" prompts, all of the player's responses are terminated by pressing RETURN.)

Figure 7.5:   Question Editing

used to add a question to the end of the set. (See Figure 7.6.) The program, which knows the number of questions already in the set, prompts with the next higher question number. The B command is used to insert a question into the question set. (See Figure 7.7.) At this point, the program does not know where the question is to be inserted, so it begins by asking the player for the number of the question before which the new question is to be placed. For example, if the player specifies insertion before question two, then the new question becomes question two, the original question two becomes question three, three becomes four, and so on. With either the A or the B

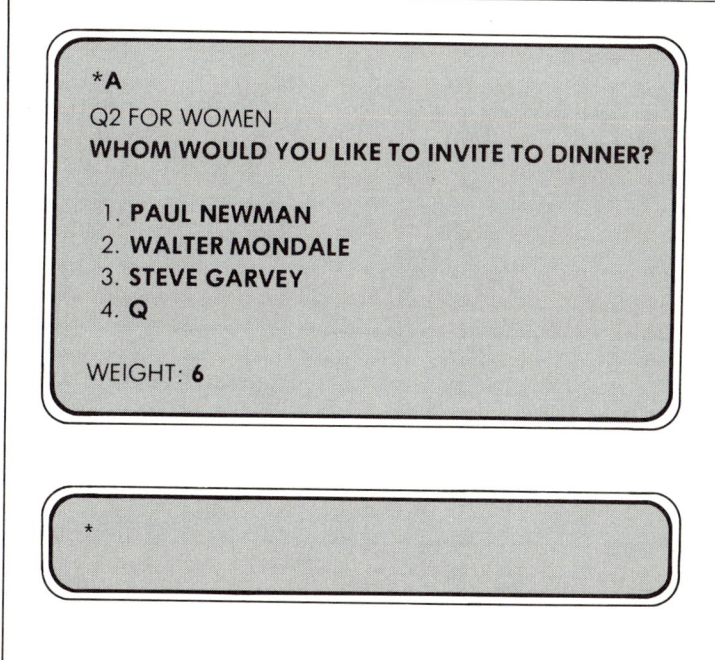

```
*A
Q2 FOR WOMEN
WHOM WOULD YOU LIKE TO INVITE TO DINNER?

1. PAUL NEWMAN
2. WALTER MONDALE
3. STEVE GARVEY
4. Q

WEIGHT: 6
```

```
*
```

   This sample dialog shows an example of the use of the "A" command to add a question to the set of questions entered in Figure 7.2 and edited in Figure 7.5. The "*" prompt at the top of this figure could be the same one shown at the bottom of Figure 7.5.
   The prompt "Q2 FOR WOMEN" is printed by the program, because this will be the second group 2 question, and an "A" command always causes the new question to be added to the end of the set of questions. If the player wishes to place the new question somewhere other than at the end of the set, the "B" command must be used.

Figure 7.6: Adding a Question

```
*Q
```
(Player terminates group 2 editing.)

```
:: 1
```
(Player initiates group 1 editing.)

```
*B
AHEAD OF QUESTION NUMBER: 1
Q1 FOR MEN
HOW OLD ARE YOU?

  1. UNDER 25
  2. 25-38
  3. 39
  4. OLD ENOUGH TO KNOW BETTER
  5. AT THE DANGEROUS AGE

WEIGHT: 9
```

```
*
```

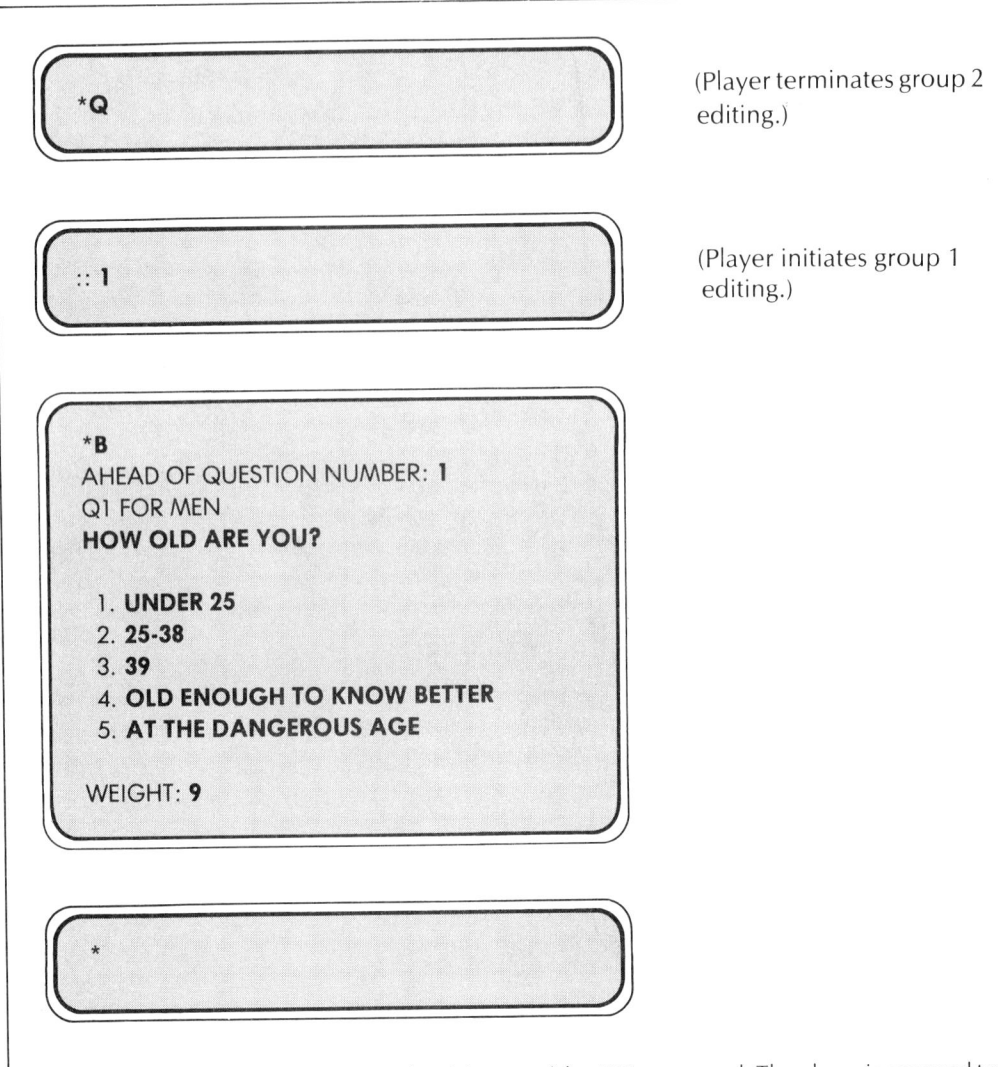

This sample dialog shows an example of the use of the "B" command. The player is assumed to have been editing group 2 questions; e.g., the "*" at the top could be the same as the final "*" of Figure 7.6. The player uses the "Q" command to return to the "::" prompt, and then the "1" command to generate the "*" for group 1 editing.

If the group 1 question file is the same as the one entered in Figure 7.2, then the questions one and two entered there will now become questions two and three, and the question entered here becomes question one.

Figure 7.7: **Inserting a Question**

command, once the question number has been determined, new information is entered in exactly the same way as it was for the I command (Figure 7.2).

The use of the letter B (rather than I) to specify the question insertion command deserves explanation. One reason for using B instead of I is that if the player were to forget to use the 1 or 2 commands, and were to enter the insertion command in response to the "::" prompt (an easy mistake to make), then "B" would result in a harmless rejection of the command (since there is no B response to the "::" prompt). On the other hand, "I" would result in an inadvertent invocation of the initialization command, causing the game to begin again from scratch. The second reason for using B instead of I is that the mnemonic value of B (for "before") helps the player remember that the new question will be placed before the question specified in the command. It is easy to become confused about whether the insertion is to be before or after the specified question, especially if you have used an editing system that follows a different convention for insertions.

The D Command

The final command used to edit the question file is the D command, for delete (see Figure 7.8). The player specifies the number of the question to be deleted, and the program does the rest. The question is removed from the set, and each question with a higher number than the deleted one has its number reduced by 1. Thus, if question four is deleted, then question five becomes the new question four, question six becomes question five, and so on.

```
*D
QUESTION NUMBER: 2
```

```
*
```

This is a sample of the dialog required to delete a question from the question set. If the above interaction occurs immediately following that of Figure 7.7, then the question "WHOM DO YOU MOST RESPECT?", which was entered as question one in Figure 7.2 and became question two after the insertion shown in Figure 7.7, will be deleted from the group 1 question set. The question "WHAT DO YOU WANT MOST FROM LIFE?" will become question two again.

Figure 7.8: Deleting a Question

This completes our discussion of the set of commands that are used to edit the group 1 and group 2 question files. These commands are a complete set of editing functions (in the sense that everything needed is provided), but there are several very desirable features that are not present. Some of these features are suggested as improvements at the end of the chapter.

The Playing Phase

The next phase of the game is the playing phase, which is arrived at by entering the Q (quit) command in response to the "::" prompt. In this phase the program displays the "[ ]" prompt, and will accept single-character commands, which are used to establish and alter player entries and to select compatible pairs of players. Figure 7.9 shows the commands that can be entered in response to the "[ ]" prompt.

Command	Action
E	*Edit:* The current contents of the specified player's entry are displayed, and the player is allowed to make alterations and additions or deletions.
L	*Load:* This command is not implemented in the version of the program shown in this book. A set of player entries for one of the groups is loaded from cassette or disk.
N	*New Player:* A new entry is added to the specified group. The entry consists of the player's name, the player's answers to the questions pertaining to the given group, and the player's preferences for answers by members of the other group. All of this information is entered by the new player in response to the prompting of the program.
P	*Pair:* Two sets of members of the other group are displayed: those whose answers best match the given player's preferences, and those whose preferences the given player's answers best match.
Q	*Quit:* The game is over. Return to the "::" prompt to start a new game.
S	*Save:* This command is not implemented in the version of the program shown in this book. The player entries for a specified group are saved on cassette or disk.

Figure 7.9: *Responses to the "[ ]" Prompt*

The N Command

Figures 7.10 and 7.11 show an example of the dialog that occurs when the N command is invoked. The program begins by displaying the names of the two groups, and by asking the player to specify one of them by entering either a 1 or a 2. Then the program asks for the player's name. After the name has been given, the program presents the player with the questions pertaining to the player's specified group. The player is asked to choose one answer to each of the questions. After the player's name and choices have been given (Figure 7.10), the player is allowed to specify preferences that will be used in scoring the compatibility of members of the other group with the player's wishes. In the example shown in Figure 7.11, the player can assign one possible answer to each of the values 2, 1, −1 and −2. The number of values and the values themselves are defined in DATA statements in the initialization routine of the program, so this aspect of the game can be easily altered. The program uses the format

VALUE n   CHOICE:

to ask for the choice, if any, to which the player wishes to assign the value n.

In order to illustrate the scoring method used, Figures 7.12 and 7.13 show another example of the dialog that follows an N command. This time the player belongs to the other group. This player's answers (Figure 7.12) are scored against the first player's preferences (Figure 7.11). The figures illustrate the scoring method used. To score player B's answers against player A's preferences, a score is assigned to each of the questions answered by player B, and these scores are added to obtain a total score. The score assigned to a given question is found by multiplying the value that player A previously specified for the choice now given by player B times the weight assigned to the question during the game-building phase of the game.

The P Command

The P command is used to obtain a display of two lists of names (see Figure 7.14):

— those members of the other group whose answers best matched the player's wishes

— those members of the other group whose wishes the player's responses best matched.

THE MATCH-UP GAME

```
[ ] N
  1. MEN
  2. WOMEN
GROUP: 1
```
(Player is in the group MEN.)

```
NAME: JOHN
```
(Player's name is JOHN.)

```
HOW OLD ARE YOU?
  1. UNDER 25
  2. 25-38
  3. 39
  4. OLD ENOUGH TO KNOW BETTER
  5. AT THE DANGEROUS AGE
CHOICE: 2
```
(Player's age is 25-38.)

```
WHAT DO YOU WANT MOST FROM LIFE?
  1. FAME
  2. FORTUNE
  3. HAPPINESS
CHOICE: 3
```
(Player desires HAPPINESS.)

This example shows the initial portion of the new player entry dialog. First the names of the two groups are displayed, and the player specifies that he is in the group called MEN. Then the program asks the player's name. The response is JOHN. Next the program steps through the group 1 questions, and JOHN chooses an answer to each one. From this point the dialog proceeds to the point at which JOHN is asked to supply information about how a compatible member of the group called WOMEN should answer the group 2 questions. This dialog is shown in Figure 7.11.

Figure 7.10: JOHN's Answers

INSIDE BASIC GAMES   175

WHO WOULD YOU MOST WISH TO BE?

1. BO DEREK
2. BETTY FRIEDAN
3. ELIZABETH HOLTZMAN
4. YOURSELF

VALUE   2   CHOICE: **4** ← (Player assigns + 2 to YOURSELF.)
VALUE   1   CHOICE: **3** ← (Player assigns + 1 to ELIZABETH HOLTZMAN.)
VALUE −1   CHOICE: (Player presses RETURN.)
VALUE −2   CHOICE: (Player presses RETURN.)

WHOM WOULD YOU LIKE TO INVITE TO DINNER?

1. PAUL NEWMAN
2. WALTER MONDALE
3. STEVE GARVEY

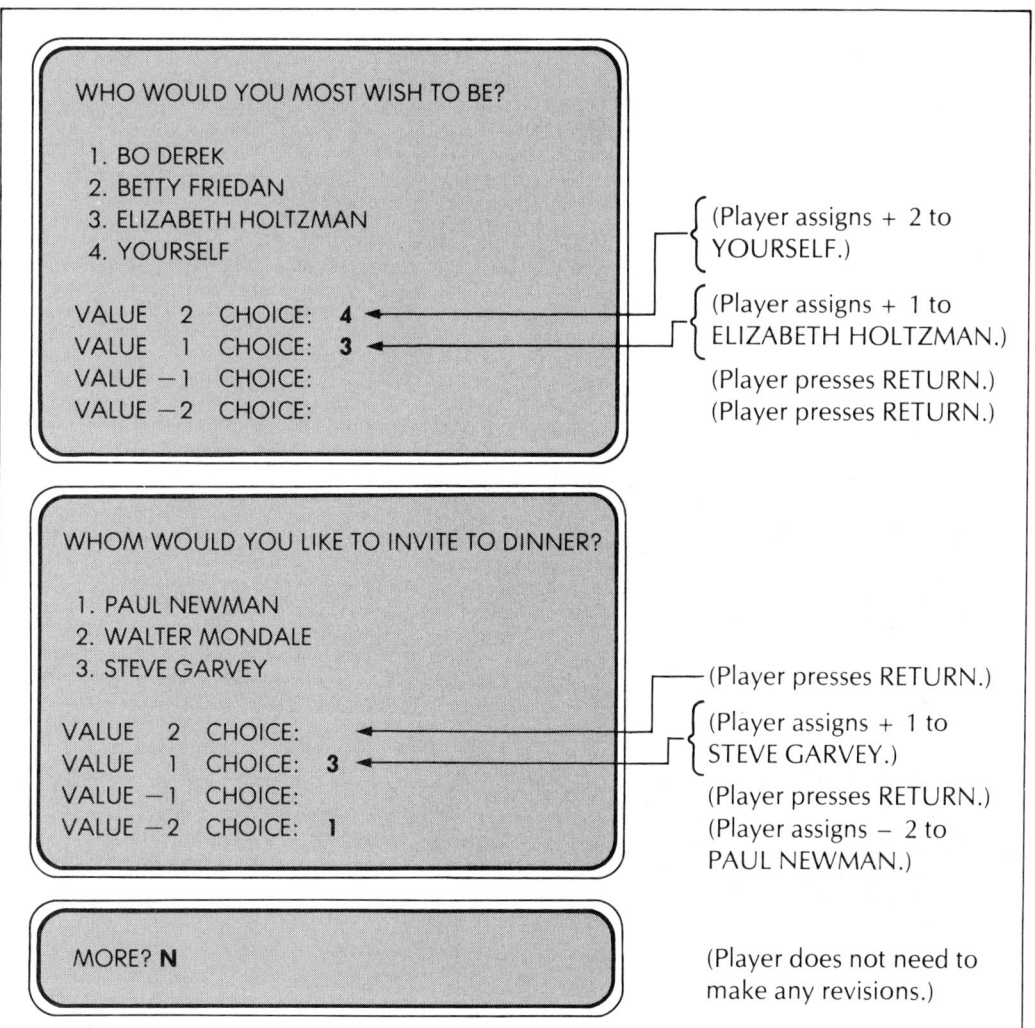

VALUE   2   CHOICE: (Player presses RETURN.)
VALUE   1   CHOICE: **3** ← (Player assigns + 1 to STEVE GARVEY.)
VALUE −1   CHOICE: (Player presses RETURN.)
VALUE −2   CHOICE: **1** (Player assigns − 2 to PAUL NEWMAN.)

MORE? **N**   (Player does not need to make any revisions.)

  This sample shows the portion of the new player entry dialog that picks up where Figure 7.10 left off. JOHN, who is a member of group 1 (MEN) is shown the questions to be asked members of group 2 (WOMEN), and is allowed to assign positive or negative values to choices for each of these questions. In this example, there are four values available (2, 1, − 1, − 2), and JOHN can assign a different choice to each of them. The number of values (4) and the numbers 2, 1, − 1, − 2 appear in DATA statements in the initialization routine. Choices assigned no value receive a value of zero by default. These values are used in measuring the compatibility of each group 2 player's answers with JOHN's wishes.

Figure 7.11: JOHN's Preferences

176 THE MATCH-UP GAME

[ ] N
1. MEN
2. WOMEN
GROUP: **2**

NAME: **SUSAN**

WHO WOULD YOU MOST WISH TO BE?

1. BO DEREK
2. BETTY FRIEDAN
3. ELIZABETH HOLTZMAN
4. YOURSELF

CHOICE: **2**

WHOM WOULD YOU LIKE TO INVITE TO DINNER?

1. PAUL NEWMAN
2. WALTER MONDALE
3. STEVE GARVEY

CHOICE: **1**

This is the first part of the new player entry for a member of group 2. Here is how SUSAN's answers are scored in the light of JOHN's preferences:

$$\begin{aligned}
&\text{Question 1: (VALUE: } 0) \times \text{(WEIGHT: } 7) = 0 \\
&\text{Question 2: (VALUE: } -2) \times \text{(WEIGHT: } 6) = -12 \\
&\text{TOTAL SCORE:} -12
\end{aligned}$$

In Figure 7.13, SUSAN's preferences are used to score JOHN's answers.

Figure 7.12: SUSAN's Answers

INSIDE BASIC GAMES 177

HOW OLD ARE YOU?

1. UNDER 25
2. 25-38
3. 39
4. OLD ENOUGH TO KNOW BETTER
5. AT THE DANGEROUS AGE

VALUE   2   CHOICE: **2**
VALUE   1   CHOICE: **3**
VALUE −1   CHOICE: **4**
VALUE −2   CHOICE: **5**

(Player assigns +2 to 25-38.)
(Player assigns +1 to 39.)
(Player assigns −1 to KNOW BETTER.)
(Player assigns −2 to DANGEROUS AGE.)

WHAT DO YOU WANT MOST FROM LIFE?

1. FAME
2. FORTUNE
3. HAPPINESS

VALUE   2   CHOICE: **3**
VALUE   1   CHOICE:
VALUE −1   CHOICE: **2**
VALUE −2   CHOICE: **1**

(Player assigns +2 to HAPPINESS.)
(Player presses RETURN.)
(Player assigns −1 to FORTUNE.)
(Player assigns −2 to FAME.)

MORE? **N**

In this sample dialog, SUSAN has entered preferences for answers made by members of group 1 (MEN). Here is how JOHN's answers are scored in the light of SUSAN's preferences:

```
Question 1: (VALUE:   2) × (WEIGHT:   9)   =    18
Question 2: (VALUE:   2) × (WEIGHT:   6)   =    12
TOTAL SCORE:                                    30
```

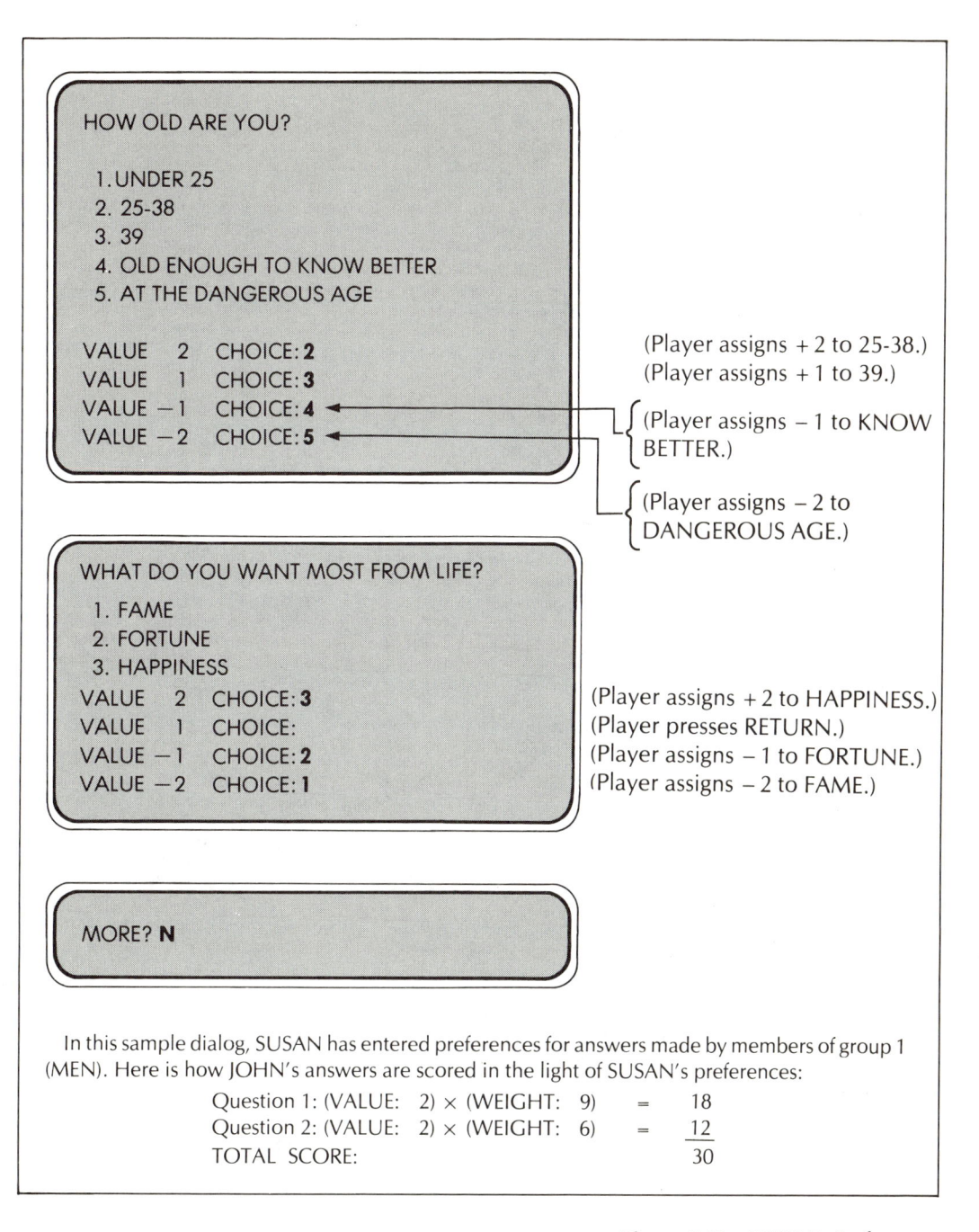

Figure 7.13: SUSAN's Preferences

178 THE MATCH-UP GAME

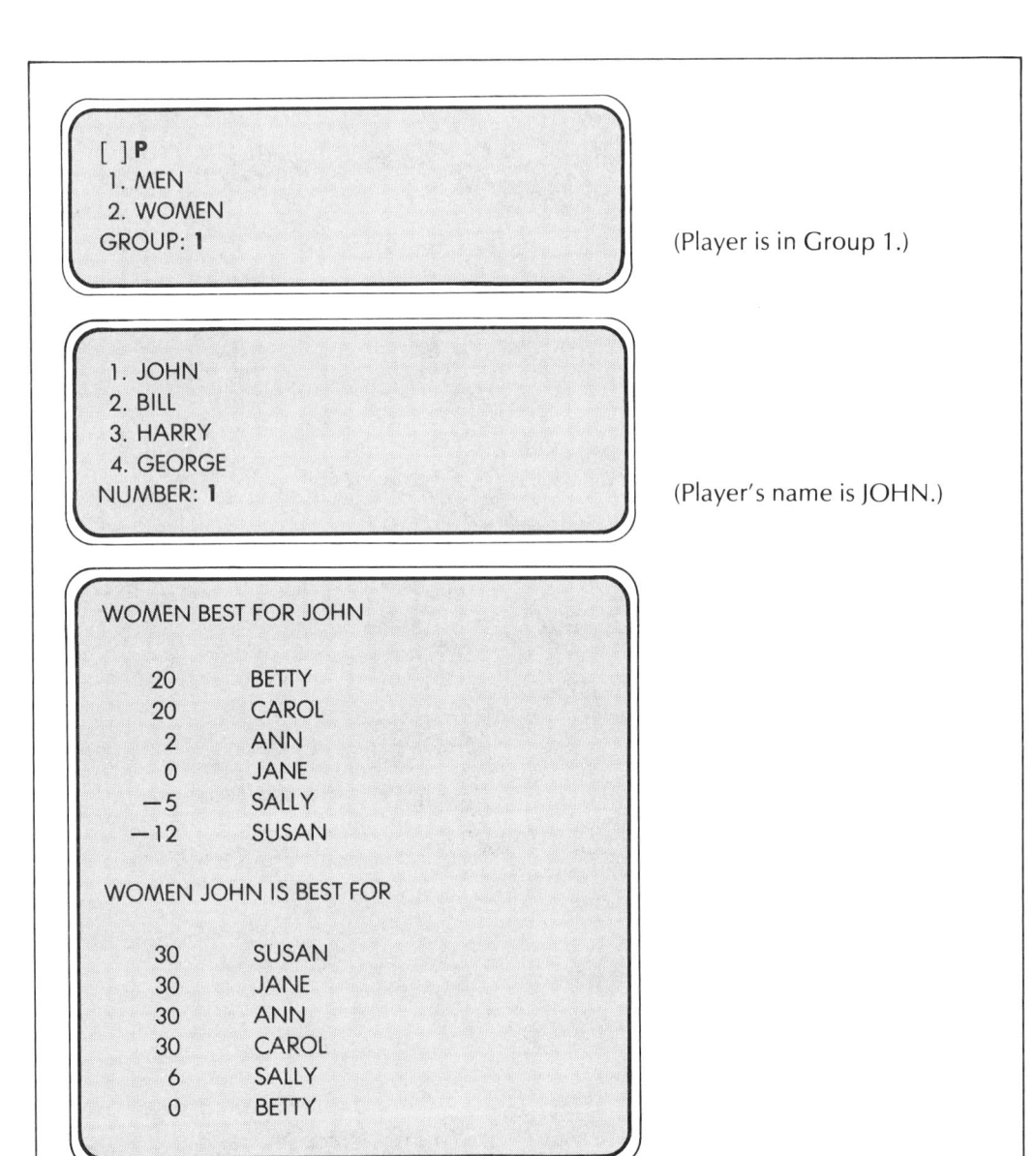

```
[ ] P
1. MEN
2. WOMEN
GROUP: 1
```
(Player is in Group 1.)

```
1. JOHN
2. BILL
3. HARRY
4. GEORGE
NUMBER: 1
```
(Player's name is JOHN.)

```
WOMEN BEST FOR JOHN

    20      BETTY
    20      CAROL
     2      ANN
     0      JANE
    −5      SALLY
   −12      SUSAN

WOMEN JOHN IS BEST FOR

    30      SUSAN
    30      JANE
    30      ANN
    30      CAROL
     6      SALLY
     0      BETTY
```

This example shows the use of the "P" command to obtain two lists: the players in the other group whose answers corresponded most closely with JOHN's wishes, and the players in the other group whose wishes corresponded most closely with JOHN's answers. The number to the left of each name is the actual score.

Figure 7.14: Pairing

INSIDE BASIC GAMES   179

```
JOHN
NAME:
```
(Player presses RETURN.)

```
HOW OLD ARE YOU?
  1. UNDER 25
  2. 25-38
  3. 39
  4. OLD ENOUGH TO KNOW BETTER
  5. AT THE DANGEROUS AGE
CHOICE: 2: 5
```
(Player changes age from 25-38 to DANGEROUS AGE.)

```
WHAT DO YOU WANT MOST FROM LIFE?
  1. FAME
  2. FORTUNE
  3. HAPPINESS
CHOICE: 3: 1
```
(Player changes goal from HAPPINESS to FAME.)

```
WHO WOULD YOU MOST WISH TO BE?
  1. BO DEREK
  2. BETTY FRIEDAN
  3. ELIZABETH HOLTZMAN
  4. YOURSELF
VALUE   2   CHOICE: 4:
VALUE   1   CHOICE: 3: D
VALUE  -1   CHOICE: 2
VALUE  -2   CHOICE: Q
```
(Player presses RETURN, so YOURSELF still scores + 2.)

(ELIZABETH HOLTZMAN no longer scores + 1.)

(BETTY FRIEDAN is now assigned a score of − 1.)

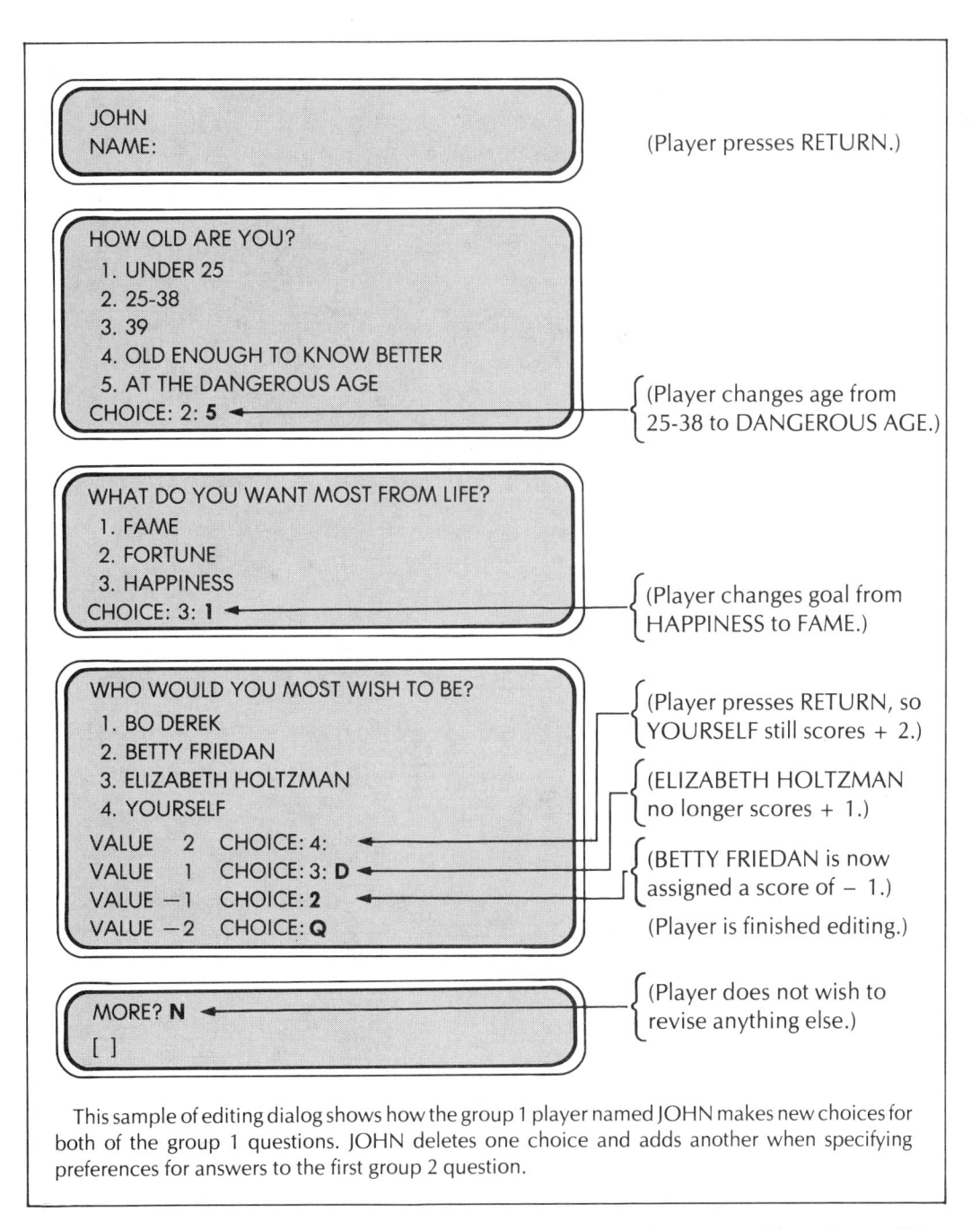

(Player is finished editing.)

```
MORE? N
[ ]
```
(Player does not wish to revise anything else.)

This sample of editing dialog shows how the group 1 player named JOHN makes new choices for both of the group 1 questions. JOHN deletes one choice and adds another when specifying preferences for answers to the first group 2 question.

Figure 7.15: Player Entry Editing

The E Command

The E command can be used in response to the "[ ]" prompt to alter an existing player entry. In order to do this, the player's identity must first be determined. This is done in the same way as for the P command (Figure 7.14): first the player's group is requested and all of the names in that group are displayed, then the player's number is requested. In the example shown, there are only a few names, but there could be more names than will fit on a screen at once. In this case, a partial list is printed and the "NUMBER: " question is asked. If RETURN or any non-numeric entry is made, the next screenload will be displayed.

After the player's identity has been determined, another pass over the player's entry dialog is made. In each case, current information, if any, is displayed, and then changes are accepted. This is the same editing function that would be entered if the answer to the "MORE?" question (see Figures 7.11 and 7.13) were "Y". A sample of this editing dialog is shown in Figure 7.15.

This concludes our presentation of the instructions for playing Match-up. While the instructions are long and involved, the playing of the game is actually quite simple.

The Match-up Program

The Match-up game is implemented by the program shown in Figures 7.16 through 7.59. There are more than forty subroutines (even excluding the "stubs" for external storage functions), and over 225 lines of BASIC instructions, derived from more than 500 lines of Free BASIC description. Because of its many subroutines, the Match-up program provides a good example of the kind of subroutine argument passing that is unique to BASIC: the use of global "implicit" arguments. This convention is extremely inflexible and error-prone, but if you are going to program in BASIC, then you must learn to use it and try to minimize the possibility of errors.

The Match-up program is large enough to be called a "real" program; it illustrates many of the techniques and principles of program structuring and design that are often difficult to see in smaller examples. Here are some of the aspects of computer programming that you can learn about by studying the Match-up program:

- General design concepts:
  - Modularity
  - Top-down structure
  - Isolation of function
  - Provision for change
  - Use of stubs to define the program's structure

- General programming techniques:
  - Design of user and programmer documentation
  - Design of user interaction
  - Error handling

- Specific programming techniques:
  - Design of editing functions
  - Insertion and deletion algorithms
  - Design of file maintenance functions
  - Space-saving

We shall now discuss the points shown above in more detail.

General Design Concepts

The first category in the list above is called General Design Concepts. Under this heading we shall discuss aspects of program design that apply to all programs.

Modularity. The term "modularity" has no widely accepted definition. Here, the term is used to call attention to the fact that the Match-up program has been divided into small program units (as evidenced by its description in over forty self-contained figures). Each of these units has a clearly identifiable job to do, and well-defined ways of interacting with other program units. The adherence to well-defined interfaces with other program units is especially important in BASIC, since every program can refer to every variable.

Top-Down Structure. Top-down structure is illustrated by the arrangement of the Match-up program. The "main" program appears

THE MATCH-UP GAME

```
#Match-up game
GOSUB init                                  #set up arrays, constants
repeat {
    GOSUB create                            #make or edit question file
    repeat {                                #play the game
        GOSUB clearscreen
        PRINT "[ ]";                        #display prompt
        GOSUB onech: PRINT X$               #get and echo command
        IF case
            X$ = "E" THEN GOSUB editplayer  #edit an existing player entry
            X$ = "N" THEN GOSUB newplayer   #add a new player
            X$ = "P" THEN GOSUB pairup      #display pairings for a player
            X$ = "L" THEN GOSUB loadplayers #load a player file
            X$ = "S" THEN GOSUB saveplayers #save a player file
            X$ = "Q" THEN {                 #end the game
                PRINT "REALLY? "
                GOSUB onech
                IF X$ = "Y" THEN
                    break
            }
        }
    }
}
100   GOSUB2260
110   GOSUB1420
120   GOSUB2200:PRINT"[ ]";:GOSUB2180:PRINTX$
130   IFX$="E"THENGOSUB210:GOTO120
140   IFX$="N"THENGOSUB260:GOTO120
150   IFX$="P"THENGOSUB970:GOTO120
160   IFX$="L"THENGOSUB2230:GOTO120
170   IFX$="S"THENGOSUB2240:GOTO120
180   IFX$<>"Q"THEN120
190   PRINT "REALLY? ";:GOSUB2180
200   IFX$="Y"THEN110
205   GOTO120
```

This is the principal loop of the Match-up game. The *create* subroutine contains a similar command selection for building a file of questions. The inner loop of the program shown here has commands for "playing" the game—building the file of players and ordering pairup printouts.

Figure 7.16: Match-up

first; after that come the subroutines called by the main program, then the subroutines called by the subroutines of the main program, and so on. Naturally, since the structure we just described is a two-dimensional branching structure (a "tree"), while the pages of a book follow one another in linear order, the actual arrangement only approximates the one described.

#Edit an existing player entry—"E" command

editplayer	GOSUB identity	#ask for (PP,GP)
	IF XX = 0 THEN	#can escape if mistake
	RETURN	
	GOSUB othergroup	#GC = "other" group for GP
	GOSUB loadwork	#load entry into workspace
	repeat {	
	GOSUB changes	#get changes to file
	GOSUB askok	#ask if more to do
	} until (OK = 1 OR OK = −1)	#update or abort
	IF OK = 1 THEN	
	GOSUB update	#update entry from workspace
	RETURN	

```
210    GOSUB1310:IFXX=0THENRETURN
220    GOSUB1290:GOSUB340
230    GOSUB460:GOSUB510:IFNOT(OK=1OROK=−1)THEN230
240    IFOK=1THENGOSUB380
250    RETURN
```

The *editplayer* routine is the "edit" routine for a player's entry. The *identity* routine gets a player index PP and the player's group index GP by conducting a dialog with the player. The routine *othergroup* computes the index GC of the group being matched with the player's group. The *loadwork* routine fetches the array elements corresponding to player (PP,GP), and places them into a workspace where they can be manipulated without changing the actual array entries. The *changes* routine makes a pass over the information in the workspace, allowing it to be edited. The *askok* routine is then called. It uses the variable OK to specify: "changes OK, update the file" (OK = 1), "more editing needed" (OK = 0), or "abort the editing process" (OK = −1). Finally, the *update* routine writes the edited workspace contents back into the array locations from which they came. If the process is aborted at any time before the call to *update* is made, there will be no change to the actual array entries.

Figure 7.17: Editing a Player Entry

184 THE MATCH-UP GAME

This type of program arrangement is helpful to the reader, since it makes the program easier to understand. Each "level" of the program is constructed from calls to lower level programs that perform progressively more specialized tasks. The degree of abstraction is reduced at each level of the program: at the highest level, the program handles abstract concepts like adding players or finding pairs of compatible players; at the lowest level, specific BASIC instructions are used to manipulate specific variables.

The term "top-down," however, usually refers to more than just the structure of a program listing. Instead, this term refers to the order in which the parts of the program are developed. In fact, the Match-up program was developed in almost exactly the same order as that in which the program units appear in this book.

#Make a new player entry

newplayer GOSUB askgroup   #get the new player's group (GP)
            GOSUB othergroup   #GC is the other group
            GOSUB clearwork   #initialize the work area
            repeat {   #obtain new information, then revisions
                GOSUB changes
                GOSUB askok
            } until (OK = 1 OR OK = −1) #until player is happy or gives up
            IF OK = 1 THEN   #if happy,
                GOSUB storenew   # then create an array entry
            RETURN

```
260  GOSUB1260:GOSUB1290:GOSUB300
270  GOSUB460:GOSUB510:IFNOT(OK=1OROK=−1)THEN270
280  IFOK=1THENGOSUB390
290  RETURN
```

The *newplayer* routine creates a new player entry. The *askgroup* routine finds out the player's group (GP), and *othergroup* computes the other group number (GC). Then the *clearwork* routine initializes the work area (i.e., fills it with zeroes and null strings). The loop of calls to *changes* and *askok* is identical to the one in *editplayer*. This is because *changes* and its subroutines are designed to edit existing entries or to create new ones. If the player is satisfied with the contents of the workspace, then the *storenew* routine is called to assign array indices to the entry and store it.

Figure 7.18: Creating a Player Entry

This book has provided many examples of program design and development, but it cannot cover every aspect of this subject. For a discussion of top-down development and the advantages it offers over other development methods, you should consult the many fine works by Dijkstra, Wirth, Yourdon and others on program development techniques. *How To Manage Structured Programming* by Ed Yourdon contains an especially clear presentation of this material.

Isolation of Function. Not only is it important for programs to be broken into small, manageable units with well-defined jobs and interfaces, but it is also important for information that does not need to be widely dispersed throughout the program to be localized to the greatest extent possible.

For example, the subroutines *unpackwish* and *packwish* (Figure 7.32) are the only locations in the program where specific use is made of the actual structure of the single-number encoding of the player's assignment of choices to preference values (see Figures 7.11 and 7.13). Outside of these routines, the encoded number is treated

---

#Clear the workspace for an entry for group GP (other is GC)

clearwork WN$ = ''''   #clear name
              FOR QX = 1 TO NQ(GP)   #clear answers to GP questions
                  WA(QX) = 0
              NEXT QX
              FOR QX = 1 TO NQ(GC)   #clear wishes for GC answers
                  WW(QX) = 0
              NEXT QX
              RETURN

```
300   WN$=''''
310   FORQX=1TONQ(GP):WA(QX)=0:NEXTQX
320   FORQX=1TONQ(GC):WW(QX)=0:NEXTQX
330   RETURN
```

The *clearwork* routine clears the workspace, which consists of the player name WN$, the array of player answers WA, and the array of player wishes WW.

Figure 7.19: Clearing the Workspace

as indecipherable, while the player's assignment of choices to preference values is represented as an array of choice numbers corresponding to the array of preference values. The routine *packwish* takes such an array and generates a single encoded number from it; the routine *unpackwish* takes an encoded number and generates an array of choice numbers. The translation functions that pass back and forth between these two forms of the player's assignment of choices are isolated in the two routines in Figure 7.32. This is a good example of isolation of function. We shall see another example in the isolation of the inter-player scoring function to the *score* routine (see Figure 7.34).

Provision for Change. Provision for change is another general design concept that is illustrated by the Match-up program. When designing computer programs, it is extremely important to recognize the following corollary of Murphy's Law:

> If any aspect of the program is absolutely immutable, someone will have a compelling reason for changing it, no matter what the original specification said.

#Load the workspace from player entry (PP,GP); other is GC.
loadwork WN$ = NM$(PP,GP)   #load the name
            FOR QX = 1 TO NQ(GP)   #load answers to GP questions
                WA(QX) = A(QX,PP,GP)
                NEXT QX
            FOR QX = 1 TO NQ(GC)   #load wishes for GC answers
                WW(QX) = W(QX,PP,GP)
                NEXT QX
            RETURN

    340   WN$=NM$(PP,GP)
    350   FORQX=1TONQ(GP):WA(QX)=A(QX,PP,GP):NEXTQX
    360   FORQX=1TONQ(GC):WW(QX)=W(QX,PP,GP):NEXTQX
    370   RETURN

The *loadwork* routine loads the workspace from the player name array NM$, the player answer array A, and the player wish array W. The player and group index values are given by PP and GP.

Figure 7.20: Loading the Workspace

Or to put it more succinctly:

If anything can't change, it will have to.

Without further explanation of why provision for change is desirable, let's examine some of the general approaches used in the Match-up program. The first and most important technique for providing for change is to write clear, well-structured and well-documented programs.

The next most important technique for accommodating future changes is to avoid the use of numeric constants in programs. In most BASIC programs, the greatest opportunity to apply this technique is in the specification of loop limits and array dimensions. In the Match-up program, array dimensions and upper loop limits are almost always variables; on the few occasions when they are not variables, they are parameters (i.e., symbolically named constants). Lower loop limits, on the other hand, are almost invariably specified by constants (usually 1 or 0). Such usage is easier to understand than the use of variables for lower limits, and a lower limit of 1 or 0 seldom needs to be changed.

Another technique for accommodating change is to identify those aspects of a program that are likely to change, and to assure that such changes can be made in one place. For example, in the Match-up

---

#Update the player entry (PP,GP) from the workspace; other is GC.

update   UP = PP: UG = GP: UC = GC   #arguments for storework
            GOSUB storework
            RETURN

  380   UP=PP:UG=GP:UC=GC:GOSUB420:RETURN

Update is the companion routine to *loadwork*. The sequence of editing is: GOSUB *loadwork*, perform editing in the workspace, GOSUB *update*. The routine *storework* (even though its name makes it appear to be a companion to *loadwork*), uses the arguments UP,UG for the player and UC for the "other" group. The reason for this is that *storework* is called with different arguments by *storenew*; if PP and GP were the arguments used by *storework*, then *update* would have to change the value of PP. Since PP has meaning within the routine calling *update*, the change to PP would be a *side effect* of the call to update. Side effects of subroutines are hard to avoid in BASIC, but every effort should be made to do so.

Figure 7.21: Updating a Player Entry from the Workspace

program, the maximum number of choices associated with any question is represented by the variable MC. This variable is set in the initialization subroutine (Figure 7.59) by the pair of instructions:

    READ MQ, MC, MP
    DATA maxquestions, maxchoice, maxplayers

(In the actual BASIC program, numeric values, rather than parameters, appear in the DATA statement.) This is the only place where a change must be made if the maximum number of choices is to be changed, since throughout the program, MC has been used in dimension statements and loop limits. If constants had been used in each of these places, they would all have had to be found and changed. Furthermore, if the program were full of numeric constants and if the maximum number of choices were to change, say, from 3 to 5, the 3s that we would need to change would first have to be distinguished from the 3s that had another meaning and didn't have to change. Also, you may have written

    IF X > 2

instead of

    IF X >= 3

and in that case changing the maximum number of choices from 3 to 5 would mean changing the 2 to a 4. If the statement had contained

    IF X > MC − 1

then the change would still occur automatically.

The initialization routine of Figure 7.59 provides for change in another way. All of the variables whose values are expected to change have been set with DATA statements. In contrast, note that the variable GM, which represents the number of groups of players, is set by the instruction

    GM = 2

An assignment statement is used instead of a DATA statement, and the constant "2" is used instead of a parameter like "maxgroup." This is done because the number of groups cannot be changed without changing the entire structure of the game. We write the variable GM throughout the program instead of "2" because we wish the program to be clear and easily understood, not because we think that the value of GM can ever change.

We have now sufficiently noted the provisions made for change in the Match-up program. As we describe the program in more detail, you will see many cases in which variables have been used for this purpose. In contrast, there is one constant that is used throughout the program: the string constant "Q". You might find it an interesting exercise to try to change the abort character "Q" (used at many places in the dialog) to something else. Do you know why a constant was used instead of a string variable?

Using Stubs to Define the Program's Structure. The last of the general design concepts that we shall discuss here is the use of stubs to define the structure. This is simply another aspect of the top-down development process, which requires definition and checkout of the interactions of higher level modules before the coding and checkout of lower level modules. In order to test the interactions of modules

```
#Store a new entry for group GP
storenew   UG = GP: UC = GC              #use current GP, GC
           IF NP(GP) < MP THEN {         #if room for another player,
               NP(GP) = NP(GP) + 1       # add it
               UP = NP(GP)
               GOSUB storework
               }
           else                          #otherwise,
               PRINT "SORRY, NO ROOM."   # don't
           RETURN
   390  UG=GP:UC=GC
   400  IFNP(GP)<MPTHENNP(GP)=NP(GP)+1:UP=NP(GP):GOSUB420:RETURN
   410  PRINT"SORRY,NO ROOM.":RETURN
```

The *storenew* routine is called after the creation of a new player entry. It assigns the next available player index to the new player and calls *storework* to save the player information that has been built up in the workplace.

This routine illustrates a common design flaw. The programmer has failed to anticipate that a test will be needed to determine whether or not there is room to store the new entry. (This test belongs at the beginning of the *newplayer* routine.) Thus, the player may go through the entire process of answering questions and stating preferences, only to be told afterwards what the program could have known all along—SORRY, NO ROOM.

Figure 7.22: A Flawed Program to Store a Player Entry

190 THE MATCH-UP GAME

that have not yet been completely coded and checked, the programmer replaces the incomplete portions of these modules by *stubs*—small programs whose interactions are like those of the incomplete portions of the modules but whose functional behavior is rudimentary.

For example, the programs concerned with external storage of question or player information are simply dummy programs that say "EXTERNAL STORAGE IS NOT IMPLEMENTED" whenever they are called. More elaborate stubs—potentially much more useful in checkout of the remaining system—might have created a question or player file in memory whenever a load was requested. During the actual checkout process, a different approach was taken. The programmer began by writing stubs for the *create* portion of the program (Figures 7.45 - 7.56). These stubs allowed keyboard entry of question and player information with minimal editing. Then, after the playing portion of the program had been perfected, the editing stubs were expanded into the versions shown in Figures 7.45 - 7.56. Since a means of entering question and player information was available early in the checkout process, stubs for the loads from external storage were not expanded to create dummy files.

#Store workspace contents into player entry (UP,UG); other is UC.

storework   NM$(UP,UG) = WN$ #store name
               FOR QX = 1 TO NQ(UG)   #store answers
                   A (QX,UP,UG) = WA(QX)
                   NEXT QX
               FOR QX = 1 TO NQ(UC)   #store wishes
                   W(QX,UP,UG) = WW(QX)
                   NEXT QX
               RETURN

```
420    NM$(UP,UG)=WN$
430    FORQX=1TONQ(UG):A(QX,UP,UG)=WA(QX):NEXTQX
440    FORQX=1TONQ(UC):W(QX,UP,UG)=WW(QX):NEXTQX
450    RETURN
```

The *storework* routine stores the workspace contents into the player name array NM$, the answer array A, and the wish array W. The player and group index values are given by UP and UG.

Figure 7.23: Storing Workspace Contents

#Ask for changes or additions to workspace for a group GP player; other is GC.

changes   GOSUB changename   #player name
             FOR QX = 1 TO NQ(GP)   #player answers
                GOSUB changechoice
                IF X$ = "Q" THEN   #Q = No more changes
                    break
                NEXT QX
             IF X$ <> "Q" THEN
                GOSUB clearscreen
                PRINT "preference request"; GN$(GC)
                GOSUB onech
                FOR QX = 1 TO NQ(GC)   #player wishes
                    GOSUB changewish
                    IF X$ = "Q" THEN   #Q = No more changes
                      break
                  NEXT QX
             RETURN

```
460   GOSUB550
470   FORQX=1TONQ(GP):GOSUB680:IFX$<>"Q"THENNEXTQX
480   IFX$="Q"THENRETURN
485   GOSUB2200:PRINT"NOW GIVE YOUR PREFERENCES FOR ANSWERS BY ";GN$(GC)
490   GOSUB2180:FORQX=1TONQ(GC):GOSUB880:IFX$<>"Q"THENNEXTQX
500   RETURN
```

The *changes* routine asks the player for changes or additions to the existing entry in the workspace. (When first called from *newplayer* the routine can only ask for additions, since there is nothing in the workspace to change.) The structure of the routine parallels those of *clearwork, loadwork,* and *storework*. The routine *changename* is called for the player name; *changechoice* is called for the player's answers to the group GP questions; *changewish* is called for the player's preferences for group GC answers.

Figure 7.24: Asking for Changes to Player Entry

THE MATCH-UP GAME

General Programming Techniques

We have been discussing design concepts that apply to all programming situations. There are also several programming techniques illustrated in the Match-up program that apply to a broad range of programs.

Design of User and Programmer Documentation. You may be surprised to learn that the design of user and programmer documentation is one of the first tasks that you should perform when creating a program. Unfortunately, this task is often left until last, and then done half-heartedly.

#Ask whether player wishes to make another pass over the workspace

askok   GOSUB clearscreen
          PRINT "MORE? "
          GOSUB onech   #single-character reply
          IF case
              X$ = "Q" THEN OK = −1   #Q = abort: don't store workspace
              X$ = "Y" THEN OK = 0   #Y = yes, make another pass
          else
              OK = 1   #otherwise, OK; no more changes
          RETURN

```
510   GOSUB2200:PRINT"MORE? ";:GOSUB2180:OK=1
520   IFX$="Q"THENOK=−1
530   IFX$="Y"THENOK=0
540   RETURN
```

The *askok* routine asks the player whether or not more editing is desired. If the player answers "Y", this routine returns with OK = 0, signaling the calling program that another call to *changes* is desired. If the player answers "Q", this routine returns with OK = −1, signaling that the entire process has been aborted and that the workspace contents should be abandoned without being written into the player entry. Any other reply causes a return with OK = 1. In this case, the calling program will *update* or *storenew* to save the workspace contents in the player entry.

The simple set of choices that this routine provides is essential to the smooth operation of any editing cycle.

Figure 7.25: Asking for More

INSIDE BASIC GAMES 193

Figure 7.26 shows the steps followed in the creation and documentation of Match-up. Notice the way that programming and documentation proceed concurrently and are kept consistent with one another.

The main reason for the development of Free BASIC was to free the BASIC programmer from the use of line numbers. The process shown in Figure 7.26 shows how important this goal is and how successfully it has been achieved. This process calls for further changes to the program after the figures have been constructed. These changes can be made easily, since the BASIC program can be renumbered and the BASIC instructions appearing in the figures can be replaced with no effect on anything that appears elsewhere in the figures or in any of the accompanying text. The program design, development and documentation have been completely freed from the use of BASIC line numbers. Line numbers are used only in the translation from Free BASIC to BASIC, a mechanical process that might ultimately be performed by a computer program. The purpose of the actual BASIC

Steps in the Creation and Documentation of the Match-up Game

- The author conceived and refined the original idea for the program. The final form of the program's behavior was decided upon before any program development began.

- The Free BASIC program descriptions that appear in Figures 7.16-7.59 were first written with much simpler "stubs" for the question-creation and editing routines of Figures 7.45-7.56.

- A manual translation from Free BASIC into BASIC was made.

- The program was debugged, and improvements were made to the screen formats and dialog. Real editing programs replaced the stubs. All changes were first made to the Free BASIC descriptions, then incorporated into the BASIC program.

- A BASIC listing was made, and the Free BASIC and BASIC were combined into the Figures 7.16-7.59. At that time, the descriptive material in the figures was written, and this close examination led to discovery of the need for small changes. These were made first to the Free BASIC and then to the BASIC.

- A sample run was made, during which Figures 7.1-7.15 and the accompanying text were prepared. This process led to further changes to the programs, and Figures 7.16-7.59 were altered accordingly.

Figure 7.26: Creation and Documentation of Match-up

instructions that appear in Figures 7.16 - 7.59 is to provide concrete examples of the process of translation from Free BASIC into BASIC. Otherwise, these BASIC instructions, like the object code that can accompany a FORTRAN or PL/I listing, are only needed to facilitate certain seldom-needed forms of debugging.

The discussion above emphasized the importance of the programming language in the process of documentation. Just as user documentation is most effectively built around actual examples of program behavior, programmer documentation is best conveyed by the actual

---

#Name-editing routine

changename GOSUB clearscreen
                IF WN$ = '''' THEN {   #if no name on file,
                    PRINT "NAME: ";   # get one
                    repeat {
                        GOSUB stringin
                        WN$ = SS$
                    } until (WN$ <> '''')
                }
                else   #if name on file,
                    PRINT WN$ # display it
                    PRINT "NAME: ";
                    GOSUB stringin   #get new name
                    IF SS$ <> '''' THEN
                        WN$ = SS$ #"RETURN" = don't change
                }
                RETURN

  550   GOSUB2200:IFWN$<>''''THEN590
  560   PRINT"NAME: ";
  570   GOSUB610:WN$=SS$:IFWN$=''''THEN570
  580   RETURN
  590   PRINTWN$:PRINT"NAME: ";:GOSUB610:IFSS$<>''''THENWN$=SS$
  600   RETURN

The *changename* routine shows the player the current string that it has on file for the player's name, and accepts a replacement (or RETURN for "no change").

Figure 7.27: Editing the Player's Name

```
#String input routine—SS$ = all characters typed, up to a RETURN
Stringin    SS$ = ""                              #initialize SS$
            repeat {
                GOSUB onech                       #wait for character
                IF X$ = "return" THEN break       #done if RETURN
                else IF X$ <> "delete" THEN {     #if not a rubout,
                    PRINT X$;                     #  echo
                    SS$ = SS$ + X$                #  add to SS$
                }
                else IF LEN(SS$) <> 0 THEN {      #if one available,
                    PRINT "delete string";        #  remove from screen
                    SS$ = LEFT$(SS$,LEN(SS$) – 1)  #  remove from SS$
                }
            }
            IF SS$ = "Q" THEN                     #remove Q from screen
                PRINT "delete string";
            else IF SS$ <> "" THEN PRINT          #echo RETURN if not null line
            RETURN

610    SS$=""
620    GOSUB2180:IFX$=CHR$(13)THEN660
630*   IFX$<>CHR$(20)THENPRINTX$;:SS$=SS$+X$:GOTO620
640**  IFLEN(SS$)<>0THENPRINTX$;:SS$=LEFT$(SS$,LEN(SS$)–1):GOTO620
660*   IFSS$="Q"THENPRINTCHR$(20);:RETURN
665    IF SS$<>""THENPRINT
670    RETURN
```

The *stringin* routine is similar in effect to the instruction LINE INPUT SS$.

When handling the delete character (RUBOUT, back arrow, DEL), the program must check to see whether there are any characters in SS$ to be deleted. This section is more complicated in the Apple version, since LEFT$(SS$,0) is illegal in Apple BASIC. (See Figure 4.10.)

The routine has two features that make its behavior different from that of LINE INPUT. If SS$ = "Q", then the Q is removed from the screen. If SS$ = "", then no echo of the RETURN character occurs; otherwise a PRINT instruction is used to start a new line.

---

*In the Apple version, CHR$(8) represents the delete character, and CHR$(8);" ";CHR$(8) represents the delete string. In the TRS-80 version, CHR$(8) represents the delete string.
**See notes on Figure 4.10 for Apple version of this line.

Figure 7.28: String Input

196 THE MATCH-UP GAME

```
#Answer-editing routine for a group GP member; GC = other group.

changechoice  FOR QX = 1 TO NQ(GP)
                  GOSUB clearscreen
                  GQ = GP: GOSUB showquestion        #display question
                  PRINT "CHOICE: ";
                  IF WA(QX) <> 0 THEN                #display choice, if any
                      PRINT WA(QX); ": "
                  repeat {                           #get Q, D, RETURN or
                      GOSUB onech                    # valid choice number
                      IF case
                          X$ = "Q" or "return" THEN  #Q = no more editing
                              break                  #RETURN = next question
                          X$ = "D" THEN {            #D = remove choice
                              WA(QX) = 0
                              break
                              }
                          else {                     #valid choice number =
                              NN = VAL(X$)           # select that choice
                              IF 1 <= NN <= NC(QX,GP) THEN {
                                  WA(QX) = NN
                                  break
                                  }
                              }
                          }
                  IF X$ = "Q" THEN
                      break
                  NEXT QX
              RETURN
```

The *changechoice* routine presents the player's current answer, if any, for each question. It then asks for input. A RETURN means no change; a D means remove the current answer, a Q means no more changes, and a valid choice number means selection of that choice.

Figure 7.29: Changing Player Choices

program listings. Since BASIC has no way in which to convey information about program structure and does not provide easily understood source programs, the Free BASIC program description technique has been given the role of the source programming language in this book. Thus, all changes were first made to the Free BASIC description, then to the actual BASIC program. This means that the programmer documentation, which is built around the Free BASIC descriptions, always reflects the actual state of the program. This is an advantage that a "self-documenting" programming language has over description tools like flowcharts: any change must first be made to the documentation before it can be made to the program.

Design of User Interaction. The next point mentioned under general programming techniques is the design of user interaction. Since the mid-1960's, when the use of interactive computer systems began to increase, the design of the interaction that occurs between user and program has been widely neglected. Many programs that are otherwise well-designed have user interactions that were thrown in haphazardly by the programmer as a by-product of the processing portions of the programs. Other programs have interactions that are modeled after the kinds of input and output appropriate to non-interactive environments. Still others are designed for use with teletypes; the properties of the screen-display devices with which they will actually be used are ignored.

One common fault seen in user interactions designed by conscientious and well-meaning programmers is over-helpfulness. Wordy prompts, offers of help, menus and other such distractions are useful to the first-time player, but these features stand between the player and the fast, undistracted action of the game. Their effect on a game

```
680    FORQX=1TONQ(GP):GOSUB2200
690*   GQ=GP:GOSUB1240:PRINT"CHOICE:";:IFWA(QX)<>0THENPRINTWA(QX);":   ";
700    GOSUB2180
710    IFX$="Q"ORX$=CHR$(13)THEN750
720    IFX$="D"THENWA(QX)=0:GOTO750
730    NN=VAL(X$):IF1<=NNANDNN<=NC(QX,GP)THENWA(QX)=NN:GOTO750
740    GOTO700
750    IFX$="Q"THENRETURN
760    NEXTQX:RETURN
```

*The Apple version of this line differs only in the spacing inside the string constants.

Figure 7.29a: BASIC for Changing Choices

198 THE MATCH-UP GAME

```
#Ask for player's preferences—unpacked in FT array

askwish  FOR FX = 1 TO NF
            PRINT: PRINT "VALUE"; FV(FX); " CHOICE ";
            IF FT(FX) <> 0 THEN
                PRINT FT(FX);
            PRINT ": ";
            repeat {
                GOSUB onech
                IF X$ = "Q", "N" or "return" THEN
                    break
                else IF X$ = "D" THEN {            #D = delete preference
                    PRINT X$;                      #echo D
                    FT(FX) = 0
                    break
                }
                else {                             #number is a preference
                    NN = VAL(X$)
                    IF 1 <= NN <= MC THEN {
                        PRINT X$;                  # echo choice
                        FT(FX) = NN
                        break
                    }
                }
            }
            IF X$ = "Q" or "N" THEN                #Q = done editing
                break                              #N = next question
        NEXT FX
        RETURN
```

The *askwish* routine displays the current set of preferences from the FT array, and accepts changes or additions. Q, D and RETURN are interpreted as for *changechoice*; N means go on to the next question.

Figure 7.30: Asking Player for Preferences

is like that of leaving the house lights on in a theatre so that members of the audience can consult their programs during the performance.

In studying the Match-up program, you should notice the frequent use of brief prompts, screen clearing and single-character input. Also notice the *stringin* routine (Figure 7.28), and the way that it is used in the program's interactions. The principal purpose of this routine is to make up for the lack of a LINE INPUT instruction in most home computer BASIC systems. It does, however, have two special features that are experimental in nature:

— Removal of a single-character "Q" from the screen.
— Omission of any echoing on null-string input (i.e., a RETURN typed with no preceding characters is not echoed).

In screen interaction, information must not disappear from the screen before the player has a chance to read it. This can occur when more than one screenload of information must be output by the program (see Figure 7.44, the *identity* program), or when information output by the program is erased because the screen is cleared prior to entry of a new command. Figures 7.18 and 7.22, the *newplayer* and *storenew* routines, and Figure 7.33, the *pairup* routine, provide examples of this problem.

Another important feature of interactive programs, used when editing and updating files, is the ability to back out of a transaction, either because it was mistakenly begun, because errors occurred along the way, or simply because the player decided not to complete it. For example, notice that the *qnum* routine (Figure 7.56) can return the value QQ = 0 to indicate to the calling program that the player wishes to terminate the action. To see the importance of this feature,

```
770*  FORFX=1TONF:PRINT:PRINT"VALUE";FV(FX);" CHOICE";
780   IFFT(FX)<>0THENPRINTFT(FX);
790   PRINT": ";
800   GOSUB2180:IFX$=CHR$(13)ORX$="Q"ORX$="N"THEN850
810   IFX$="D"THENPRINTX$;:FT(FX)=0:GOTO850
820   NN=VAL(X$)
830   IF1<=NNANDNN<=MCTHENPRINTX$;:FT(FX)=NN:GOTO850
840   GOTO800
850   IFX$="Q"ORX$="N"THEN870
860   NEXTFX
870   RETURN
```

*The Apple version of this line differs only in the spacing inside the string constants.

Figure 7.30a: BASIC for Asking for Preferences

consider an alternative design for *qnum* that would ask repeatedly for the question number until a valid one was given. This design would be very frustrating to the player who had inadvertently typed "D" for "delete a question" and was then faced with the prompt "QUESTION NUMBER:". As *qnum* is designed, the player need only press RETURN and the deletion process will be aborted.

The Match-up program uses several different prompts. Thus, for example, a player answering questions and requesting a pair-up has no access to the question-editing commands, and a player editing questions for one group has no access to the editing commands for the other group. These features protect against inadvertent invocation of commands outside of a special set. However, there is no protection against intentional invocation of any command. Any player can move from one prompt to another at any time, since the commands to do so are not secret, and are not protected in any special way. In one case (see Figure 7.16), the program asks "REALLY?" and waits for an answer of "Y" before responding to a potentially destructive command.

#Preference-editing routine for group GP member; other = GC

changewish   FOR QX = 1 TO NQ(GC)
        GOSUB clearscreen
        GQ = GC: GOSUB showquestion   #print question and choices
        WC = WW(QX): GOSUB unpackwish   #current wishes to FT
        GOSUB askwish   #add/change
        GOSUB packwish: WW(QX) = WC   #put back into WC
        IF X$ = "Q" THEN   #Q = no more editing
           break
        NEXT QX
    RETURN

```
880   FORQX=1TONQ(GC):GOSUB2200
890   GQ=GC:GOSUB1240:WC=WW(QX):GOSUB920:GOSUB770:GOSUB950:WW(QX)=WC
900   IFX$<>"Q"THENNEXTQX
910   RETURN
```

The *changewish* routine steps through the group GC questions, displaying the current player's preferences, if any, from the WW array in the workspace, and accepting changes and additions. The subroutines *unpackwish* and *packwish* are called to translate between the packed format of preferences used in the WW array, and the unpacked preferences manipulated in the FT array by *askwish*.

Figure 7.31: Editing a Player's Preferences

Error Handling. The error handling that occurs in the Match-up program is done subtly, without glaring error messages to disrupt the flow of the game. Most erroneous single-character input is merely ignored and not echoed. Figure 7.27 (*changename*) gives an example of how an illegal null input is handled (and illustrates why *stringin* does not echo the RETURN on a null input). Figure 7.22 (*storenew*) gives an example of the wrong way to handle a common error.

Specific Programming Techniques

In addition to the general programming techniques discussed in the previous section, several specific programming techniques are

---

#Routines to translate between a packed wish WC and the array FT

unpackwish FOR FX = 1 TO NF
                FT(FX) = WC mod (MC + 1)
                WC = (WC − FT(FX))/(MC + 1)
                NEXT FX
                RETURN

packwish   WC = 0: FM = 1
                FOR FX = 1 TO NF
                WC = WC + FT(FX)*FM
                FM = FM*(MC + 1)
                NEXT FX
                RETURN

```
920  FORFX=1TONF:FT(FX)=WC−(MC+1)*INT(WC/(MC+1))
930  WC=(WC−FT(FX))/(MC+1):NEXTFX:RETURN
950  WC=0:FM=1:FORFX=1TONF:WC=WC+FT(FX)*FM:FM=FM*(MC+1):NEXT FX
960  RETURN
```

The *unpackwish* and *packwish* routines pass between the packed and unpacked formats of a set of preferences for a question. In the unpacked format, each preference is a number between 0 and MC. A zero means no preference, a value of n (1 <= n <= MC) means choice n. There are *numfit* preferences, and each has a value; these values are stored in the FV array. When computing the numeric value of a match-up, each answer receives a score equal to the question weight from the WT array times the value from the FV array corresponding to the first position at which the answer appears in the array of preferences. If the answer isn't one of the preferences, it gets a score of zero.

In the packed format, the preferences, which are numbers between 0 and MC, are treated as digits in a number in the base MC + 1 representation. That is, each preference is multiplied by a power of MC + 1, and the results are added together.

Figure 7.32: Packing and Unpacking Preferences

THE MATCH-UP GAME

#Find the top matches for a player

pairup	GOSUB identity	#player is (PP,GP)
	GOSUB othergroup	#other group = GC
	GOSUB clearscreen	
	GOSUB clearscore	#initialize "top scores"
	PW = PP: GW = GP: GA = GC	#players' wishes, others' answers
	FOR PX = 1 TO NP(GC)	
	PA = PX: GOSUB score	#match (PW,GW) and (PA,GA)
	GOSUB savescore	#save top scores
	NEXT PX	
	PRINT GN$(GC); " BEST FOR "; NM$(PP,GP)	
	GS = GC: GOSUB printscores	#print top scores
	GOSUB clearscore	#initialize "top scores"
	PA = PP: GA = GP: GW = GC	#player's anwers, others' wishes
	FOR PX = 1 TO NP(GC)	
	PW = PX: GOSUB score	#match (PW,GW) and (PA,GA)
	GOSUB savescore	#save top scores
	NEXT PX	
	PRINT: PRINT GN$(GC); " THAT "; NM$(PP,GP); " IS BEST FOR"	
	GS = GC: GOSUB printscores	#print top scores
	GOSUB onech: RETURN	#leave on screen until key pressed

```
970  GOSUB1310:GOSUB1290:GOSUB2200:GOSUB1110
980  PW=PP:GW=GP:GA=GC
990  FORPX=1TONP(GC):PA=PX:GOSUB1050:GOSUB1140:NEXT PX
1000 PRINT GN$(GC);" BEST FOR ";NM$(PP,GP):GS=GC:GOSUB1120
1010 GOSUB1110:PA=PP:GA=GP:GW=GC
1020 FORPX=1TONP(GC):PW=PX:GOSUB1050:GOSUB1140:NEXT PX
1030 PRINT:PRINTGN$(GC);" THAT ";NM$(PP,GP);" IS BEST FOR":GS=GC:GOSUB1120
1040 GOSUB2180:RETURN
```

The *pairup* routine finds the members of the other group best matching this player's preferences and the members of the other group whose preferences this player best matches. The *score* routine computes the score associated with each pairing, the *savescore* routine keeps a sorted array of the top scores, and the *printscores* routine prints the names of the players in the other group whose scores are in the array of top scores.

Figure 7.33: Pairing Players

INSIDE BASIC GAMES 203

#Compute score SC of player (PA,GA)'s answers vs. (PW,GW)'s wishes

score   SC = 0   #initialize score
        FOR QX = 1 TO NQ(GA)   #step thru group GA questions
            WC = W(QX,PW,GW)   #(PW,GW)'s wish
            AV = A(QX,PA,GA)   #(PA,GA)'s answer
            GOSUB fit
            SC = SC + FT*WT(QX,GA)   #update score
        NEXT QX
        RETURN

```
1050  SC=0:FORQX=1TONQ(GA)
1060  WC=W(QX,PW,GW):AV=A(QX,PA,GA):GOSUB1080
1070  SC=SC+FT*WT(QX,GA):NEXTQX:RETURN
```

The *score* routine repeatedly calls *fit* to match answers against preferences, and accumulates a score SC by adding the weighted fits together.

Figure 7.34: Scoring Answers vs. Preferences

#Measure fit FT of answer AV vs. coded wishes WC

fit   FT = 0   #initialize fit
        GOSUB unpackwish   #translate WC into FT array
        FOR FX = 1 TO NF   #step thru FT, FV arrays
            IF FT(FX) = AV THEN {   #if AV has been assigned a value,
                FT = FV(FX)   # then FT = that value
                break
            }   #otherwise FT is as initialized above
        NEXT FX
        RETURN

```
1080  FT=0:GOSUB920
1090  FOR FX=1TONF:IFFT(FX)=AVTHENFT=FV(FX):RETURN
1100  NEXTFX:RETURN
```

The *fit* routine finds the first occurrence, if any, of the answer AV in the preferences encoded in WC. The encoded preference WC is translated by *unpackwish* into the array FT, and then AV is compared against the array elements in turn.

Figure 7.35: Scoring One Answer

also illustrated by the Match-up program. These techniques find application in many different programs.

Design of Editing Functions. Two styles of editing are used in the Match-up game. The first style is illustrated by the commands shown in Figure 7.4. This is a standard approach to editing sets of information consisting of numbered items (e.g., a "page" of text consisting of n lines, numbered 1 through n for editing purposes). The essential functions are to add, delete or modify an item. Adding an item is accomplished by one of two separate functions: inserting an item or adding an item onto the end. This separation into two functions occurs because of the problem of how to specify where the inserted item is to go. If the insertion command requires specification of the item before which the new item is to be placed, then a different means of specifying "insert at the end" must be found. (Do you see why?) On the other hand, if the insertion command required specification of an item after which the new item is to be placed, a problem would arise when trying to insert an item before the first item. The easiest solution is the one used in the Match-up game.

#Initialize the array of top scores

clearscores SX = 0: RETURN   #set array index to zero

   1110 SX=0:RETURN

The *clearscores* routine initializes the array pair TS, TI by zeroing the index SX of the "last" element.

Figure 7.36: Clearing Top Scores

#Display the top scores—group GS—from TS,TI arrays

printscores   IF SX <= 0 THEN PRINT "NONE"   #if array empty,
                                                    # don't print any
            else
                  FOR SZ = 1 TO SX   #step through arrays
                       PRINT TS(SZ), NM$(TI(SZ),GS) #print score and player name
                  NEXT SZ
            RETURN

   1120 IFSX<=0THENPRINT"NONE":RETURN
   1130 FORSZ=1TOSX:PRINTTS(SZ),NM$(TI(SZ),GS):NEXTSZ:RETURN

The *printscores* routine steps through the array pair (TS,TI) printing the score from TS and the player name using the index from TI and the group number GS.

Figure 7.37: Displaying Top Scoring Players

The other kind of editing used in the Match-up game is exemplified by the dialog shown in Figure 7.15. If you wish to edit your choices and preferences, the program will display them one at a time, showing you what it currently has on file at each point, and allowing you to alter it if you wish. If "no answer" is a legal possibility, then altering an answer can also mean removing a given answer without replacing it. This removal is the function of the "D" input shown in Figure 7.15. This type of deletion is essentially different from that associated with the "D" command used in question editing. When editing questions you are free to modify the structure of the file of questions as well as its contents. In editing your choices and preferences, you have no control over the structure; you can only affect the contents. This type of editing is often used with interactive debuggers—programs that allow you to examine and alter the contents of the computer's memory, since you cannot alter the amount or the addresses of memory locations.

This type of editing can only be used for a small file of items, since you must go through the items one by one until you locate the item that you wish to change. The player must be able to advance quickly through the items, or the process will be tedious. This quick advance is made possible by single-character input in the Match-up game. You only need to press RETURN to allow a choice or preference to stand and then move on to the next. You simply type the single-character choice number (there can't be more than nine choices per question) to alter a choice or preference. You type the single character "Q" when you have finished.

Insertion and Deletion Algorithms. Figures 7.52 and 7.55 show examples of deletions. Figures 7.40 and 7.53 show insertions. Similar algorithms for insertion and deletion are applicable whenever there is an array in which the order of the elements is significant. When an item is added to the array, it must be placed at the correct index, and all items that follow that index must be displaced. When an item is removed, all items that follow it must be moved so that no vacant space is left.

Let's look at what happens to each item of the array when insertion or deletion occurs. For insertion, each item (if any) after the inserted item gets the next higher-numbered index; for deletion, each item (if any) after the deleted item gets the next lower-numbered index.

When an item is deleted from within an array, an open slot is created in the middle, and the first item to move must be the one whose destination is the empty slot (see Figure 7.38). This move creates an empty slot, and the next item to be moved is the one headed

206 THE MATCH-UP GAME

1	1	1	1	1
2	2	2	2	2
.	.	.	.	.
.	.	.	.	.
.	.	.	.	.
k − 1	k − 1	k − 1	k − 1	k − 1
k	empty	k + 1	k + 1	k
k + 1	k + 1	empty	k + 2	k + 1
.	.	.	.	.
.	.	.	.	.
.	.	.	.	.
N − 1	N − 1	N − 1	N	N − 1
N	N	N	empty	EMPTY
EMPTY	EMPTY	EMPTY	EMPTY	

Array before deletion. | Item k is deleted, nothing has moved. | Item k + 1 is moved up, its slot becomes empty. | Each item has been moved up, item N's slot is empty. | Items are renumbered.

Slot k is the first slot to become empty. The item closest to the empty slot (item k + 1) is moved first.

Figure 7.38: Deletion of Item k

INSIDE BASIC GAMES 207

1	1	1	1	1	
2	2	2	2	2	
.	.	.	.	.	
.	.	.	.	.	
.	.	.	.	.	
k − 1	k − 1	k − 1	k − 1	k − 1	
k	k	k	empty	k	New item
k + 1	k + 1	k + 1	k	k + 1	Old item k
.	.	.	.	.	
.	.	.	.	.	
.	.	.	.	.	
N − 1	N − 1	N − 1	N − 2	N − 1	
N	N	empty	N − 1	N	
empty	empty	N	N	N + 1	Old item N
EMPTY	EMPTY	EMPTY	EMPTY	EMPTY	
Array before insertion	Next free slot is declared "empty."	Item N is moved down, its slot becomes empty.	Others are moved down. Empty slot is between k − 1 and k.	New item becomes item k. Old k through old N are renumbered.	

Slot N + 1 is the first to become empty. The item closest to the empty slot (item N) is moved first.

Figure 7.39: Insertion Ahead of Item k

for that empty slot. In this way, the empty slot moves toward the end of the array, and is finally incorporated into any free space that exists there.

When an item is to be inserted (see Figure 7.39), there is no free space where it is ultimately to be placed (unless it is being added at the end). The first unused slot (if any) at the end of the array is identified, and then the item closest to it (i.e., the final item) moves there. This causes the empty slot to move into the slot where the final item used to be, and the next-to-last item is moved down to fill that slot. In this way, the empty slot is moved to the place chosen to be the location of the new item. Then the new item is inserted. The renumbering mentioned in the captions of Figures 7.38 and 7.39 happens automatically, since the array indices belong to the slots, not to the items.

In summary, the rule to remember when inserting and deleting is: the item nearest the empty slot moves first.

File Maintenance. The next programming technique that we are going to discuss is file maintenance. This is a very broad subject, and the Match-up program only illustrates one small portion of it. In general, a file is a collection of items called records, each of which contains a specific instance of the subject of the file. For example, in a file of customer names, phone numbers and shoe sizes, a record might contain

    PAUL BUNYAN   (509) BLUE OX 7-0048   17EEEE

Files of this type are usually stored in an external storage medium like disk or tape.

In the Match-up program, the "files" are the arrays of question and player information. For example, the "record" for player 1 of group 1 consists of:

    NM$(1,1)
    A(QX,1,1)   QX = 1,...,NQ(1)
    W(QX,1,1)   QX = 1,...,NQ(2)

The string NM$(1,1) is the player's name, the number A(QX,1,1) is the player's answer to question number QX for group 1, and the number W(QX,1,1) is the encoded form of the player's preferences for possible answers to question number QX for group 2. The expressions QX = 1,...,NQ(1) and QX = 1,...,NQ(2) indicate that there is one answer for each question for group 1, and one encoded preference for each question for group 2. NQ(1) is the number of questions for group 1, NQ(2) is the number of questions for group 2.

#Insert score SC and player index PX into array pair (TS,TI)

```
savescore  IF SX = 0 THEN                              #if array is empty,
              GOSUB addon: RETURN                      #  insert "at end"
           FOR SZ = 1 TO SX
              IF SC > TS(SZ) THEN {                    #insert ahead of SZ
                 IF SX < MT THEN
                    SX = SX + 1                        #room to add one
                 SE = SX − 1                           #highest index to move
                 IF SE >= SZ THEN
                    FOR SY = SE TO SZ STEP − 1         #move highest index first
                       TS(SY + 1) = TS(SY)             #each moves up one
                       TI(SY + 1) = TI(SY)
                    NEXT SY
                 TS(SZ) = SC                           #insert new item
                 TI(SZ) = PX
                 RETURN
                 }
              NEXT SZ                                  #don't insert ahead of SZ
           IF SX < MT THEN                             #if any room left,
              GOSUB addon                              #  insert at the end
           RETURN
```

```
1140   IFSX=0THENGOSUB1230:RETURN
1150   FORSZ=1TOSX:IFSC<=TS(SZ)THEN1210
1160   IFSX<MTTHENSX=SX+1
1170   SE=SX−1:IFSE<SZTHEN1200
1180   FORSY=SETOSZSTEP−1
1190   TS(SY+1)=TS(SY):TI(SY+1)=TI(SY):NEXTSY
1200   TS(SZ)=SC:TI(SZ)=PX:RETURN
1210   NEXTSZ:IFSX<MTTHENGOSUB1230
1220   RETURN
```

The *savescore* routine steps through the TS array, comparing the score SC with the elements of TS. Since TS is in descending order of magnitude, the first position at which SC exceeds the TS entry is the index at which SC and PX will be inserted into TS and TI.

Figure 7.40: Keeping Sorted Scores

210 THE MATCH-UP GAME

One of the principal design objectives for file maintenance routines is the assurance of the integrity of the information in the files. That is, file maintenance routines must see that the right information is put into the right place and that wrong information goes nowhere. Naturally, the file routines can have no way of knowing whether PAUL BUNYAN's shoe size is realy 17EEEE, but they can at least provide a way to assure that the person entering that information is satisfied with it before it is placed into the files. This is the purpose of the workspace used for editing player information in the Match-up program.

The *editplayer* routine shown in Figure 7.17 shows how the workspace is used: the *loadwork* subroutine fetches the "record" for the given player (i.e., the above-mentioned items from the NM$, A and W arrays) and places that player information into the workspace, the *changes* subroutine allows the player to alter the information in the workspace (the original array contents are not changed during this editing) and finally, the *update* subroutine replaces the original array contents with the edited information from the workspace. The *update* routine is called only if the player has answered "N" to the "MORE?" question. An answer of "Q" aborts the entire edit, while any other answer allows the player to make another pass through the editing cycle.

Make sure that you understand the editing process that we have just finished discussing. Every good editing routine must allow you to throw up your hands and abandon an already-begun edit without damaging the original information. Making changes to a "working copy" is the principal technique used for this purpose. (Can you think of another way to do this?)

As you study the Match-up program, you should pay attention to the way in which question file editing is handled. Ask yourself how it differs from the techniques that we have just discussed, and why.

---

addon   SX = SX + 1   #make the list longer by one
           TS(SX) = SC   #add this entry at the end
           TI(SX) = PX
           RETURN
      1230 SX=SX+1:TS(SX)=SC:TI(SX)=PX:RETURN

The *addon* routine is a small subroutine called only from the *savescore* routine.

Figure 7.40a: *A Subroutine of Savescore*

```
#Display question (QX,GQ) and its choices
showquestion   PRINT Q$(QX,GQ)
               FOR CX = 1 TO NC(QX,GQ)
                   PRINT CX; " "; C$(CX,QX,GQ)
               NEXT CX
               RETURN


1240  PRINTQ$(QX,GQ)
1250  FORCX=1TONC(QX,GQ):PRINTCX;" ";C$(CX,QX,GQ):NEXTCX:RETURN
```

The *showquestion* routine displays one question and its choices. The question index is QX, the group number is GQ.

Figure 7.41: Displaying a Question

```
#Ask for the player's group
askgroup   PRINT                              #new line
           FOR GX = 1 TO 2                    #print group names
               PRINT GX; " "; GN$(GX)
           NEXT GX
           PRINT "GROUP: ";                   #ask for player's group
           repeat {
               GOSUB onech
               GP = VAL(X$)
           } until (GP = 1 OR GP = 2)
           PRINT X$;                          #echo player's answer
           RETURN


1260  PRINT:FORGX=1TO2:PRINTGX;" ";GN$(GX):NEXTGX:PRINT"GROUP: ";
1270  GOSUB2180:GP=VAL(X$):IFNOT(GP=1ORGP=2)THEN1270
1280  PRINTX$;:RETURN
```

The *askgroup* routine displays the two group names, asks for the player's group number, and accepts an answer of 1 or 2.

Figure 7.42: Which Kind Are You?

Space Saving. The final programming techniques that we wish to discuss concern space saving. This is a universal problem; there is never enough memory. From the tiniest microprocessor to the largest "mainframe" computer, sooner or later your program will run out of memory. To paraphrase a famous principle:

The program expands to fill the available memory.

Also (looking at the problem in a more positive light), if you are writing a program that will become a product, or part of a product, then a smaller program can result in a less expensive product with a wider potential market. For example, a BASIC program that runs in 8K on the XYZ computer is likely to be purchased by more XYZ owners than a program with the same features that requires 48K (unless, of course, all XYZ computers have at least 48K of memory).

There are two ways to save space in home computers:

— Compression of the BASIC program.

— Space-saving programming and data storage algorithms.

The opportunities available for compressing your BASIC program depend upon how your system works. For example, on the TRS-80, any spaces that you enter between the line number and the end of a BASIC statement become part of the program; each space (except the first) costs you one byte of memory. On the Pet, spaces between

---

#Compute number GC of "other" group for GP

othergroup IF GP = 1 THEN   #if you're 1,
               GC = 2   # the other is 2
               else   #and vice versa
               GC = 1
               RETURN

1290 IFGP=1THENGC=2:RETURN
1300 GC=1:RETURN

The *othergroup* routine takes a group number GP and computes the "other" group number GC. Note that this simple relationship cannot be represented straightforwardly by a BASIC function. (You could "cheat": DEF FNOT(X) = 3 − X.)

Figure 7.43: Which Kind You Aren't

the line number and the first non-blank character of the statement are ignored (Pet supplies one space in the listing), but all other spaces in the line become part of the program. On the Apple, any spaces that you type are ignored; the Apple provides its own spacing in listings. (On all of these systems, each space occurring within a string constant— e.g., "HOW ARE YOU"— costs you one byte, even on the TRS-80, which has single-character codes (ASCII 192 through 255) for up to 63 consecutive blanks.)

The Match-up program was first developed on an 8K Pet, and the BASIC instructions that appear in Figures 7.16 through 7.59 have almost no extraneous spaces. Naturally, this decreases their readability slightly, but this is not important, since the actual BASIC instructions are not intended for human eyes. Only the Free BASIC description is intended to be read.

For the same reason, the lack of REM statements among the BASIC instructions is not a problem, since the Free BASIC and BASIC forms of the program are paired in small modules in the figures comprising the Match-up program. In reality, since you must translate from Free BASIC into BASIC by hand, and you won't have a program to make this pairing for you, it might be helpful to begin each subroutine with a REM statement containing the subroutine's name. That way, if you can display all lines containing a specified program fragment, you can generate a "symbol table" that relates line numbers to subroutine names. For example, on a Pet with the Nestar "Programmer's Toolkit," you can enter the command

    FIND REM

to produce a display that looks like

    210   REM EDITPLAYER
    260   REM NEWPLAYER
    300   REM CLEARWORK
          .
          .
          .

If you were to direct this output to cassette, you could then run it through a program that turned it into the form:

    EDITPLAYER   210
    NEWPLAYER   260
    CLEARWORK   300
          .
          .

214 THE MATCH-UP GAME

```
#Ascertain player's identity (PP,GP)

identity  GOSUB askgroup                              #get player's group
          GOSUB clearscreen
          IF NP(GP) = 0 THEN {
              PRINT "NO ";GN$(GP);" ON FILE."
              XX = 0: RETURN                          #signal "fail" return
              }
          XX = 1                                      #signal "success" return
          FOR PX 1 TO NP(GP) STEP screenload
              GOSUB clearscreen
              PT = min(PX + screenload - 1,NP(GP))    #highest index for this load
              FOR PY = PX TO PT                       #display a screenload of
                  PRINT PY; " "; NM$(PY,GP)           #  numbers and names
              NEXT PY
              repeat {
                  PRINT "NUMBER: ";:GOSUB stringin    #ask for player's number
                  PP = VAL(SS$)
                  IF 1 <= PP <= NP(GP) THEN           #valid number for this group
                      RETURN
                  else IF PP = 0 THEN                 #non-numeric
                      break
                  }
              IF X$ = "Q" THEN                        #Q aborts
                  {XX = 0: RETURN}                    #signal "fail" return
          NEXT PX
          PRINT "THAT'S ALL THE ";GN$(GP)             #all screenloads displayed
          XX = 0: RETURN                              #signal "fail" return
```

The *identity* routine asks for the player's group, then shows the names and numbers of all members of that group, one screenload at a time. After each screenload is displayed, the player is asked to enter a number. A reply of Q aborts the process. Any other non-numeric reply causes the next screen to be displayed.

Figure 7.44: What's Your Number?

```
1310 GOSUB1260:GOSUB2200:IFNP(GP)>0THEN1330
1320 PRINT"NO ";GN$(GP);" ON FILE.":XX=0:RETURN
1330 XX=1:FORPX=1TONP(GP)STEP10:GOSUB2200
1340 PT=PX+9:IFPT>NP(GP)THENPT=NP(GP)
1350 FORPY=PXTOPT:PRINTPY;" ";NM$(PY,GP):NEXTPY
1360 PRINT"NUMBER: ";:GOSUB610:PP=VAL(SS$)
1370 IF1<=PPANDPP<=NP(GP)THENRETURN
1380 IFPP<>0THEN1360
1390 IFX$="Q"THENXX=0:RETURN
1400 NEXTPX
1410 PRINT"THAT'S ALL THE ";GN$(GP):XX=0:RETURN
```

Figure 7.44a: BASIC for Identity

```
#Set up the questions and weights
create  repeat {
            GOSUB clearscreen                    #start on clear screen
            PRINT "::";:GOSUB onech: PRINT X$    #get and echo command
            IF case
                X$ = "I" THEN GOSUB newgame      #start from scratch
                X$ = "N" THEN GOSUB editnames    #edit the group names
                X$ = "1" THEN GOSUB editone      #edit group 1 questions
                X$ = "2" THEN GOSUB edittwo      #edit group 2 questions
                X$ = "L" THEN GOSUB loadin       #load from cassette/disk
                X$ = "S" THEN GOSUB writeout     #write to cassette/disk
                X$ = "Q" THEN break              #let the play begin
            }
        RETURN

1420 GOSUB2200:PRINT"::";:GOSUB2180:PRINTX$
1430 IFX$="I"THENGOSUB1510:GOTO1420
1440 IFX$="N"THENGOSUB1700:GOTO1420
1450 IFX$="1"THENGOSUB1740:GOTO1420
1460 IFX$="2"THENGOSUB1750:GOTO1420
1470 IFX$="L"THENGOSUB2210:GOTO1420
1480 IFX$="S"THENGOSUB2220:GOTO1420
1490 IFX$<>"Q"THEN1420
1500 RETURN
```

The *create* routine contains command selection for the generation and editing of questions.

Figure 7.45: Selecting the Setup Command

216 THE MATCH-UP GAME

```
#Accept data for game


newgame  FOR GX = 1 TO 2
            PRINT: PRINT "GROUP"; GX; ": "
            repeat                              #get group name
                GOSUB stringin
                until (SS$ <> "")
            GN$(GX) = SS$
            NQ(GX) = 0                          #initialize number of questions
            while (NQ(GX) < MQ) {               #ask for up to MQ questions
                GOSUB clearscreen
                QX = NQ(GX) + 1
                GOSUB inquestion                #get one question, write choices
                IF X$ = "Q" THEN                #no more questions
                    break
                else
                    NQ(GX) = QX                 #update number of questions
            }
            NEXT GX
        RETURN

1510    FORGX=1TO2
1520*   PRINT:PRINT"GROUP";GX;":   ";
1530    GOSUB610:IFSS$=""THEN1530
1540    GN$(GX)=SS$:NQ(GX)=0
1550    IFNQ(GX)>=MQTHEN1590
1560    GOSUB2200:QX=NQ(GX)+1:GOSUB1600
1570    IFX$="Q"THEN1590
1580    NQ(GX)=QX:GOTO1550
1590    NEXTGX:RETURN
```

The *newgame* routine steps through the items necessary to set up the game and asks the player to supply them.

---
*The Apple version of this line differs only in the spacing inside the string constants.

Figure 7.46: Setting Up a New Game

INSIDE BASIC GAMES 217

Finally, if you directed that output to cassette, then used it as input to a sorting program, you could obtain a useful alphabetical listing:

ASKGROUP	1260
ASKOK	510
ASKWISH	770

. . .

There are many techniques that can be used for compressing your BASIC program. Your system's BASIC manual probably contains suggestions for this. The other approach that we shall discuss here is the use of space-saving programming and data storage algorithms.

Often there is no "best" way to perform a certain computation, or to store certain information. Usually there are "trade-offs"—for example, one approach might result in a small program that takes a long time to perform the computation, while another approach might use a large program that carries out the computation quickly. There is no formula that will make it easier for you to make these trade-off decisions. In fact, short of actually carrying the project to completion in two different ways, it is usually impossible to state precisely what the relative space and time requirements of two different approaches will be.

In the Match-up program, the most obvious space vs. time trade-off occurs in the encoding of player preferences. For each question of the "other" group's set, a player can assign a choice to each of NF preference values. (NF is a variable whose value is set in a DATA statement in the *init* routine—the value is 4 in Figures 7.11, 7.13, 7.15.) The easiest way to store these preferences would be in an array whose dimensions are NF, MQ, MP, GM. That is, the choice would be indexed by group, player number, question number and preference value number. Assuming values of NF = 4, MQ = 6, MP = 12, GM = 2, this array would contain 4 × 6 × 12 × 2 = 576 items. These items would occupy 6,838 bytes of storage on an Apple or Pet, 5,474 on a TRS-80. The approach used in the Match-up program is to encode all of the preferences for a given question into one number (we'll see how this is done shortly) so that the array dimensions are MQ, MP, GM. Using the values above produces an array of 144 items, which occupies 1,376 bytes on an Apple or Pet, and 1,104 bytes on a TRS-80—a space saving of nearly 80% in either case.

Preferences are packed into one word by treating them as numbers

THE MATCH-UP GAME

```
#Get question QX for group GX
inquestion  PRINT: PRINT "Q"; QX; "FOR"; GN$(GX)    #ask for question
            repeat {
                GOSUB stringin                       #accept string
                IF SS$ = "Q" THEN {
                    X$ = "Q"                         #Q = no more questions
                    RETURN
                }
                else IF SS$ <> "" THEN {             #require non-null string
                    Q$(QX,GX) = SS$
                    break
                }
            }
            NC(QX,GX) = 0                            #initialize number of choices
            while (NC(QX,GX) < MC) {                 #allow up to MC choices
                CX = NC(QX,GX) + 1
                PRINT CX; " ";                       #ask for next choice
                repeat {
                    GOSUB stringin                   #accept string
                    IF SS$ = "Q" THEN
                        {break: break}               #Q = no more choices
                    else IF SS$ <> ""THEN {          #require non-null string
                        C$(CX,QX,GX) = SS$
                        break
                    }
                }
                NC(QX,GX) = CX                       #update choice number
            }
            PRINT: PRINT "WEIGHT: ";
            GOSUB stringin
            WT(QX,GX) = VAL(SS$)
            RETURN
```
The *inquestion* routine asks for and accepts a question and its choices. Editing functions are not performed in this routine.

Figure 7.47: Getting a Question

INSIDE BASIC GAMES 219

```
1600* PRINT:PRINT"Q";QX;"FOR ";GN$(GX)
1610  GOSUB610:IFSS$="Q"THENX$="Q":RETURN
1620  IFSS$=""THEN1610
1630  Q$(QX,GX)=SS$:NC(QX,GX)=0
1640  IFNC(QX,GX)>=MCTHEN1690
1650  CX=NC(QX,GX)+1:PRINTCX;" ";
1660  GOSUB610:IFSS$="Q"THEN1690
1670  IFSS$=""THEN1660
1680  C$(CX,QX,GX)=SS$:NC(QX,GX)=CX:GOTO1640
1690  PRINT:PRINT"WEIGHT: ";:GOSUB610:WT(QX,GX)=VAL(SS$):RETURN
```

*The Apple version of this line differs only in the spacing inside the string constants.

Figure 7.47a: BASIC for Inquestion

```
#Edit the group names

editnames   FOR GX = 1 TO 2
                PRINT GX; " "; GN$(GX)         #display group name
                GOSUB stringin                  #accept change
                IF SS$ <> "" THEN               #RETURN = "no change"
                    GN$(GX) = SS$
                NEXT GX
            RETURN


1700  FORGX=1TO2:PRINTGX;" ";GN$(GX)
1710  GOSUB610:IFSS$<>""THENGN$(GX)=SS$
1720  NEXTGX:RETURN
```

The *editnames* routine displays the group name for each of the two groups, and allows the player to enter a change. A RETURN signals "no change"; otherwise the name is replaced by the string typed.

Figure 7.48: Editing Group Names

to the base MC + 1, where MC is the variable containing the maximum choice number allowed. This value is set in a DATA statement; it cannot exceed 9.

You probably know that if b is any whole number, then any other whole number N can be written in exactly one way as

$$N = a_0 + a_1 \times b + a_2 \times b^2 + \ldots + a_k \times b^k$$

where $a_0, a_1, \ldots, a_k$ are whole numbers in the range 0 to b − 1. For example (using b = 10),

$$1980 = 0 + 8 \times 10 + 9 \times 100 + 1 \times 1000;$$

that is,

$$1980 = 0 + 8 \times 10 + 9 \times 10^2 + 1 \times 10^3$$

There is no other way to break 1980 up into ones, tens, hundreds, and thousands with each of the coefficients (i.e., the 1, 9, 8, 0) in the range 0 to 9.

To illustrate the packing of preferences, let's assume that each question can have up to 9 choices, that is, MC has the value 9. Then the packed preferences are represented by a number to the base 10 (MC + 1). Each digit is the choice number between zero and nine (zero means no choice) corresponding to one of the preference values. In Figure 7.13, SUSAN's preferences for question one would be encoded as 1432; SUSAN's preferences for question two would be encoded as 1203.

With small numbers of preference values and question choices, this scheme allows a player's preferences for one question to be encoded into one number. This works until the number becomes so large that it is "rounded off." The limit at which this "rounding off" occurs varies from system to system. On an Apple or Pet, it is nine digits; with TRS-80 it is six digits, unless a "double precision" variable is used. In that case the 6 × 12 × 2 array that took 1,104 bytes would take 2,196 bytes instead. The maximum size that an encoded preference can attain can be computed from the following formula:

$$M = (MC + 1) \uparrow NF - 1$$

For example, if MC has the value 9 and NF (the number of preference values) has the value 4, then M = 9999.

#Question-editing dispatch routines

editone GX = 1: GOSUB edit
RETURN

edittwo GX = 2: GOSUB edit
RETURN

edit   repeat {
        GOSUB clearscreen   #start on clear screen
        PRINT "*";   #prompt for input
        GOSUB onech: PRINT X$ #get the command
        IF case
            X$ = "E" THEN GOSUB modques   #edit a question
            X$ = "D" THEN GOSUB delques   #delete a question
            X$ = "A" THEN GOSUB addques   #add a question (at the end)
            X$ = "B" THEN GOSUB insertques   #insert a question
            X$ = "Q" THEN break   #no more editing
        }
        RETURN

```
1740 GX=1:GOSUB1760:RETURN
1750 GX=2:GOSUB1760:RETURN
1760 GOSUB2200:PRINT"*";:GOSUB2180:PRINTX$:IFX$="Q"THENRETURN
1770 IFX$="E"THENGOSUB1990:GOTO1760
1780 IFX$="D"THENGOSUB1850:GOTO1760
1790 IFX$="A"THENGOSUB1820:GOTO1760
1800 IFX$="B"THENGOSUB1910:GOTO1760
1810 GOTO1760
```

This is the common code for handling the two edit commands "1" and "2". In either case, a "*" is displayed as a prompt. Then commands are accepted to allow modifying, deleting, adding or inserting a question. These are the fundamental editing functions that need to be provided in any text-editing situation.

Figure 7.49: Selecting the Question-Editing Function

222 THE MATCH-UP GAME

#Add a question to group GX

addques   IF NQ(GX) >= MQ THEN {
           PRINT "NO ROOM"
           RETURN
           }
           QX = NQ(GX) + 1　　　　　　　#index of the new question
           GOSUB inquestion
           IF X$ <> "Q" THEN　　　　　　#Q aborts the process
              NQ(GX) = QX　　　　　　　#update the number of questions
           RETURN

```
1820 IFNQ(GX)>=MQTHENPRINT"NO ROOM":RETURN
1830 QX=NQ(GX)+1:GOSUB1600:IFX$<>"Q"THENNQ(GX)=QX
1840 RETURN
```

This routine adds a question onto the end of the list of questions for group GX. The mistake made in the *storenew* routine (Figure 7.22) is not repeated here. Thus, the process is terminated immediately if there is no room.

Note the handling of the variable QX and the array entry NQ(GX). QX is set immediately to the new number, but if the question-entry process is aborted (*inquestion* returns with X$ = "Q"), then NQ(GX) is not changed.

Figure 7.50: Adding a Question

#Delete a question from group GX

delques   GOSUB qnum　　　　　　　　#get question number QQ
           IF QQ <> 0 THEN　　　　　　　　#QQ = 0 aborts
              GOSUB qout
           RETURN

```
1850 GOSUB2160:IFQQ<>0THENGOSUB1870
1860 RETURN
```

This routine deletes a question from group GX. The actual mechanics of deletion are implemented in the *qout* routine.

Figure 7.51: Deleting a Question

We have seen how the encoding of preferences can save large amounts of memory. The price paid for this memory saving is seen in the *pairup* program (Figure 7.33), where 2* NP(GC) calls to *score* (Figure 7.34), which calls *fit* (Figure 7.35), result in

$$NP(GC)*NQ(GC) + NP(GC)*NQ(GP)$$

calls to *unpackwish*. All of these calls could be eliminated if the much larger arrays that we considered earlier had been used, but many small computer systems would be unable to allocate this much memory to that array. However, there is an intermediate trade-off

---

#Delete question QQ from group GX

qout   NQ(GX) = NQ(GX) − 1   #one less question
        IF QQ > NQ(GX) THEN   #nothing to do if
            RETURN   # deleting final question
        FOR QX = QQ TO NQ(GX)   #otherwise, move each
            Q$(QX,GX) = Q$(QX + 1,GX)   # question up by one
            IF NC(QX + 1,GX) > 0 THEN   #move choices, if any
                FOR CX = 1 TO NC(QX + 1,GX)
                      C$(CX,QX,GX) = C$(CX,QX + 1,GX)
                NEXT CX
            NC(QX,GX) = NC(QX + 1,GX)   #move number of choices
            WT(QX,GX) = WT(QX + 1,GX)   #move weight
            NEXT QX
        RETURN

```
1870 NQ(GX)=NQ(GX)−1:IFQQ>NQ(GX)THENRETURN
1880 FORQX=QQTONQ(GX):Q$(QX,GX)=Q$(QX+1,GX):IFNC(QX+1,GX)<=0THEN1900
1890 FORCX=1TONC(QX+1,GX):C$(CX,QX,GX)=C$(CX,QX+1,GX):NEXTCX
1900 NC(QX,GX)=NC(QX+1,GX):WT(QX,GX)=WT(QX+1,GX):NEXTQX:RETURN
```

The *qout* routine implements the mechanics of deleting question QQ from group GX. The question string, the choices, the number of choices and the weight must all be moved, as each question beyond the one deleted is moved up one in the list. The case in which a question has no choices is provided for—even though it shouldn't happen.

Figure 7.52: *Mechanics of Deleting a Question*

224 THE MATCH-UP GAME

```
#Insert a question into group GX
insertques   IF NQ(GX) >= MQ THEN {          #if no room,
                PRINT "NO ROOM"              # say so and exit
                RETURN
             }
             PRINT "AHEAD OF ";              #find out where to put the question
             GOSUB qnum
             IF QQ = 0 THEN RETURN           #QQ = 0 if player aborts the process
             FOR QX = NQ(GX) TO QQ STEP −1   #move last question first
                Q$(QX + 1,GX) = Q$(QX,GX)
                IF NC(QX,GX) > 0 THEN        #move choices, if any
                   FOR CX = 1 TO NC(QX,GX)
                      C$(CX,QX + 1,GX) = C$(CX,QX,GX)
                   NEXT CX
                NC(QX + 1,GX) = NC(QX,GX)    #move number of choices
                WT(QX + 1,GX) = WT(QX,GX)    #move weight
             NEXT QX
             NQ(GX) = NQ(GX) + 1             #update number of questions
             QX = QQ: GOSUB inquestion       #get the question
             IF X$ = "Q" THEN                #Q means abort
                GOSUB qout                   #give back the space
             RETURN

1910  IFNQ(GX)>=MQTHENPRINT"NO ROOM":RETURN
1920  PRINT"AHEAD OF ";:GOSUB2160:IFQQ=0THENRETURN
1930  FORQX=NQ(GX)TOQQSTEP−1:Q$(QX+1,GX)=Q$(QX,GX)
1940  IFNC(QX,GX)<=0THEN1960
1950  FORCX=1TONC(QX,GX):C$(CX,QX+1,GX)=C$(CX,QX,GX):NEXTCX
1960  NC(QX+1,GX)=NC(QX,GX):WT(QX+1,GX)=WT(QX,GX):NEXTQX
1970  NQ(GX)=NQ(GX)+1:QX=QQ:GOSUB1600:IFX$="Q"THENGOSUB1870
1980  RETURN
```

The *insertques* routine inserts a question into group GX. The player is asked to specify the number of the question ahead of which the next question is to be inserted. Room for the new question is provided by moving each question, starting from the specified number, to the next higher-numbered position. For insertion, it is essential that each question moves into its new space before making its old space available to the next lower-numbered question.

Figure 7.53: Inserting a Question

that can be made. In the first half of the *pairup* program, the given player's preferences are checked against the answers of each of the players in the other group, so that each of the player's encoded preferences is unpacked NP(GC) times. If, instead, these preferences were all unpacked once and saved in a two-dimensional version of the FT array, then the first term in the sum shown above could be reduced from NP(GC)*NQ(GC) to NQ(GC). The two-dimensional FT array would have dimensions NF and MQ. Using the same numbers that

---

#Edit a question in group GX

modques GOSUB qnum   #question number
         IF QQ = 0 THEN
             RETURN
         PRINT Q$(QQ,GX)   #display question
         GOSUB stringin   #accept change
         IF SS$ = "Q" THEN   #Q = no change and quit
             RETURN
         else IF SS$ <> "" THEN   #RETURN = no change
             Q$(QQ,GX) = SS$ #store new string
         GOSUB choices   #edit choices
         PRINT: PRINT "WEIGHT: "; WT(QQ,GX); ": ";   #display weight
         GOSUB stringin   #accept change
         IF VAL(SS$) <> 0 THEN   #non-numeric or zero
             WT(QQ,GX) = VAL(SS$)   # means no change
         RETURN

```
1990  GOSUB2160:IFQQ=0THENRETURN
2000  PRINTQ$(QQ,GX):GOSUB610:IFSS$="Q"THENRETURN
2010  IFSS$<>""THENQ$(QQ,GX)=SS$
2020  GOSUB2050:PRINT:PRINT"WEIGHT: ";WT(QQ,GX);": ";:GOSUB610
2030  IFVAL(SS$)<>0THENWT(QQ,GX)=VAL(SS$)
2040  RETURN
```

The *modques* routine allows the player to edit a question in group GX. Editing of the question string is done here. Editing of the choices is done in the *choices* routine.

Figure 7.54: Editing a Question

226 THE MATCH-UP GAME

```
#Edit the choices for question (QQ,GX)
choices   CX = 1                              #initialize choice index
          while (CX <= NC(QQ,GX)) {           #for each existing choice,
              PRINT CX; " "; C$(CX,QQ,GX)     # display number and text
              GOSUB stringin                  # and accept change
              IF SS$ = "Q" THEN RETURN        #Q = quit
              else IF SS$ = "D" THEN {        #D = delete the choice
                  NC(QQ,GX) = NC(QQ,GX) - 1   #one less choice
                  IF CX <= NC(QQ,GX) THEN     #if not deleting last choice,
                      FOR CY = CX TO NC(QQ,GX)  # move following choices up
                          C$(CY,QQ,GX) = C$(CY + 1,QQ,GX)
                      NEXT CY
              }
              else {                          #not Q or D
                  IF SS$ <> "" THEN           #if not the null string,
                      C$(CX,QQ,GX) = SS$      # replace choice with SS$
                  CX = CX + 1
              }
          }
          while (NC(QQ,GX) < MC) {            #while there is still room
              CX = NC(QQ,GX) + 1              # set choice index and
              PRINT CX; " ";: GOSUB stringin  # ask for choice
              IF SS$ = "" or "Q" THEN RETURN  #RETURN or Q ends it all
              else {
                  C$(CX,QQ,GX) = SS$          #set choice to SS$
                  NC(QQ,GX) = CX              #count it
              }
          }
          RETURN
```

The *choices* routine allows the player to edit the set of choices for the question whose indices are (QQ,GX). First the current choices are stepped through, and the player is allowed to edit or delete each one. (D = delete, RETURN = no change.) Then new choices can be added at the end. In the addition phase, a Q or a RETURN terminates the process.

In the first portion of the program, CX is controlled explicitly. A FOR...NEXT loop could not be used, since deletion of choices could change the upper limit NC(QQ,GX).

Figure 7.55: Editing the Choices

```
2050  CX=1
2060  IFCX>NC(QQ,GX)THEN2130
2070  PRINTCX;" ";C$(CX,QQ,GX):GOSUB610:IFSS$="Q"THENRETURN
2080  IFSS$<>"D"THEN2110
2090  NC(QQ,GX)=NC(QQ,GX)-1:IFCX>NC(QQ,GX)THEN2060
2100  FORCY=CXTONC(QQ,GX):C$(CY,QQ,GX)=C$(CY+1,QQ,GX):NEXTCY:GOTO2060
2110  IFSS$<>""THENC$(CX,QQ,GX)=SS$
2120  CX=CX+1:GOTO2060
2130  IFNC(QQ,GX)=>MCTHENRETURN
2140  CX=NC(QQ,GX)+1:PRINTCX;" ";:GOSUB610:IFSS$=""ORSS$="Q"THENRETURN
2150  C$(CX,QQ,GX)=SS$:NC(QQ,GX)=CX:GOTO2130
```

Figure 7.55a: BASIC for Choices

```
#Get a question number QQ for group GX; QQ = 0 if invalid


qnum    PRINT "QUESTION NUMBER: ";
        GOSUB stringin
        QQ = VAL(SS$)
        IF QQ < 0 OR QQ > NQ(GX) THEN
            QQ = 0
        RETURN




2160  PRINT"QUESTION NUMBER: ";:GOSUB610:QQ=VAL(SS$):
      IFQQ<0ORQQ>NQ(GX)THENQQ=0
2170  RETURN
```

The *qnum* routine asks for a question number and checks its validity.

Figure 7.56: Getting a Question Number

we used earlier, this would mean that the FT array would be enlarged from 4 items to 4 × 6 = 24 items and the number of calls to *unpackwish* would be reduced from 12 × 6 + 12 × 6 = 144 to 6 + 12 × 6 = 78. This seems like a small increase in memory use in exchange for a small reduction in execution time, but if the number of players in each group increased from 12 to 24, then there would be no further increase in memory use, and the number of calls would be reduced from 24 × 6 + 24 × 6 = 288 to 6 + 24 × 6 = 150.

This completes our discussion of space saving. (You might wish to carry out the change that we described above.) Many interesting questions arise about how best to organize the communication among *pairup, score, fit* and *unpackwish* for the new structure. For example, should *unpackwish* be rewritten to unpack all of a player's wishes at once? If not, will there be two separate versions of FT—one with one dimension, one with two? How will you measure the actual time savings resulting from this change?

This concludes our discussion of the Match-up program. You should read through the program listings, recalling what has been discussed, and try to see what else you can learn.

#Some old favorites—Pet versions

onech   repeat
                    GET X$
                until (X$ <> '''')
            RETURN

clearscreen PRINT "clr";
            RETURN

    2180 GETX$:IFX$=""THEN2180
    2190 RETURN
    2200 PRINTCHR$(147);:RETURN

The Apple and TRS-80 versions of these routines are as shown in Figures 2.11 and 2.12.

Figure 7.57: Single Character Input and Screen Clearing

Changes and Improvements

After (or in the course of) your study of the Match-up program, you may wish to implement some of the following changes and additions:

- Implement the external storage features that are provided for in the program.
- If external storage features are implemented, or if the game building phase is re-entered after player files have been built up, the player and question files can become incompatible. Consider the following features:

#External storage routines

loadin   GOSUB copout
                RETURN

writeout   GOSUB copout
                RETURN

loadplayers GOSUB copout
                RETURN

saveplayers GOSUB copout
                RETURN

copout   PRINT "EXTERNAL STORAGE IS NOT IMPLEMENTED"
                RETURN

   2210 GOSUB2250:RETURN
   2220 GOSUB2250:RETURN
   2230 GOSUB2250:RETURN
   2240 GOSUB2250:RETURN
   2250 PRINT "NOT IMPLEMENTED":RETURN

These routines are "stubs" that indicate the place in the program at which code to implement the loading and saving of files on external storage media will eventually be added.

Figure 7.58: External Storage Facility

THE MATCH-UP GAME

```
#Initialization
init  GM = 2                              #two groups allowed
      READ MQ,MC,MP                       #max. questions, choices, players
      DATA maxquestions, maxchoice, maxplayers
      DIM NQ(GM), NP(GM),                 #no. of questions and players;
          GN$(GM), NM$(MP,GM),            #group names, player names
          Q$(MQ,GM), WT(MQ,GM),           #questions, weights,
          NC(MQ,GM), C$(MC,MQ,GM),        #no. of choices, choices
          A(MQ,MP,GM), W(MQ,MP,GM)        #player answers, wishes
      READ MT: DATA maxtop                #size of "top scores" array
      DIM TS(MT),TI(MT)                   #top scores and player numbers
      READ NF: DATA numfit                #size of "wishes" array
      DIM FV(NF),FT(NF)                   #wish value and wish expansion arrays
      FOR FX = 1 TO NF                    #read wish values
         READ FV(FX)
         NEXT FX
      DATA highest wish value,...,lowest wish value
      RETURN


2260*  GM=2:READMQ,MC,MP:DATA2,5,2
2270   DIMNQ(GM),NP(GM),GN$(GM),NM$(MP,GM),Q$(MQ,GM)
2280   DIMWT(MQ,GM),NC(MQ,GM),C$(MC,MQ,GM)
2290   DIMA(MQ,MP,GM),W(MQ,MP,GM)
2300   READMT:DATA3
2310   DIMTS(MT),TI(MT)
2320   READNF:DATA4
2330   DIMFV(NF),FT(NF)
2340   FORFX=1TONF:READFV(FX):NEXTFX
2350   DATA 2,1,−1,−2
2360   RETURN
```

The data statements allow the setting of array sizes. For a small home computer (8K of RAM), array sizes must be set to very small values—like 2 questions, 5 choices, and 2 players per group. Much larger values can be used on larger systems.

---

*The values of *maxquestions*, *maxchoice* and *maxplayers* shown in this line are small enough to allow the program to run on an 8K Pet. Much larger values are possible on larger systems. On a 16K TRS-80, for example, values of 10, 5, 30 can be used. Remember, *maxchoice* must not exceed 9.

Figure 7.59: Initialization

- Automatically making changes to the player files when questions are added to or deleted from the question files.

- Automatically re-encoding the player preferences if MC or NF values change.

- Automatically, or by command, clearing the player files under certain circumstances.

Implement any of these features that you feel are appropriate. Which problems won't these changes solve?

- Provide editing commands for the player file that will allow you to specify the point in the editing cycle at which you wish to start.

- Provide single character commands that will allow you to skip to the start of preference editing, or allow you to go on to the next question without repeatedly using RETURN to step through all of the preference values for the question that you're on.

- Correct the problem inherent in *storenew* (Figure 7.22).

- Provide a mass pairing function that generates a list of couples in such a way as to maximize the total of all preference scores of paired players. If one group is larger than the other, assign more than one member of the larger group to certain members of the smaller group.

Summary

The Match-up game uses sets of multiple-choice questions to pair members of two groups. The program implementing this game illustrates general design concepts and general and specific programming techniques. In particular, the general design concepts discussed are modularity, top-down structure, isolation of function, provision for change and the use of stubs to define the structure. General programming techniques covered include the design of user and programmer documentation, and the design of user interaction and error handling. Specific programming techniques described are the design of editing functions, insertion and deletion algorithms, the design of file maintenance functions and techniques for saving space.

CHAPTER 8
Craps

Craps is a home computer version of the popular casino game. Some of the fast action and excitement of the casino have been recreated by the use of simple programming and design techniques.

Instructions for Craps

In its most fundamental form, Craps is a simple dice game. Each player, in turn, receives the dice and makes a series of throws. When your turn comes you begin by making a bet. If your series of throws results in a win, then you collect your bet, make another bet, and begin another series of throws. If your series of throws results in a loss, then you pay your bet and pass the dice to the next player.

If your first throw in a series is a 7 or an 11 (the numbers refer to the sum of the dots on the two die faces), then you win immediately. If your first throw is 2, 3 or 12, then you lose immediately. If your first throw is any of the other six possibilities (4, 5, 6, 8, 9, 10), then that number becomes your "point." You must continue to throw the dice until you throw a 7, in which case you lose, or you throw your point, in which case you win. Thus, if your point is 8, you continue to throw, hoping for an 8, until either you throw an 8 and win, or you throw a 7 and lose. Any other numbers that you throw during this time, including 2, 3, 11, and 12 are ignored. For example, the following sequence of throws results in a win:

8, 5, 11, 6, 9, 3, 8

Only this fundamental form of craps has been implemented here. At an actual casino craps table, the large layout facilitates additional betting over and above the player's main bet. For example, you might bet that the next throw will be a 3. You do this by placing your chip at a certain place on the layout. If the next throw is not a 3, your chip is taken by the house; if it is a 3, a payment of 12 chips is placed beside your chip. If you don't intend to bet all 13 chips on a 3 on the next throw, then you must pick them up quickly—action is fast at a craps table, and there are so many bets made that it is easy to forget where your chips are placed.

234   CRAPS

```
WON    BET    THIS RUN    BEST RUN
 0      0        0            0
```

Initial display.

Player types N.

```
MAY I HAVE YOUR NAME PLEASE? JOHN
THANK YOU.
```

Program requests and player enters name.

```
WON    BET    THIS RUN    BEST RUN
 0      0        0            0
                JOHN
```

Initial display with name.

The player has responded to the initial display by typing N (the "new player" command). The program requests the player's name, and the player responds JOHN. The program says THANK YOU. Then, after a short pause, it returns to the initial display, which now includes the player's name.

Figure 8.1: JOHN is First

None of this additional betting has been included in this version of the game because representing the layout on the screen would be difficult, and because there are no separate input/output facilities for each player. Nevertheless, some additional betting could be introduced. Additional betting features are included in the suggested improvements that appear at the end of this chapter.

The program for our version of Craps begins with the start-of-game display (see Figure 8.1). There are several commands that can be entered at this point. The game can be played by a single player, or by several players. If there are several players, the N command can be used to enter each player's name. This is illustrated in Figure 8.1. Figure 8.2 shows how the program announces each player's place in the rotation. (Note: since the input of names uses the INPUT instruction, two-word names (e.g., JOHN SMITH) must be entered in quotes.)

Let's suppose that MARY and SUSAN have been added, so that there are now three players. It is still JOHN's turn. The next step is for JOHN to enter a bet. This is illustrated in Figure 8.3. After entering a bet of 25, JOHN begins to throw the dice by pressing the space bar once.

Figure 8.4 shows what happens next. First, a picture of a pair of dice appears. Then the result of the throw is displayed in words (e.g., SORRY, YOU LOSE or POINT IS 5). After a short pause, the dice picture disappears, but the result remains. Then, if the result is a win,

```
MAY I HAVE YOUR NAME PLEASE? MARY
THANK YOU. YOU WILL FOLLOW JOHN.
```

In response to another N command, the program has asked for the player's name. This player has entered the name MARY, and the program has placed MARY after JOHN in the rotation.

Figure 8.2: MARY Follows JOHN

JOHN's initial display reappears; if it is a loss (as shown in Figure 8.4), the next player's initial display appears (in this case, MARY's). If the result is neither a win nor a loss, a line appears stating the point and the number of throws.

This last case is what happens to MARY (Figure 8.5). MARY enters a bet of 50 and presses the space bar to throw the dice. A 6 and a 2 appear, then the result (POINT IS 8), and, finally, the "game line" (POINT: 8 THROWS: 1). The dice disappear, and the program awaits

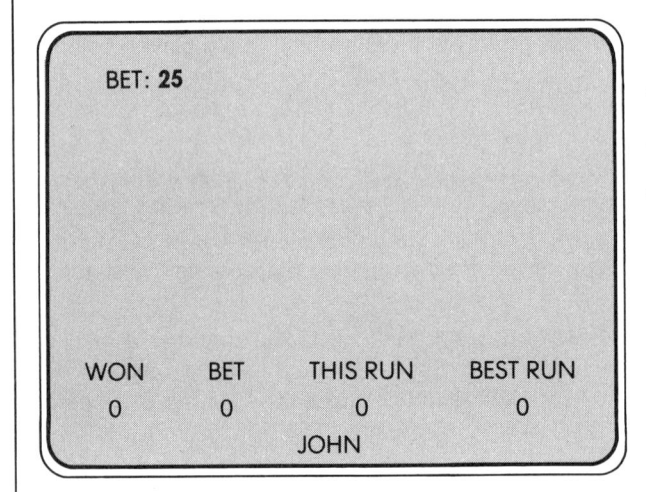

BET: **25**

WON	BET	THIS RUN	BEST RUN
0	0	0	0

JOHN

JOHN has typed B.

The program asked for a bet amount, and JOHN entered 25.

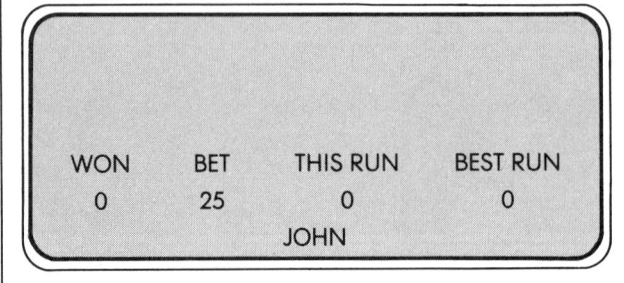

WON	BET	THIS RUN	BEST RUN
0	25	0	0

JOHN

Now the display shows JOHN's bet amount. The bet amount will not change unless JOHN changes it.

The player has now entered a bet. This is the last of the preliminaries—the action can begin.

Figure 8.3: JOHN's Bet

INSIDE BASIC GAMES   237

the next throw. Mary waits for the magic moment, presses the space bar in a special way, and two 4s appear—MARY wins. MARY's initial display appears again. The WON, THIS RUN and BEST RUN entries have been updated, while the BET entry remains unchanged.

Now that you have a general idea of how the game is played, let's look at a list of the commands that you can enter when the initial display appears. This list appears in Figure 8.6. We shall now discuss each of these commands.

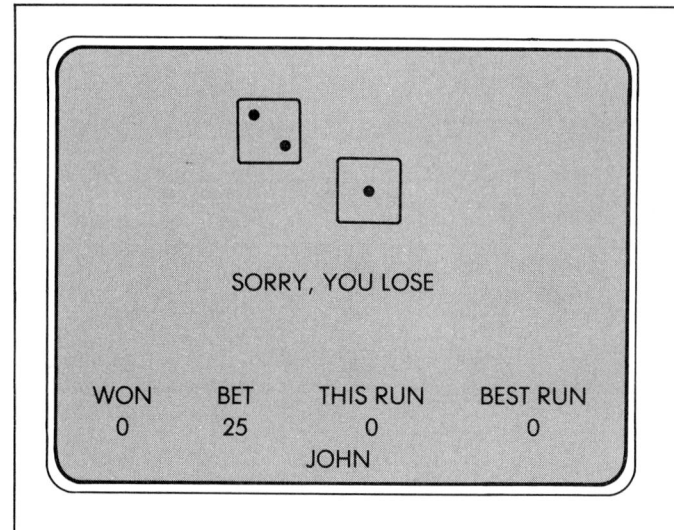

```
                  ┌───┐
                  │ • │
                  │  •│
                  └───┘ ┌───┐
                        │ • │
                        └───┘

              SORRY, YOU LOSE


   WON    BET   THIS RUN   BEST RUN
    0     25       0          0
                 JOHN
```
JOHN presses the space bar.
First the dice appear.
Then the message:
SORRY, YOU LOSE.

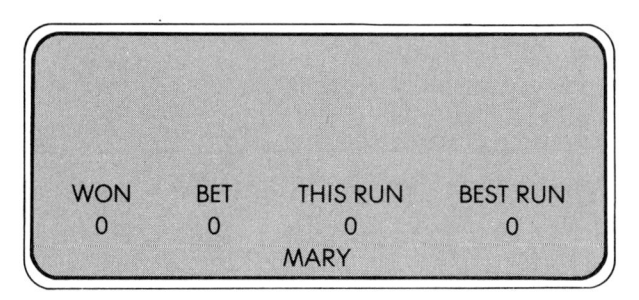

```



   WON    BET   THIS RUN   BEST RUN
    0     0        0          0
                 MARY
```
After a brief pause, MARY's initial display appears.

JOHN's turn consists of a single throw. A throw of 3 makes JOHN lose immediately, causing the dice to pass to the next player.

Figure 8.4: JOHN Throws the Dice

238 CRAPS

POINT IS 8 POINT: 8 THROWS: 1 WON BET THIS RUN BEST RUN 0 50 0 0 MARY	First, the dice picture appears. Then "POINT IS 8" Then "POINT: 8 THROWS: 1"
POINT IS 8 POINT: 8 THROWS: 1 WON BET THIS RUN BEST RUN 0 50 0 0 MARY	Next, the dice disappear. The program is waiting for MARY to throw again, (i.e., press the space bar).
YOU WIN! POINT: 8 THROWS: 2 WON BET THIS RUN BEST RUN 0 50 0 0 MARY	MARY throws another 8.
WON BET THIS RUN BEST RUN 50 50 1 1 MARY	The initial display appears reflecting MARY's win.

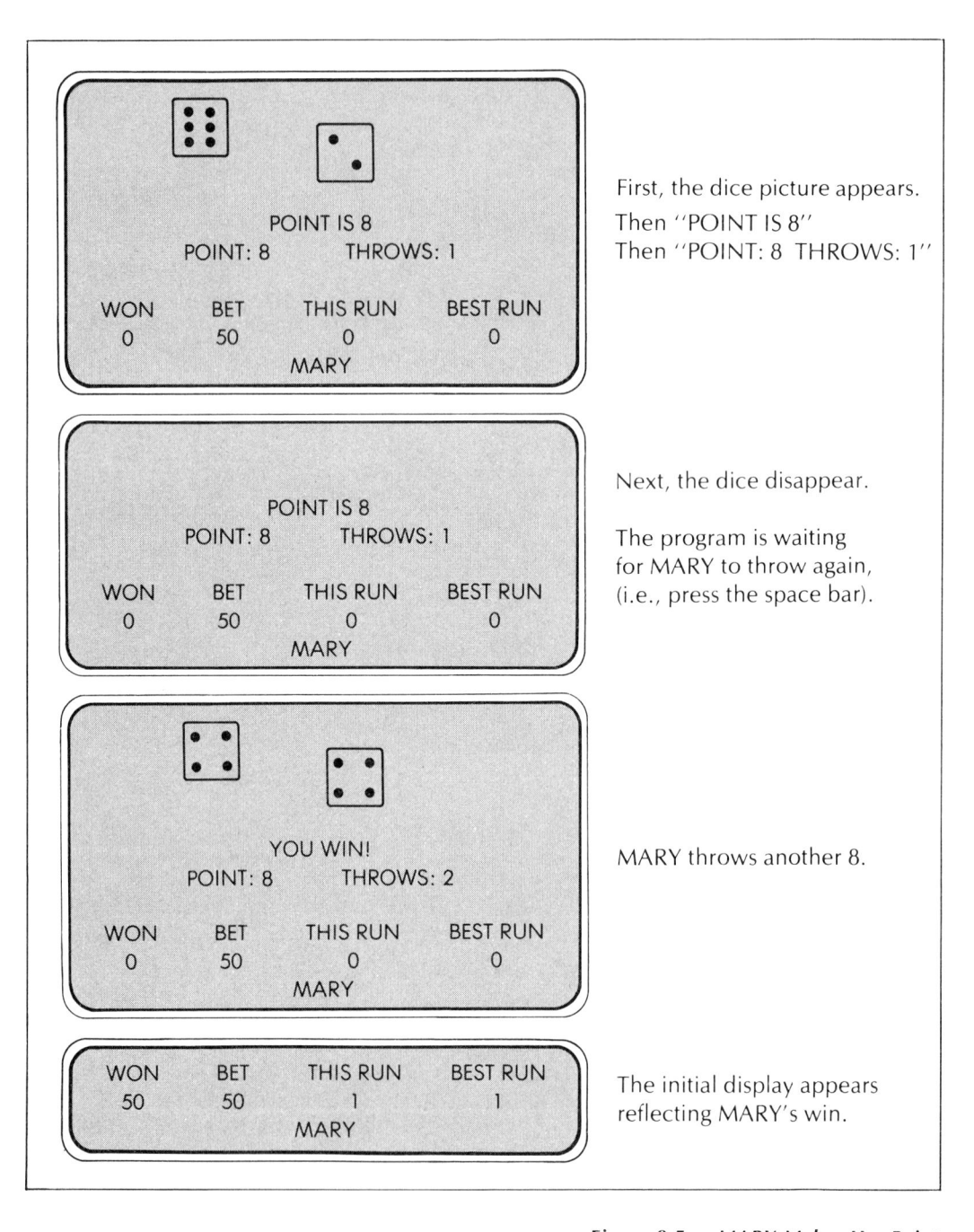

Figure 8.5:   MARY Makes Her Point

The Commands

The N command (see Figure 8.1) is used for entering the name of a new player. N causes a corresponding entry to be made in the statistics array, which is used to remember the information (WON, BET, etc.) from the bottom line of your initial display while other players are playing. Thus, when your turn comes again, the program will know as much about you as it did when your last turn ended.

The P command allows you to pass the dice to the next player without waiting until you lose. (Normally, you keep the dice as long as you continue to win.) Your statistics are saved, and then the next player's statistics are recalled.

Command	Meaning
N	New player. The program asks for a name and places the player's name into the rotation.
P	Pass the dice. The next player's initial display appears.
Q	Quit. A player is allowed to leave the game. A list of players with numbers is displayed. The number of the quitting player is entered, and that player is removed from the game (entering zero aborts the process with no one removed). Each player's winnings/losses are also displayed in the list.
S	Sidebets. Nothing happens—this feature has not been implemented.
space	First throw. The player's first throw is made, starting the current series.
B	Bet. The player is allowed to enter a numeric bet amount.
D	Double. The bet amount is doubled.
L or W	Losses/winnings. The total amount won/lost is bet.
O	Original. The last amount entered through a B command becomes the bet amount.
R	Ride. The bet amount becomes the amount of the last bet, plus the amount won. This command has the same effect as D.

Figure 8.6:   Commands Available at Initial Display

The Q command allows a player (any player—not just the one whose initial display appears on the screen) to leave the game. Figure 8.7 illustrates the dialog involved. A list of player names and numbers appears. Any player can then quit, simply by entering the appropriate number.

WON	BET	THIS RUN	BEST RUN	
25	25	0	2	
		JOHN		

JOHN's initial display

SUSAN wants to quit. Player enters Q.

NUMBER	WON/LOST	NAME
1	25	JOHN
2	−700	MARY
3	−20	SUSAN

WHAT IS YOUR NUMBER, PLEASE? **3**

SUSAN is removed.

WON	BET	THIS RUN	BEST RUN
25	25	0	2
		JOHN	

JOHN's initial display appears again.

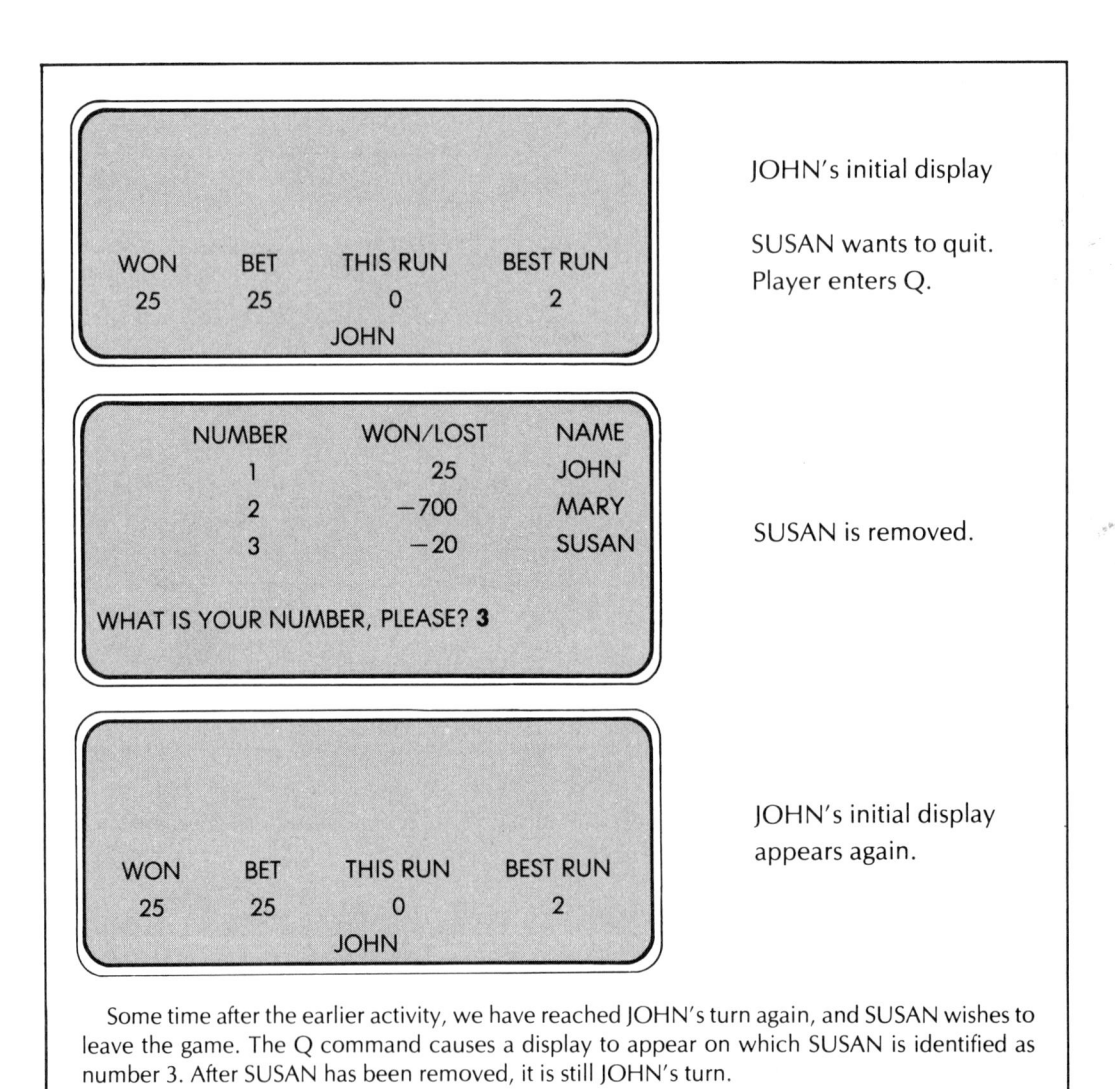

Some time after the earlier activity, we have reached JOHN's turn again, and SUSAN wishes to leave the game. The Q command causes a display to appear on which SUSAN is identified as number 3. After SUSAN has been removed, it is still JOHN's turn.

Figure 8.7: Dropping SUSAN

The S command doesn't do anything. (When we discuss the program, we shall explain why it is included.)

The space bar is used to start the game. The player's first throw appears on the screen, and play proceeds as described above.

The betting commands allow various alterations of the player's bet. Figure 8.3 shows how the B command is used to enter a numeric bet amount. The other bet commands (D, L, O, R, W) alter the bet value on the statistics line but do not affect the remainder of the display.

The Craps Program

The program that implements Craps appears in Figures 8.8 through 8.39. Many aspects of this program have been discussed in previous chapters and will not be discussed explicitly here. These include:

— Modular, top-down structure

— Isolation of function (e.g., *throw* and *draw*)

— Menu selection for player identification (Figures 8.31 and 8.32)

— Mechanics of deletion from an array (Figure 8.32)

— File handling functions (Figure 8.34).

We shall confine our discussion to several new points:

— Use of the display screen for a mixture of constant and varying information

— Attainment of smooth, fast game action

— Context switching

— Randomizing.

Use of the Display Screen

The Craps program presented here depends upon the use of a display screen. Much of the character of the game would be destroyed if it were converted to run on a hardcopy device like a teleprinter. At the start of each set of throws, the screen is cleared. The various aspects of the game are then displayed at different places on the screen: the dice appear in one place, the statistics in another, the throw result and the "game line" in others. Each of these aspects of the game manages its own screen space, blanking out or overwriting old information when necessary.

The separate management of screen areas relies upon a cursor-positioning routine (see Figure 8.27) that allows the cursor to be set to a specified line and column position. The assignment of line and column positions to the elements of the display is done in the *initialize* routine (Figure 8.35).

The display of the dice is handled by the *draw* and *cleardice* routines (Figures 8.38 and 8.39). The simple *draw* routine in Figure 8.38 relies on the Pet's cursor-moving characters. After positioning the cursor for each die picture, *draw* first displays the string BX$, which results in the appearance of a box on the screen. This box surrounds a square area that will contain the "dots," and causes the cursor to be positioned at the upper left corner of that square area. The Pet version of the string BX$ contains cursor left, cursor down and cursor up characters, intermixed with the horizontal and vertical line characters (shift $, shift #, shift % and shift ') that make up the box. After the box is in place, one of six die face strings from the array D$ is printed. These strings combine spaces, circular dots (shift Q) and cursor-moving commands to fill the nine positions inside the box.

This display can also be achieved on other systems. The Apple version (not shown here) uses a two-dimensional array D$(6,5) instead of the one-dimensional D$(6) used for the Pet version. In the Apple version, there is no BX$ string for the box. Rather, each picture consists of five lines, including both the box outlines and the dots. Thus, on the Apple, the drawing of the first die picture is accomplished by the following instructions:

```
FOR  LX = L1 TO L1 + 4
     LL = LX: CC = C1: GOSUB cursor
     PRINT D$(D1, LX − L1 + 1);
NEXT LX
```

A similar sequence of instructions is used to erase the pattern in the Apple version of *cleardice*. The box sides are made from hyphens and colons on the Apple, and the dots are asterisks. The TRS-80 version (referred to in the footnotes to the figures) is similar to the Apple version, but a TRS-80 version similar to the Pet version is also possible.

The display and clearing of the bet, result, game and statistics lines are easy to understand. Clearing is done with strings of blanks. See the routines *clearbet* (Figure 8.23), *clresult* (Figure 8.24) and *clstat* (Figure 8.25). No clearing of the game line is necessary. The line is merely overwritten, and the new copy always completely covers the old (see Figure 8.20).

#Craps

```
        GOSUB initialize              #set up constants and formats
        repeat {
            GOSUB start               #clear away old game, show stats
            GOSUB onech               #get an input character
            GOSUB incode              #decode it (return IC)
            ON IC GOSUB
                askbet,               #establish bet amount
                game,                 #play one game
                new,                  #new player joins the game
                quit,                 #player leaves the game
                sidebet,              #any bets aside from main bet
                pass                  #pass the dice
        }
```

```
100    GOSUB 1950
130*   GOSUB 960:GOSUB 2340:GOSUB 160
140    ON IC GOSUB 560,230,1400,1690,1820,1560
150    GOTO 130
```

This is the principal loop of Craps. Each iteration of the loop is for one game. The player begins by entering a single-character input that determines what happens next. A space starts the game and causes the first throw of the dice. Other inputs are used to allow players to enter or leave the game, or to make bets other than the player's main bet. The player can pass the dice or alter the bet amount. Bets can be entered numerically, doubled, set to the total amount already won or lost, or returned to the "old" (last numerically entered) amount. Winnings can also be allowed to "ride."

The routine *start* displays the current player's name and statistics. As soon as the player loses, the dice are passed to the next player.

---

*Because of the length of the initialization routine in the Apple and TRS-80 versions, the routines shown in Figures 8.37, 8.38 and 8.39 have different line numbers in the Apple and TRS-80 versions. The correspondence is:

Routine	Pet Line	Apple/TRS-80 Line	Figure
draw	2240	2500	8.38
cleardice	2300	2580	8.39
onech	2340	2660	8.37

Thus, the "GOSUB 2340" in line 130 becomes "GOSUB 2660" in the Apple and TRS-80 versions.

Figure 8.8: *Craps*

Smooth, Fast Game Action

The achievement of smooth, fast game action is an extremely important aspect of game design. The Craps program assures fast game action by:

— Single-character input, without intrusive prompts.

— Automatic disappearance of the dice and some of the messages from the screen after a fixed (adjustable) interval of time.

— The existence of several display areas that can change with each keystroke.

#Analyze input codes

incode   IF case
        X$ = " " THEN IC = 2   #space bar to play
        X$ = "N" THEN IC = 3   #new player
        X$ = "Q" THEN IC = 4   #player quits
        X$ = "S" THEN IC = 5   #side bets
        X$ = "P" THEN IC = 6   #pass the dice
   else
        IC = 1   #default: treat as bet code
   RETURN

```
160   REM INCODE
170   IF X$=" " THEN IC=2:RETURN
180   IF X$="N" THEN IC=3:RETURN
190   IF X$="Q" THEN IC=4:RETURN
200   IF X$="S" THEN IC=5:RETURN
210   IF X$="P" THEN IC=6:RETURN
220   IC = 1:RETURN
```

The *incode* routine analyzes the first input character of the game. If the character is one that the routine recognizes, then the routine sets the appropriate value into IC, which is used in the ON...GOSUB construction in the main routine. If the character is not recognized, IC is set to the default value, which will cause the code to be treated as a bet-setting command by the *askbet* routine.

Figure 8.9: Decoding the First Input Character

Context Switching

Our version of Craps provides a simple example of context switching, which is a very important aspect of more sophisticated programs, such as time-sharing systems. To apply the language of computer science to the craps table, we say that each player represents a *process*. Your process consists of your actions during your turn: betting and throwing the dice. The other players' processes consist of their actions during their turns.

In a casino, the dice table, the dice, the position on the betting layout where the bet from the player with the dice is placed—these are all *shared resources*. Only one process at a time can use these resources (i.e., only one player can be throwing the dice); the other processes are said to be *suspended* while this happens.

In our Craps program, the keyboard, screen display and controlling programs are the shared resources, and the processes are the various players' interactions with these resources. The only items in the program that vary from player to player are the player's name and statistics. These items constitute the *context* in which the programs are executing when a given player's process is active. Specifically, the context consists of the variables WN (winnings/losses), BT (bet amount), BR (best run of wins), OB (original bet) and CP (current player number—used as an index to the name array, and as one of the indices to the statistics array).

The context switching occurs in the *pass* routine in Figure 8.30. In this example, most of the machine and program are shared—the only separate resources of the processes are their slots in the statistics and name arrays.

This simple switch is the basis of any kind of *multiprocessing* program. In a more complex program, like a timesharing system, each process might have its own display and keyboard as well. In that case, the context switching would occur much more frequently—whenever the program was waiting for input from the keyboard of any process, the next process would be allowed to run. You would never notice the fact that the program was processing someone else's betting and dice throwing. In fact, you would have the illusion that you had the machine entirely to yourself. When you were waiting for the program's response, it would be giving you its full attention, but when you thought the program was waiting for you, it would actually be paying no attention to you at all. (This is similar to the practice of the great singer Chaliapin who claimed that by breathing when he wasn't expected to and by not breathing when he was expected to, he was able to give the impression that he never breathed at all.)

There is a great deal that we haven't covered about timesharing and other multiprocessing systems, but the simple context switching illustrated by the Craps program does give you an idea of how such systems work.

```
# Play one game

game    GOSUB throw                              #initial throw
        IF D = 2, 3 or 12 THEN                   #2, 3 or 12 loses
            {GOSUB lose: RETURN}
        else IF D = 7 or 11 THEN                 #7 or 11 wins
            {GOSUB win: RETURN}
        else {
            PT = D: TH = 1                       #set point, number of throws
            RS$ = "POINT IS " + STR$(PT)         #announce "result" of throw
            GOSUB result
            }
        GOSUB gamestats                          #display point and throws
        repeat {                                 #try to make the point
            GOSUB onech                          #wait for player to press key
            GOSUB throw: TH = TH + 1             #generate a throw
            GOSUB gamestats                      #display point and throws
            IF D = PT THEN
                {GOSUB win: break}               #making point wins
            else IF D = 7 THEN
                {GOSUB lose: break}              #7 loses
            else {
                RS$ = STR$(D) + "— THROW AGAIN"
                GOSUB result
                }
            }
        RETURN
```

The *game* routine goes through one complete set of throws. An initial 7 or 11 causes an immediate win; a 2, 3, or 12, an immediate loss. Any other throw becomes the "point," and subsequent throws are made until either the point or 7 is thrown.

Figure 8.10: One Set of Throws

INSIDE BASIC GAMES 247

```
230   REM GAME
240   GOSUB 370
250   IF D=2 OR D=3 OR D=12 THEN GOSUB 490:RETURN
260   IF D=7 OR D=11 THEN GOSUB 420:RETURN
270   PT=D:TH=1
280   RS$="POINT IS "+STR$(PT):GOSUB 1020
290   GOSUB 980
300*  GOSUB 2340:GOSUB 370
310   TH=TH+1:GOSUB 980
320   IF D=PT THEN GOSUB 420:GOTO 360
330   IF D=7 THEN GOSUB 490:GOTO 360
340   RS$=STR$(D)+" — THROW AGAIN":GOSUB 1020
350   GOTO 300
360   RETURN
```

*In the Apple and TRS-80 versions, 2340 is replaced by 2660 (see Figure 8.8).

Figure 8.10a: BASIC for Game

```
#Throw the dice

throw    D1 = INT(6*RND(1) + 1)      #first die value
         D2 = INT(6*RND(1) + 1)      #second die value
         GOSUB draw                   #display picture of dice
         D = D1 + D2                  #total value of throw
         RETURN

   370   REM THROW
   380*  D1=INT(6*RND(1)+1)
   390*  D2=INT(6*RND(1)+1)
   400** GOSUB 2240
   410   D=D1+D2:RETURN
```

The *throw* routine generates two random numbers between one and six—the values on the faces of the two dice. The subroutine *draw*, which displays the dice, appears at the end of the program, with *cleardice*, *wait*, and *initialize*. These four routines have different implementations on different home computers. The remainder of the program is essentially identical in the Pet, Apple and TRS-80 versions.

*In the TRS-80 version, RND(1) is replaced by RND(0).
**In the Apple and TRS-80 versions, 2240 is replaced by 2500 (see Figure 8.8).

Figure 8.11: One Throw

Randomizing

It is important that the dice values be unpredictable. Unfortunately, since a given computer and program may behave identically each time power is turned on, the sequence of numbers generated by the random number function, RND, is predictable. The process by which the programmer conceals this predictability from the player is called *randomization*. The technique used in the *onech* routine in Figure 8.37 to achieve randomization is to call upon the RND (random number) function an unpredictable number of times. The number of calls is unpredictable because the call occurs as part of a loop that waits for the player to press a key. The longer the player waits, the more times the RND function is called. The duration of the loop is so short (a few thousandths of a second at most) that the player really can't control the number of iterations.

#Process a win

win	WN = WN + BT	#increase winnings
	RD = 2*BT	#amount riding
	PS = PS + 1	#another consecutive win
	BR = max(BR,PS)	#BR = best pass ever
	ZP = CP: GOSUB savestats	#update stats array
	RS$ = "YOU WIN!": GOSUB result	#announce result
	GOSUB delay	#give player time to read it
	RETURN	

```
420  REM WIN
430  WN=WN+BT
440  RD=2*BT:PS=PS+1
450  IF PS>BR THEN BR=PS
460  ZP=CP:GOSUB 1890
470  RS$="YOU WIN!":GOSUB 1020
480  GOSUB 1380:RETURN
```

When the player wins, the *win* routine is called to update the statistics and announce the result.

Figure 8.12: Housekeeping After a Win

This technique works well on the Pet and TRS-80 systems, but it does not work for an Apple, since the Apple single-character input waits until a character has been entered before returning. Therefore, you only go through the loop once. In order to randomize in a way that the player can't control, you must use the Apple " game paddles." The following sequence will work:

1. Tell the player to press the paddle button.
2. Wait until the button is not being pressed.
3. Repeatedly call RND and check the button, until it is being pressed.

Step 2 prevents the player from controlling the number of calls on RND by pressing the button before the program starts checking.

#Process a loss

lose WN = WN − BT   #decrease winnings
      RD = OB   #amount riding = "old bet"
      PS = 0   #streak has ended
      ZP = CP: GOSUB savestats   #update stats array
      RS$ = "SORRY, YOU LOSE": GOSUB result   #announce result
      GOSUB delay   #give player time to read it
      GOSUB pass   #pass dice to next player
      RETURN

```
490  REM LOSE
500  WN=WN−BT
510  RD=OB:PS=0
520  ZP=CP:GOSUB 1890
530  RS$="SORRY, YOU LOSE":GOSUB 1020
540  GOSUB 1380:GOSUB 1560
550  RETURN
```

When the player loses, the *lose* routine is called to update the statistics, announce the result, and pass the dice to the next player.

Figure 8.13: Housekeeping After a Loss

250 CRAPS

```
#Ask for the bet


askbet    repeat {
              GOSUB betcode                        #evaluate input character
              ON B GOSUB
                  double,                          #double bet
                  ride,                            #let winnings ride
                  won,                             #bet amount won/lost
                  old,                             #return to "old" bet
                  number                           #enter a numeric bet
              IF BT <= 0 THEN {                    #invalid bet code
                  LL = BL
                  CC = BC
                  GOSUB cursor
                  PRINT "SORRY, PLEASE..."
                  GOSUB onech                      #get another input char
              }
              else IF BT > MX THEN {               #bet exceeds limit
                  LL = BL
                  CC = BC
                  GOSUB cursor
                  PRINT "HOUSE LIMIT IS"; MX       #mention house limit
                  BT = MX                          #set bet to maximum
                  IF OB > MX THEN
                      OB = MX
                  GOSUB delay                      #give player time to read
              }
          } until (0 < BT <= MX)
          RETURN
```

The *askbet* routine processes the single-character bet command passed in X$ from the main routine. It calls one of the routines *double, ride, won, old* or *number,* then evaluates the bet amount BT returned.

Figure 8.14: Routing the Bet Command

INSIDE BASIC GAMES   **251**

```
560   REM ASK FOR BET
570   GOSUB 700
580   ON B GOSUB 860,880,900,920,770
590   IF 0<BT AND BT<=MX THEN 690
600   IF BT<=MX THEN 650
610   LL=BL:CC=BC:GOSUB 1270
620*  PRINT "HOUSE LIMIT IS";MX;" ";
630   BT=MX:IF OB>MX THEN OB=MX
640   GOSUB 1380:GOTO 690
650   LL=BL:CC=BC:GOSUB 1270
660   PRINT "SORRY, PLEASE ENTER A NEW BET CODE";
670** GOSUB 2340
680   GOTO 570
690   RETURN
```
*The Apple version of this line differs only in the spacing in the string constant.
**See note in Figure 8.8.

Figure 8.14a: BASIC for Askbet

#Interpret the single-character bet code

betcode   IF case
          X$ = "D"   THEN B = 1   #double bet amount
          X$ = "R"   THEN B = 2   #let last winnings ride
          X$ = "W" or "L" THEN B = 3   #bet = total won/lost
          X$ = "O"   THEN B = 4   #bet = original amount
          X$ = "B"   THEN B = 5   #player will enter a number
      else   B = 4   #default = original
      RETURN

```
700   REM BETCODE
710   IF X$="D" THEN B=1:RETURN
720   IF X$="R" THEN B=2:RETURN
730   IF X$="W" OR X$="L" THEN B=3:RETURN
740   IF X$="O" THEN B=4:RETURN
750   IF X$="B" THEN B=5:RETURN
760   B=4:RETURN
```

    The *betcode* routine takes the hand-me-downs from *incode* and sorts through them. If this routine also fails to recognize the command, then the command is treated like the "old bet" command. This is a relatively harmless default action. Control returns to the main program, so that the player who enters *askbet* accidentally is not forced to enter a valid bet code.

Figure 8.15: Decoding the Bet Command

252 CRAPS

```
#Accept a number

number      AA = 0                              #initialize number
            GOSUB betprompt                     #ask for input
            repeat {
                GOSUB onech                     #next character
                NN = ASC(X$) − ASC("0")         #make into a digit,
                IF NN < 0 OR NN > 9 THEN        #if not a digit, then done
                    break
                PRINT X$;                       #echo digit
                AA = 10*AA + NN                 #accumulate the number
            }
            BT = AA                             #set bet amount
            OB = BT                             #set "original" bet
            RETURN

#Prompt for input

betprompt   GOSUB clearbet
            PRINT "BET: ";
            RETURN


770   REM NUMBER
780   AA=0:GOSUB 940
790*  GOSUB 2340
800   NN=ASC(X$)−ASC("0")
810   IF NN<0 OR NN>9 THEN 840
820   PRINT X$;:AA=10*AA+NN
830   GOTO 790
840   BT=AA:OB=BT
850   RETURN

940   REM BETPROMPT
950   GOSUB 1180:PRINT "BET: ";:RETURN
```

The *number* routine and its subroutine *betprompt* provide an example of poor program design. (Can you explain why this is so?) An alternative routine is shown in Figure 8.17.

---
*In the Apple and TRS-80 versions, 2340 is replaced by 2660 (see Figure 8.8).

Figure 8.16: Accepting Numeric Bet Input

#Alternate version of numeric bet input

number repeat {
        GOSUB clearbet　　　　　　　　　　#clear bet line
        PRINT "BET: ";: GOSUB stringin　　#get input up to a RETURN
        IF XX$ = "" THEN　　　　　　　　　#escape valve
            RETURN
        else
            NN = VAL(XX$)　　　　　　　　#convert input to a number
       } until (NN <> 0)　　　　　　　　　#don't settle for 0 or non-numeric
       BT = NN: OB = BT　　　　　　　　#set bet amount, "original" bet
       RETURN

This routine avoids reinventing the wheel by using the *stringin* routine of the Match-up program, and the VAL function that is built into BASIC.

Figure 8.17: Better Numeric Bet Input

#Bet-specifying subroutines

double BT = 2*BT: RETURN　　　　　#double the bet

ride　　BT = RD: RETURN　　　　　　#let winnings ride

won　　BT = ABS(WN):RETURN　　　　#bet the total amount won or lost

old　　 BT = OB: RETURN　　　　　　#return to the original bet

```
860  REM DOUBLE
870  BT=2*BT:RETURN

880  REM RIDE
890  BT=RD:RETURN

900  REM WON
910  BT=ABS(WN):RETURN

920  REM OLD
930  BT=OB:RETURN
```

These are the small routines called from *askbet* to implement the various bet-setting options.

Figure 8.18: Setting the Bet

254 CRAPS

#Start the game

start   GOSUB clearscreen   #clear the screen
            GOSUB stats   #display the stats line and player name
            RETURN

    960   REM START
    970   GOSUB 1250:GOSUB 1050:RETURN

The *start* routine clears the screen and displays the basic information about the player—name, winnings, amount bet, and so forth. No information concerning any specific turn appears.

Figure 8.19: Preparing to Play

#Print game statistics line

gamestats   LL = GL: CC = GC: GOSUB cursor
                PRINT "POINT: "; PT; " THROWS: "; TH;
                RETURN

    980   REM GAMESTATS
    990   LL=GL:CC=GC:GOSUB 1270
    1000* PRINT "POINT:";PT;" THROWS:";TH;
    1010 RETURN

The *gamestats* routine displays the line containing the player's point and the number of throws already made. *Gamestats* is called from *game*.

First *gamestats* positions the cursor, then it prints the information. Since the value of PT (the point) does not vary during a game, the only part of the display that appears to change is the value of TH, the number of throws.

---

*The Apple version of this line differs only in the spacing inside the string constants.

Figure 8.20: Displaying Point and Throws

INSIDE BASIC GAMES 255

```
#Display the result of the throw — RS$

result    GOSUB clresult           #blank out the result line
          PRINT RS$                #display the result
          GOSUB delay              #give player time to look
          GOSUB cleardice          #rake in the dice
          RETURN


   1020  REM RESULT
   1030  GOSUB 1150
   1040* PRINT RS$;:GOSUB 1380:GOSUB 2300:RETURN
```

The *result* routine displays the result of the throw. The string RS$ is passed to this routine by the calling program (*game, win* or *lose*). RS$ says things like "YOU WIN" or "POINT IS 8" or "6 — THROW AGAIN." The blanking out of the result line, in addition to assuring that the previous result is completely obliterated, causes a noticeable blinking of the line, even if the current result and the previous result are identical. This blinking lets the player know that the program has finished processing the throw.

After the result line is displayed, the program delays briefly (the duration of the delay is set in the initialization routine by specifying a value for the variable DC); then the dice picture is cleared from the screen.

Notice how this short routine results in three distinct, noticeable screen changes:

— the blanking of the result line
— the new result display
— the clearing of the dice.

This sequenced set of changes holds the player's attention and makes the game more absorbing.

---
*In the Apple and TRS-80 versions, 2300 is replaced by 2580 (see Figure 8.8).

Figure 8.21: Displaying the Throw Result

```
#Display player statistics

stats    GOSUB clstat                              #blank out stats area
         IF WN < 0 THEN                            #say WON or LOST
             PRINT "LOST",
         else
             PRINT "WON",
         PRINT "BET", "THIS RUN", "BEST RUN"       #display remainder of title
         PRINT ABS(WN), BT, PS, BR; "    ";        #display values
         IF CP > 0 THEN {                          #if any player, display name
             CC = NC: LL = NL: GOSUB cursor
             PRINT NM$(CP);
         }
         RETURN

    1050 REM STATS
    1060 GOSUB 1210
    1070* IF WN<0 THEN PRINT "LOST",:GOTO 1090
    1080* PRINT "WON",
    1090* PRINT "BET","THIS RUN","BEST RUN"
    1100* PRINT ABS(WN),BT,PS,BR;"    ";
    1110 IF CP=0 THEN 1140
    1120 CC=NC:LL=NL:GOSUB 1270
    1130 PRINT NM$(CP);
    1140 RETURN
```

The *stats* routine displays the player's statistics—amount won or lost, the bet, the number of consecutive wins (the current "run") and the most consecutive wins ever achieved by this player (best "run"). The values printed are from the global variables WN, BT, PS, BR; these variables are loaded from the player's slots in the stats array by the *pass* routine when this player's turn comes around.

One nice touch is the printing of WON or LOST, depending upon whether WN has a positive or negative value, rather than simply displaying something like "WINNINGS: −100" when the player is losing.

---

*In the Apple version of this routine, the spacing of the four displayed statistics and their captions is accomplished with TAB(10), TAB(20) and TAB(30) entries in the PRINT lists, rather than through the use of commas. The reason for this difference is that the use of commas in Apple BASIC PRINT statements leads to the appearance of three items per line rather than four.

Figure 8.22: Displaying the Player Statistics

```
#Clear for the bet input

clearbet   GOSUB start                           #display starting screen
           CC = BC: LL = BL: GOSUB cursor        #position cursor
           RETURN

    1180  REM CLEARBET
    1190  GOSUB 960
    1200  CC=BC:LL=BL:GOSUB 1270:RETURN
```

The *clearbet* routine prepares for bet input. Since nothing important (other than the player statistics) is on the screen, *clearbet* takes the shortcut of simply clearing the screen and redisplaying the stats (all done in *start*), then positioning the cursor properly.

Figure 8.23: Preparing for Bet Input

```
#Clear the result line

clresult   CC = RC: LL = RL: GOSUB cursor        #position cursor
           PRINT RC$;                            #blank and reposition
           RETURN

    1150  REM CLRESULT
    1160  CC=RC:LL=RL:GOSUB 1270
    1170  PRINT RC$;:RETURN
```

The *clresult* routine blanks out the result line and repositions the cursor for a display of the new result. The repositioning of the cursor is accomplished by the inclusion of cursor-moving characters in the string RC$. This makes the routine specific to the Pet, but the Apple and TRS-80 version is only slightly different. On those systems, RC$ contains no cursor-moving characters, and the first line is repeated after the PRINT RC$;. Since the difference is only minor, this routine is not included with the other configuration-specific routines at the end of the program. In fact, the Apple and TRS-80 version would work perfectly well on the Pet, with no loss of efficiency. The actual BASIC instructions for the Apple and TRS-80 are:

```
    1150  REM CLRESULT
    1160  CC=RC: LL=RL: GOSUB 1270: PRINT RC$;
    1170  CC=RC: LL=RL: GOSUB 1270: RETURN
```

Figure 8.24: Blanking the Result Line

258 CRAPS

#Clear the statistics area on the screen

clstat   CC = 0: LL = SL: GOSUB cursor   #first column of first stat line
           PRINT BL$: PRINT BL$;   #clear two lines
           CC = 0: LL = SL: GOSUB cursor   #reposition
           RETURN

```
1210 REM CLSTAT
1220 CC=0:LL=SL:GOSUB 1270
1230 PRINT BL$:PRINT BL$;
1240 CC=0:LL=SL:GOSUB 1270:RETURN
```

The *clstat* routine clears two lines of the screen (lines SL, SL + 1). The cursor is then positioned at the start of the blank area, to prepare for display of statistics by the *stats* routine.

Figure 8.25: Blanking the Statistics Area

#Clear the screen

clearscreen   PRINT "clr";
                  RETURN

```
1250 REM CLEARSCREEN
1260 PRINT CHR$(147);:RETURN
```

The *clearscreen* routine clears the screen. This is the Pet version. Apple and TRS-80 versions differ slightly—the first line is replaced by HOME on Apple, or by CLS on TRS-80. In all three cases, the screen is cleared, and the cursor moves to the upper left position. (See Figure 2.12.)

Figure 8.26: Clearing the Screen

INSIDE BASIC GAMES 259

```
#Position the cursor to line LL, column CC — Pet version
cursor   PRINT "home";
         CC = CC mod columns: LL = LL mod lines
         IF CC <> 0 THEN                          #move cursor right CC times
             FOR ZZ = 1 TO CC
                 PRINT "right";
             NEXT ZZ
         IF LL <> 0 THEN                          #move cursor down LL times
             FOR ZZ = 1 TO LL
                 PRINT "down";
             NEXT ZZ
         RETURN

1270 REM CURSOR
1280 PRINT CHR$(19);
1290 IF CC<0 THEN CC=CC+40:GOTO 1290
1300 IF CC>39 THEN CC=CC-40:GOTO 1300
1310 IF LL<0 THEN LL=LL+24:GOTO 1310
1320 IF LL>23 THEN LL=LL-24:GOTO 1320
1330 IF CC=0 THEN 1350
1340 FOR ZZ=1 TO CC:PRINT CHR$(29);:NEXT ZZ
1350 IF LL=0 THEN RETURN
1360 FOR ZZ=1 TO LL:PRINT CHR$(17);:NEXT ZZ: RETURN

The Apple and TRS-80 versions differ as noted in Figure 4.16.
```

Figure 8.27: Positioning the Cursor

```
#Delay long enough for the player to read what appears on the screen
delay    FOR ZZ = 1 TO DC
             NEXT ZZ
         RETURN

1380 REM DELAY
1390 FOR ZZ=1 TO DC:NEXT ZZ:RETURN
```

The *delay* routine runs through a do-nothing loop the number of times specified in DC, which is set in a DATA statement in the *initialize* routine. The actual length of the delay is what determines the speed of the game's action. The proper setting of DC is best determined by trial and error. A more precise alternative, not available on Apple or TRS-80 systems, would be to use the Pet TI feature. (See Figure 9.36.)

Figure 8.28: Doing Nothing for a While

260 CRAPS

```
          #Add a new player
new       IF NP >= MP THEN {                          #if no room,
              CC = BC: LL = BL: GOSUB cursor          # say so
              PRINT "NO ROOM...";
              GOSUB delay: RETURN
              }
          NP = NP + 1: GOSUB clearscreen              #one more player
          INPUT "... NAME, PLEASE"; NM$(NP)           #get name
          PRINT: PRINT "THANK YOU. ";
          IF NP > 1 THEN {                            #if not the first player
              PRINT "...FOLLOW "; NM$(NP − 1)
              ZP = CP: GOSUB savestats                #save current stats
              GOSUB zap: ZP = NP: GOSUB savestats     #give player fresh stats
              ZP = CP: GOSUB getstats                 #restore current stats
              }
          else                                        #if first, current stats
              {ZP = 1: GOSUB savestats: CP = 1}       #become player 1 stats
          GOSUB delay: GOSUB delay: RETURN            #let player read screen
1400  IF NP<MP THEN 1450
1420  CC=BC:LL=BL:GOSUB 1270
1430  PRINT "NO MORE ROOM AT THE TABLE";
1440  GOSUB 1380:RETURN
1450  NP=NP+1:GOSUB 1250
1470* INPUT "MAY I HAVE YOUR NAME PLEASE";NM$(NP)
1480  PRINT:PRINT "THANK YOU. ";
1490  IF NP<=1 THEN 1540
1500  PRINT "YOU WILL FOLLOW ";NM$(NP−1)
1510  ZP=CP:GOSUB 1890
1520  GOSUB 1860:ZP=NP:GOSUB 1890
1530  ZP=CP:GOSUB 1920
1540  CP=1:ZP=1:GOSUB 1890
1550  GOSUB 1380:GOSUB 1380:RETURN
```

The *new* routine adds a new player to the game. The player is given initialized statistics and a place in the rotation.

---
*The Apple version of this line differs only in the value of the string constant.

Figure 8.29: Adding a Player

INSIDE BASIC GAMES 261

```
#Pass the dice

pass    IF NP = 0 THEN RETURN            #if no players, don't bother
        ZP = CP: GOSUB savestats          #save current player's stats
        CP = (CP mod NP) + 1              #set to next player
        ZP = CP: GOSUB getstats           #load new player's stats
        PS = 0: RETURN                    #begin with no wins

  1560  IF NP=0 THEN RETURN
  1580  ZP=CP:GOSUB 1890
  1590  CP=CP+1:IF CP>NP THEN CP=1
  1600  ZP=CP:GOSUB 1920:PS=0
  1610  RETURN
```

The *pass* routine passes the dice from one player to the next. It is called automatically from *lose*, or by an explicit player command.

Figure 8.30: Passing the Dice

```
#Display names and winnings of all players

displayer   GOSUB clearscreen                         #clear screen
            PRINT "NUMBER", "WON/LOST", "NAME"        #display title
            PRINT
            FOR ZZ = 1 TO NP
               PRINT ZZ, ST(1, ZZ), NM$(ZZ)           #display number,
            NEXT ZZ                                   # winnings, name
            PRINT: RETURN

  1620  GOSUB 1250
  1640* PRINT:PRINT "NUMBER","WON/LOST","NAME":PRINT
  1650  FOR ZZ=1 TO NP
  1660* PRINT ZZ,ST(1,ZZ),NM$(ZZ):NEXT ZZ
  1670  PRINT
  1680  RETURN
```

The main purpose of the *displayer* routine is to reveal the numbers assigned to the players. The players can then refer to themselves by number in commands such as the Q (quit) command.

*The Apple versions of these lines use TAB(10) and TAB(20) instead of commas for spacing the output.

Figure 8.31: Giving the Players Numbers

CRAPS

```
#Player quits

quit    GOSUB displayer                          #show names and numbers
        repeat {
            INPUT "...NUMBER, PLEASE"; QP        #ask number
            IF QP = 0 THEN                       #0 = escape valve
                RETURN
            else IF 0 < QP <= NP THEN            #if in range,
                break                            # delete it
            else                                 #otherwise,
                PRINT "PLEASE REFER TO LIST"     # complain
            }
        IF QP = CP THEN                          #if current player,
            GOSUB pass                           # pass the dice first
        IF QP < NP THEN                          #if not deleting last one,
            FOR ZZ = QP TO NP - 1                # move end ones up
                FOR YY = 1 TO nstats
                    ST(YY, ZZ) = ST(YY, Z + 1)
                NEXT YY
                NM$(ZZ) = NM$(ZZ + 1)
            NEXT ZZ
        NP = NP - 1                              #one less player
        IF CP >= QP THEN                         #renumber current player
            CP = CP - 1
        IF NP = 0 THEN                           #if no players,
            GOSUB zap                            # clear stats
        RETURN
```

This routine allows a player to be removed from the rotation. All of the usual deletion principles apply. (See Figure 7.38.)

Figure 8.32: Quitting

INSIDE BASIC GAMES   263

```
1690  REM QUIT
1700  GOSUB 1620
1710* INPUT "WHAT IS YOUR NUMBER, PLEASE";QP
1720  IF QP=0 THEN RETURN
1730  IF QP<=0 OR QP>NP THEN PRINT "PLEASE REFER TO THE LIST":GOTO 1710
1740  IF QP=CP THEN GOSUB 1560
1750  IF QP=NP THEN 1790
1760  FOR ZZ=QP TO NP−1:FOR YY=1 TO 4
1770  ST(YY,ZZ)=ST(YY,ZZ+1):NEXT YY
1780  NM$(ZZ)=NM$(ZZ+1):NEXT ZZ
1790  NP=NP−1:IF CP>=QP THEN CP=CP−1
1800  IF NP=0 THEN GOSUB 1860
1810  RETURN
```

*The Apple version of this line differs only in the appearance of a question mark in the string constant.

Figure 8.32a: BASIC for Quit

#Side bets—stub

sidebet   CC = BC: LL = BL: GOSUB cursor   #go to bet line
               PRINT "SIDEBETS NOT...";   #announce inadequacy
               GOSUB delay   #let player read it
               RETURN

```
1820  REM SIDEBET
1830  CC=BC:LL=BL:GOSUB 1270
1840  PRINT "SIDEBETS NOT YET IMPLEMENTED";
1850  GOSUB 1380:RETURN
```

This "stub" informs the player that the "sidebet" command really doesn't do anything.

Figure 8.33: Side Betting

#Manipulate the stats array

zap   WN = 0   #initialize winnings,
          BT = 0   #bet,
          BR = 0   #best run
          OB = 0   #and "original" bet
          RETURN

savestats ST(1, ZP) = WN   #store player ZP's stats
          ST(2, ZP) = BT
          ST(3, ZP) = BR
          ST(4, ZP) = OB
          RETURN

getstats WN = ST(1, ZP)   #fetch player ZP's stats
          BT = ST(2, ZP)
          BR = ST(3, ZP)
          OB = ST(4, ZP)
          RETURN

```
1860 REM ZAP
1870 WN=0:BT=0:BR=0:OB=0
1880 RETURN

1890 REM SAVESTATS
1900 ST(1,ZP)=WN:ST(2,ZP)=BT:ST(3,ZP)=BR:ST(4,ZP)=OB
1910 RETURN

1920 REM GETSTATS
1930 WN=ST(1,ZP):BT=ST(2,ZP):BR=ST(3,ZP):OB=ST(4,ZP)
1940 RETURN
```

This set of routines provides the fundamental capabilities needed for maintenance of the stats array. The *zap* routine initializes the "working" stats variables; *savestats* stores the working values into the array; and *getstats* loads the working values from the array.

Figure 8.34: Statistics File Handling

#Set up constants and formats — Pet version

initialize   READ SL: DATA statline   #screen positions
        READ GL, GC: DATA gameline, gamecol
        READ RL, RC: DATA resultline, resultcol
        READ L1, C1: DATA die 1 line, die 1 col
        READ L2, C2: DATA die 2 line, die 2 col
        READ BL, BC: DATA betline, betcol
        READ NL, NC: DATA nameline, namecol
        READ DC: DATA delayconstant   #loop limit in *delay*
        READ MX: DATA houselimit   #maximum bet allowed
        FOR ZZ = 1 TO 6
            READ D$(ZZ)
            NEXT ZZ
            DATA "die face for 1"   #3 by 3 dot patterns
            DATA "die face for 2"   # (including
            DATA "die face for 3"   #   cursor-moving
            DATA "die face for 4"   #   characters)
            DATA "die face for 5"
            DATA "die face for 6"
        READ BX$: DATA "box surrounding dots"   #frame for die face
        READ BF$: DATA "blank out box & dots"   #die eraser
        READ RC$: DATA "blank out result line"   #result line eraser
        READ BL$: DATA "blank line"   #full line eraser
        WN = 0: PS = 0: BT = 0: BR = 0   #initialize variables
        NP = 0: CP = 0
        MP = maxplayers
        DIM NM$(maxplayers),   #player names
            ST(nstats, maxplayers)   #player statistics
        RETURN

   The *initialize* routine initializes the constants, variables and arrays used throughout the program. The dot pattern strings in the D$ array, and the BX$ and BF$ die box and die blanking strings all contain and depend upon the cursor-moving characters recognized by the Pet. The routines using these strings appear at the end of the program, following this routine. Versions for other systems are completely different.

Figure 8.35: Initializing

CRAPS

```
1950  REM INITIALIZE
1960  READ SL:DATA 20
1970  READ GL,GC:DATA 16,9
1980  READ RL,RC:DATA 11,10
1990  READ L1,C1:DATA 1,10
2000  READ L2,C2:DATA 3,17
2010  READ BL,BC:DATA 0,3
2020  READ NL,NC:DATA 23,14
2030  READ DC:DATA 200
2040  READ MX:DATA 500
2050  FOR XX=1 TO 6:READ D$(XX):NEXT XX
2060* DATA "(see note)"
2070* DATA "(see note)"
2080* DATA "(see note)"
2090* DATA "(see note)"
2100* DATA "(see note)"
2110* DATA "(see note)"
2120  READ BX$
2130* DATA "(see note)"
2140* READ BF$:DATA "(see note)"
2150  READ C1$:DATA "            "
2160* READ C2$:DATA "(see note)"
2170  RC$=C1$+C2$
2180  READ BL$
2190  DATA "                    "
2200  GOSUB 1860: PS=0
2210  NP=0:MP=9:CP=0
2220  DIM NM$(9),ST(4,9)
2230  RETURN
```

These are the actual BASIC instructions for the initialize routine of Figure 8.35, Pet version. The Apple and TRS-80 versions are shown in Figure 8.36a.

---

*Lines 2060-2110 contain the dice face pictures. Each consists of 3 rows of blanks and dots. The first and second rows in each case are followed by 1 "cursor down" and 3 "cursor back" characters. Thus, each of these strings consists of 17 characters.

Line 2130 contains the "box" around the dice faces. The string contains a "top line" of space, 3 underlines, space, 3 "sides" consisting of right vertical line, 3 spaces, left vertical line and a "bottom line" consisting of 1 space and 3 overlines. The top line and each of the sides is followed by 5 "cursor back" and 1 "cursor down" characters. The bottom line is followed by 3 "cursor back" and 3 "cursor up" characters (leaving the cursor in the upper left corner of the box). Thus, the string contains 55 characters.

Line 2140 contains the die-blanking string, consisting of 5 rows of 5 blanks each. The first 4 lines are each followed by 5 "cursor back" and 1 "cursor down" characters.

The string in line 2160 consists entirely of "cursor back" characters. The length of the string is the same as the length of the string of blanks in line 2150 (18 characters).

Figure 8.36: BASIC for Initializing

```
1950    REM INITIALIZE
1960*   READ SL: DATA 20
1970*   READ GL,GC: DATA 16,9
1980*   READ RL,RC: DATA 11,10
1990    READ L1,C1: DATA 1,10
2000*   READ L2,C2: DATA 3,17
2010    READ BL,BC: DATA 0,3
2020*   READ NL,NC: DATA 23,14
2030*   READ DC: DATA 400
2040    READ MX: DATA 500
2050    FOR ZZ=0 TO 6: FOR XX=1 TO 5
2060    READ D$(ZZ,XX)
2070    NEXT XX: NEXT ZZ
2080    DATA "   "," "    "," "    "," "    "," "    "
2130    DATA " --- "," ":    :"," ": * :"," ":    :"," " --- "
2180    DATA " --- "," ":*  :"," ":    :"," ":  *:"," " --- "
2230    DATA " --- "," ":*  :"," ": * :"," ":  *:"," " --- "
2280    DATA " --- "," ":* *:"," ":    :"," ":* *:"," " --- "
2330    DATA " --- "," ":* *:"," ": * :"," ":* *:"," " --- "
2380    DATA " --- "," ":* *:"," ":* *:"," ":* *:"," " --- "
2440    READ RC$: DATA "            "
2450**  READ BL$: DATA "                              "
2460    GOSUB 1860: PS=0
2470    NP=0:MP=9:CP=0
2480    DIM NM$(9),ST(4,9)
2490    RETURN
```

This is the actual BASIC listing for the Apple and TRS-80 version of Craps. The reason for the large gaps in line numbers between 2080 and 2440 is that in an earlier version of this program, each of 2080, 2130, 2180, 2230, 2280, 2330 and 2380 was replaced by five separate lines.

---

*In the TRS-80 versions of these lines, the SL value is 13, the GL value is 11, the GC value is 21, the RL value is 10, the RC value is 22, the C2 value is 25, the NL value is 15, the NC value is 26, and the DC value is 100.
**In the TRS-80 version of this line, 63 blanks appear, since the screen width is 64 columns.

Figure 8.36a: Apple and TRS-80 Versions of Initialization

#Wait for a single character and simultaneously randomize

onech repeat {
        GET X$ #check for a character
        ZZ = RND(1)   #tickle the random number maker
       } until (X$ <> '''')   #until a character is typed
    RETURN

```
2340  REM ONECH
2350* GET X$:IF X$<>"" THEN RETURN
2360* ZZ=RND(1):GOTO 2350
```

Onech is the single-character input routine that appears throughout the book, but with the addition of repeated calls on the RND function. Since the number of times that the loop is executed is likely to be large and impossible for the player to control, *onech* provides true randomization.

Because of the way the Apple GET instruction works, randomization cannot be achieved in the Apple version of this routine. The Apple version is as shown in Figure 2.11.

The TRS-80 and Apple versions of this routine occupy lines 2660-2680.

---

*In the TRS-80 versions of these lines, GET X$ is replaced by X$ = INKEY$, and RND(1) is replaced by RND(0).

Figure 8.37: Single-Character Input

#Draw the dice picture — Pet version

draw   LL = L1   #position cursor
          CC = C1
          GOSUB cursor
          PRINT BX$; D$(D1);   #draw first die
          LL = L2   #position cursor
          CC = C2
          GOSUB cursor
          PRINT BX$; D$(D2);   #draw second die
          RETURN

```
2240 REM DRAW
2250 LL=L1:CC=C1:GOSUB 1270
2260 PRINT BX$;D$(D1);
2270 LL=L2:CC=C2:GOSUB 1270
2280 PRINT BX$;D$(D2);
2290 RETURN
```

Because of the inclusion of cursor-moving characters in the BX$ and D$ array strings, the *draw* routine is extremely simple. Apple and TRS-80 versions are completely different.

The Apple and TRS-80 versions use a two-dimensional version of the D$ array. The actual BASIC instructions are:

```
2500 REM DRAW
2510 FOR LX=L1 TO L1+4
2520 LL=LX:CC=C1:GOSUB 1270
2530 PRINT D$(D1,LX−L1+1);:NEXT LX
2540 FOR LX=L2 TO L2+4
2550 LL=LX:CC=C2:GOSUB 1270
2560 PRINT D$(D2,LX−L2+1);:NEXT LX
2570 RETURN
```

Figure 8.38: Displaying the Dice

#Rake in the dice—Pet version

cleardice   CC = C1
               LL = L1
               GOSUB cursor
               PRINT BF$;
               CC = C2
               LL = L2
               GOSUB cursor
               PRINT BF$;
               RETURN

```
2300  REM CLEARDICE
2310  CC=C1:LL=L1:GOSUB 1270:PRINT BF$;
2320  CC=C2:LL=L2:GOSUB 1270:PRINT BF$;
2330  RETURN
```

The *cleardice* routine clears the dice picture (drawn by *draw*) from the screen. It is Pet-specific for the same reason that *draw* is. Apple and TRS-80 versions are completely different. Here are the actual BASIC instructions:

```
2580  REM CLEARDICE
2590  FOR LX=L1 TO L1+4
2600  LL=LX:CC=C1:GOSUB 1270
2610  PRINT D$(0,LX−L1+1);:NEXT LX
2620  FOR LX=L2 TO L2+4
2630  LL=LX:CC=C2:GOSUB 1270
2640  PRINT D$(0,LX−L2+1);:NEXT LX
2650  RETURN
```

The function provided by *cleardice* is extremely important. The dice are cleared from the screen without disturbing the remaining screen contents. The separate management of screen areas contributes to the fast action and absorbing nature of the game.

Figure 8.39: Raking in the Dice

Suggested Additions and Improvements

There are many ways to augment and improve upon this game. Here are a few suggestions.

- Prevent an overly enthusiastic player from inadvertently making the first throw of the next game by pressing the space bar too many times. This can be done by inserting

    repeat GET X$ until (X$ = '''')

    at an appropriate place in the program. Decide whether to prevent an inadvertent first throw of any game, or only of the next player's game.

- Develop a way to introduce the entire standard casino betting layout into the game. This will mean that any player (not just the player with the dice) will be able to place a bet or remove winnings between any two throws.

- Consider a different accounting scheme, in which players are issued chips, and the total number of chips that each player possesses is displayed instead of winnings and losses. (This might facilitate the introduction of the full betting layout.) Then, when players quit, their chips are cashed in and credited against their original and subsequent purchases, and a final reckoning is made.

- Add additional statistics, for example, the best run by any player (not just the current player), the most throws required to reach a decision (for any player, and for the current player), highest winnings and losses, or any others that you can think of.

- Add additional betting options, including "automatic" strategies like "doubling up." (Many such strategies are described in books on casino gambling.)

- Display the final value of each player's statistics before switching to the next player.

If you play the Craps game for a while, you will probably come up with many ideas of your own.

Summary

Craps is a computer game that implements features of the popular casino game. The program that implements Craps illustrates several points that were not brought out in the discussions of other games in this book.

Several aspects of the game are tied to independently managed screen areas. Cursor positioning makes this possible.

Smooth, fast game action is achieved by the use of single-character input, timed automatic disappearance of display elements, and the changing of the display with each keystroke.

The passing of the dice from one player to another is a simple example of context switching, a concept that is important in time sharing systems and other multiprocessing programs.

The unpredictable length of time that the program spends waiting for the player's input is used to assure unpredictable dice values. Repeated calls to the RND routine are made in the input wait loop.

Many improvements to this game are possible, expecially the inclusion of additional betting and improved accounting.

CHAPTER 9
Alien Life

Alien Life is a composite game derived from two games: Alien Encounter, which is a simple graphics-oriented game developed by the author, and Game of Life, which is a popular graphics-oriented game that has had many home computer implementations. Alien Life arises from the "grafting" of Game of Life onto Alien Encounter.

Alien Encounter

The object of Alien Encounter is to plan messages to be transmitted to alien beings many light years away from Earth. The messages that you create will be sequences of zeroes and ones that encode pictures. The Alien Encounter program helps you to create the picture and carries out the encoding. The design of the picture is left up to you.

In *Intelligent Life in the Universe*, I.S. Shklovskii and Carl Sagan show that our radio technology is capable of reliable transmission of sequences of "zeroes" and "ones" over interstellar distances. By zeroes and ones in radio transmission we mean "dits" and "dahs," like those used in Morse Code. Figure 9.1 shows how a simple 3 × 4 dot matrix picture can be encoded in a sequence of 12 dits and dahs.

Unfortunately, this decoding depends upon the fact that the receiving party knows that a 3 × 4 dot matrix picture is intended. Figure 9.2 shows three alternative two-dimensional dot matrix interpretations of the same set of dits and dahs. Figure 9.3 shows one of the three possible three-dimensional interpretations.

Since the alien beings to which we are directing our transmissions have no way of knowing our intentions, we must minimize the number of alternative interpretations of these transmissions. We can do this by transmitting sequences of N zeroes and ones (N bits, for short), where N is the product of two prime[†] numbers. If N is the product of two primes, then there are only two dot matrix interpretations of a sequence of N bits.

[†]By definition, a prime number is a positive whole number that cannot be written as the product of two other positive whole numbers. For example, 2, 3, 5 and 7 are primes, but 9 is not prime, because 9 = 3 × 3. Also by definition, 1 is not a prime.

276 ALIEN LIFE

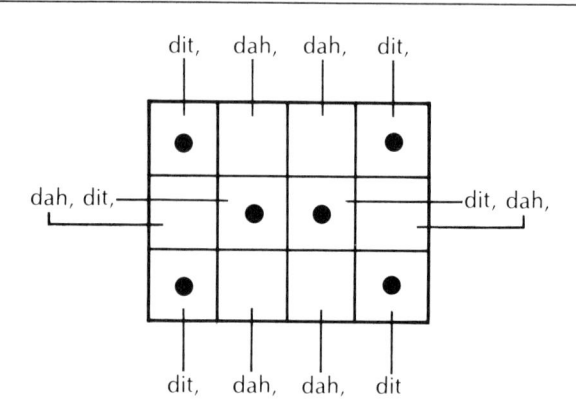

The sequence of twelve "dits" and "dahs" corresponds to the dots and blanks in a 3 x 4 "dot matrix" picture. The sequence represents a radio transmission consisting of two identifiably different kinds of signal (like Morse Code). The transmission is sent repeatedly to assure unambiguous identification of the twelve "dits" and "dahs."

Figure 9.1: **A Simple Picture Transmission**

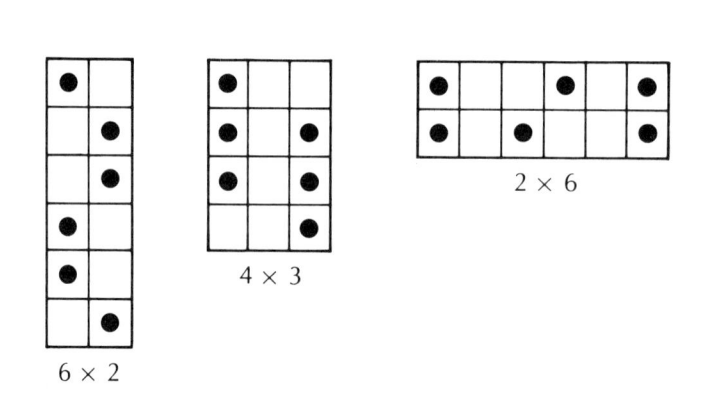

These are three alternative interpretations of the picture shown in Figure 9.1. If the receiver of the transmission did not know that it represented a 3 × 4 dot matrix, then each of these other possibilities would have to be considered.

Figure 9.2: **Other Interpretations of the Simple Picture**

INSIDE BASIC GAMES   **277**

An elaborate 29 × 19 dot matrix picture devised in 1961 by Frank Drake appears in the Shklovskii and Sagan book cited earlier. Figure 9.4 shows a 13 × 11 dot matrix picture adapted from that 29 × 19 picture. It represents the "self-portrait" of an alien, and includes a sequence of four dots by which the alien can be expected to refer to itself in future communications. Figure 9.5 shows the only other possible interpretation of the 143-bit sequence shown in Figure 9.4.

You may wish to exchange such messages with other Alien Encounter players to see whether or not another human can grasp their intended meaning. The program presented in this chapter provides no support for such exchanges, but suggestions for additions to the program are given at the end of the chapter.

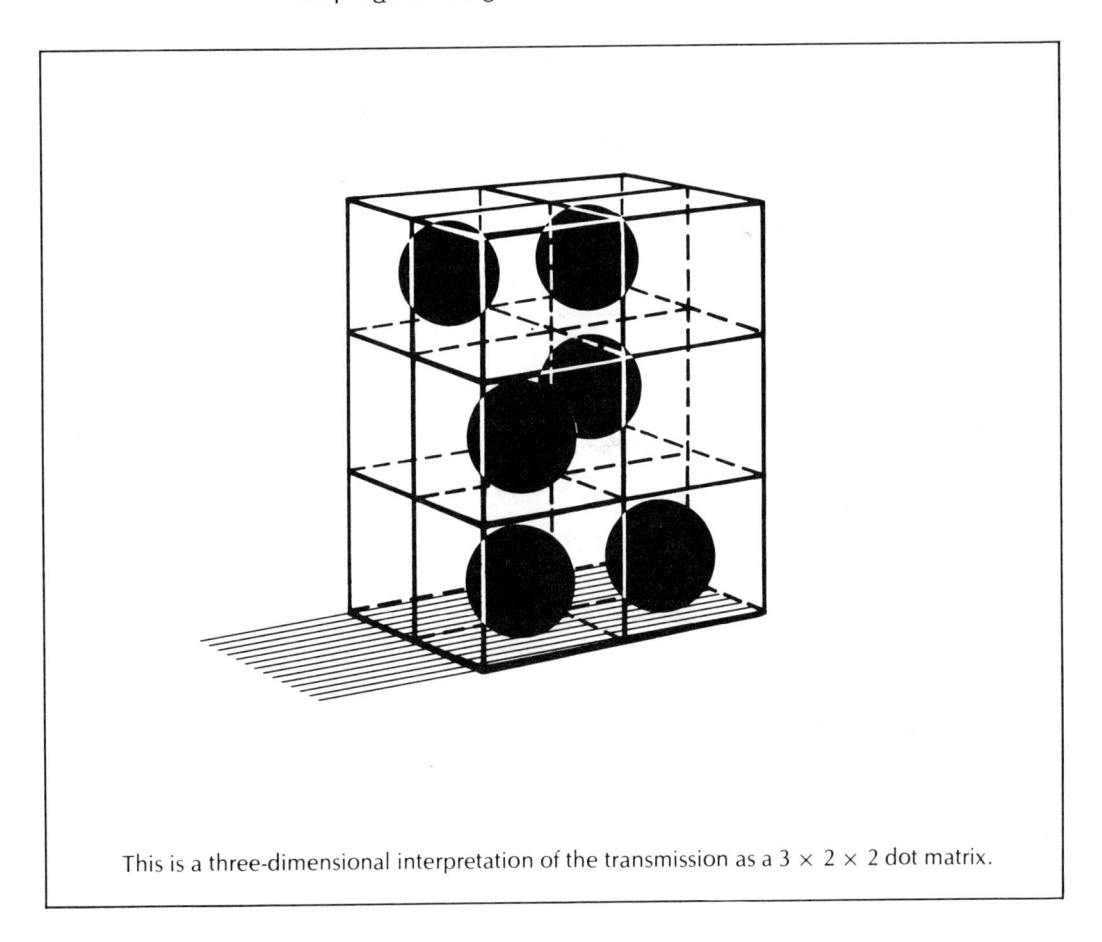

This is a three-dimensional interpretation of the transmission as a 3 × 2 × 2 dot matrix.

Figure 9.3:   A Solid Interpretation of the Simple Picture

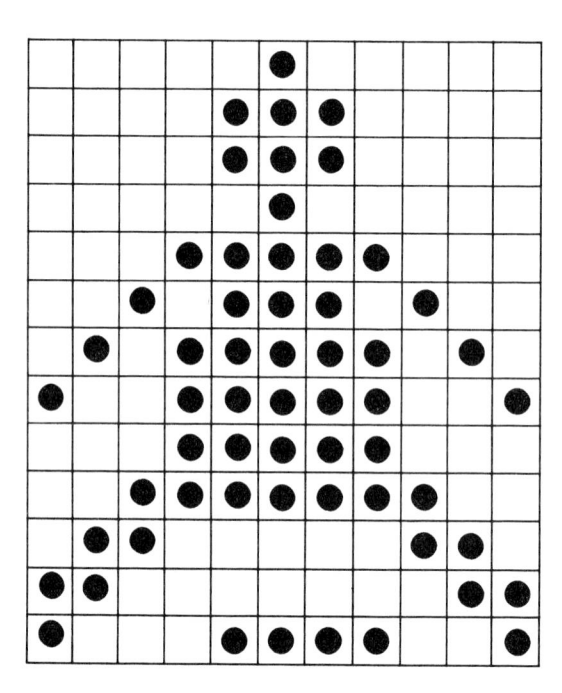

This figure shows a 13 × 11 "dot matrix" self-portrait of an alien creature. The row of four dots at the bottom is the creature's "name." That is, in future communications the alien will use a row of four dots to refer to itself. This picture has been adapted from an elaborate 29 × 19 dot matrix "first message" devised by Frank Drake in 1961.

Sent by itself, the above message would consist of the following sequence of "ones" and "zeroes":

000001000000000111000000001110000
000001000000001111100000101110100
010111110101001111100100011111000
001111111000110000011011000000011
10001111001.

That is, this sequence of "dits" and "dahs," where 0 might be a "dit" and 1 a "dah," would be sent repeatedly to assure accurate transmission of all 143 "bits." Then the scientist receiving it need only guess that it represents a 13 × 11 picture.

Figure 9.4: First Message from an Alien

INSIDE BASIC GAMES **279**

To begin the play of Alien Encounter, the program requests the dimensions of the message to be devised. Figure 9.6 illustrates this dialog. The program asks

LINES, COLUMNS:

and you respond with a pair of prime numbers separated by a comma. If you should happen to enter numbers in an unrecognized format or numbers that are larger than the maximum values allowed, the program will simply ask the question again. However, if you enter two numbers of appropriate size but not both prime, the program will ask

NON-PRIMES OK?

If you did not intend to use non-primes, then press the "N" key, and

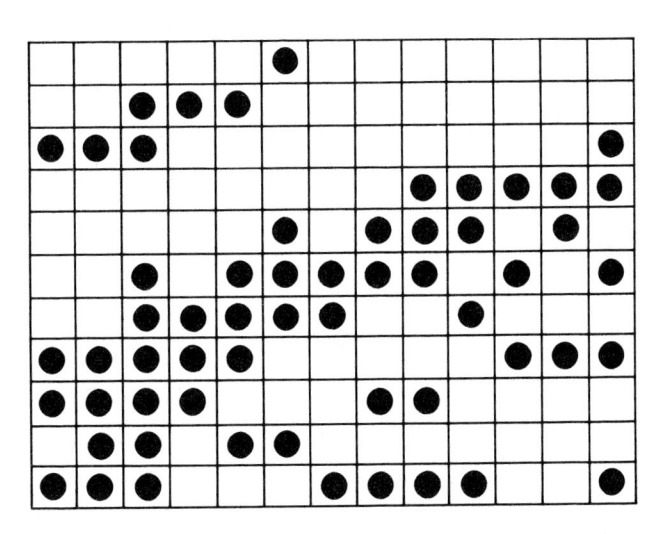

This is the picture that results from misinterpreting the alien message as an 11 × 13 dot matrix rather than as a 13 × 11 dot matrix.

Figure 9.5: Misinterpretation of the Alien Message

the program will again ask for the dimensions. If you press any other key, the program will use the specified non-prime dimensions. (There is one other possible answer to the program's request for picture dimensions. We shall discuss that response shortly.)

As seen in Figure 9.6, once the dimensions have been established, the program displays a box around the picture area, and a flashing picture cursor in the upper left corner of that area. The program then waits for you to enter single-character commands. Aside from the "E" (exit) and "L" (Game of Life) commands (which will be described later), the commands that you will enter are used first to make and remove dots, and then to move the picture cursor.

LINES, COLUMNS: **13, 11**

The program asks for picture dimensions, and the player responds "13 by 11."

The program draws a frame around a 13 × 11 area of the screen and displays a flashing "cursor" in the upper left corner.

This is the initial dialog for Alien Encounter.

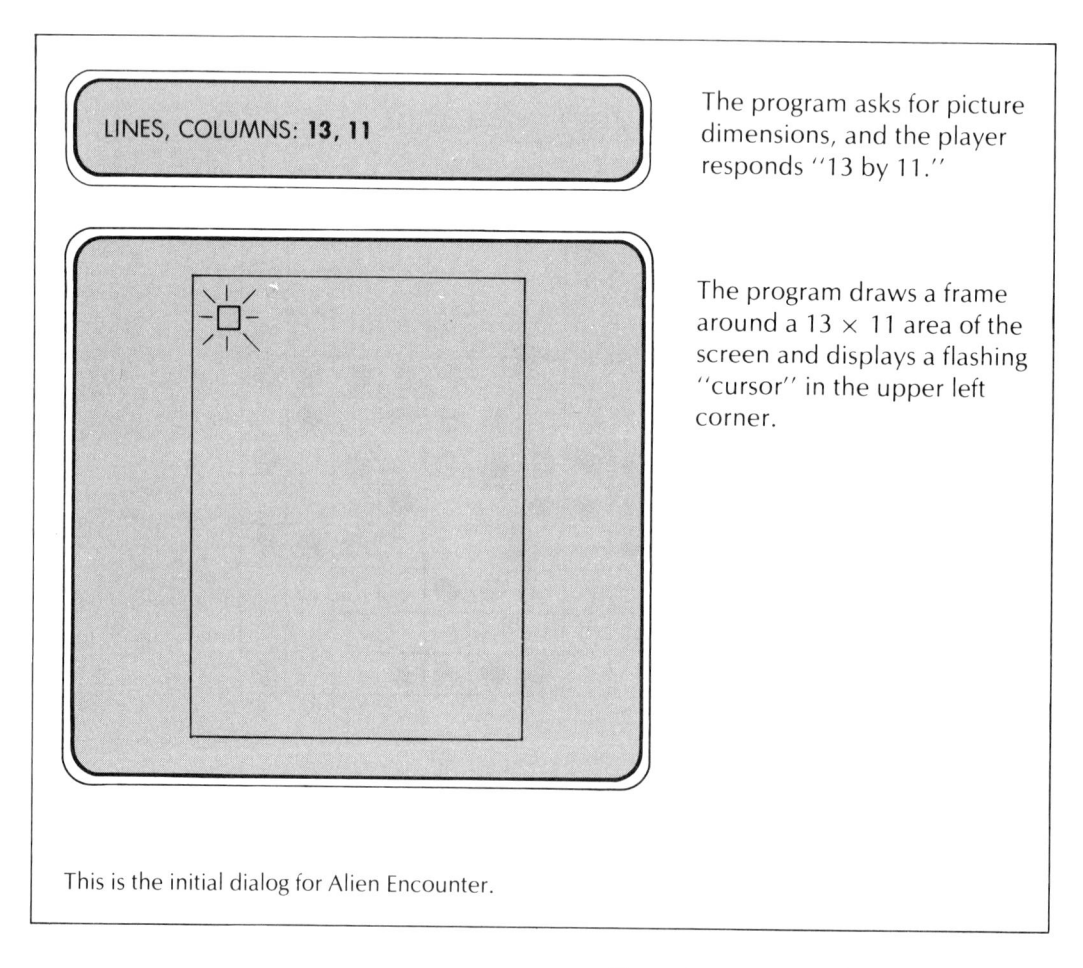

Figure 9.6: **Starting the Alien Encounter**

Each command is executed in two phases. In the first phase, one of three actions takes place:

— A dot is placed at the current picture cursor position.

— The dot (if any) at the current picture cursor position is erased.

— There is no change to the dot or blank at the current cursor position.

In the second phase, the cursor is moved in one of four directions: up, down, right or left. The move is to the next position in the given direction, but with "wraparound." (The term "wraparound" is used in situations in which the "end" and the "beginning" are considered to be adjacent positions.) For example, an attempt to move right from the rightmost column causes the picture cursor to be placed at the leftmost position of the same line.

Figure 9.7 shows the commands and the relation of the drawing commands to positions in the numeric keypad. (If your system does not have a numeric keypad, then you may wish to use other characters. We shall see how this can be done, when we discuss the Alien Life program.) In order to make the many commands easy to remember, the following approach was used:

— A set of four keys was chosen to command picture cursor motion in the four directions. Holding down the SHIFT key while pressing one of these four keys will cause a dot to be displayed at the current picture cursor position before the indicated motion occurs. Pressing the key without the SHIFT key will result in no change to the dot or blank at the current position, and the picture cursor will be moved as indicated.

— The four keys that command picture cursor motion were chosen in a four-cornered arrangement. The "up" key is the top corner, the "down" key is the bottom corner, and the "left" and "right" keys are the left and right corners.

The first of these points illustrates "orthogonality," a term taken from the theory of vector spaces. Simply put, the two functions of action selection and motion selection have been assigned to independent "coordinate axes," and any action can be combined with any motion. Unfortunately, the "axis" that we have chosen for action selection has only two "points" on it: "SHIFT key pressed" and "SHIFT key not pressed." Thus, there is no way to specify the third action (removing a dot) by pressing another key while pressing one of

Drawing Commands

- 8* Cursor up
- 2* Cursor down
- 4* Cursor left
- 6* Cursor right
- 0   Erase dot, then move cursor left
- 7   Erase dot, then move cursor up
- 5   Erase dot, then move cursor right
- 1   Erase dot, then move cursor down

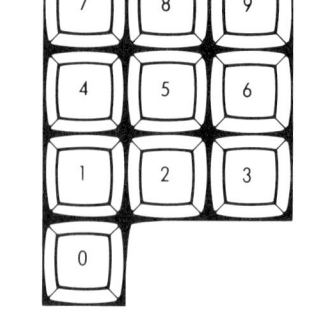

*With SHIFT key, make mark, then move "cursor" as indicated. Without SHIFT, simply move "cursor."

Numeric Keypad

Miscellaneous Commands

E Exit. Emit pattern as a sequence of ones and zeroes. (Intended for use with external storage, but in the current version, the sequence of ones and zeroes is simply displayed on the screen.)

L Life. (To be described later.)

These are the commands that can be input to the Alien Encounter program when the flashing picture cursor is being displayed. Each is a single character.

Figure 9.7: **Alien Encounter Commands**

the four direction keys. (On some home computers, a three-valued axis is provided by the SHIFT and CTRL keys. The three values are "SHIFT pressed," "CTRL pressed," "neither pressed." If you have such a system, you may wish to change the command codes to take advantage of this feature.)

Now that the drawing commands have been described, let's look at an example of their use. In Figure 9.8 you have begun to draw the picture shown in Figure 9.4. Starting from the position shown in Figure 9.6, you have used the sequence of commands shown in Figure 9.8 to perform a systematic "sweep" back and forth, completing each row of dots and passing on to the next.

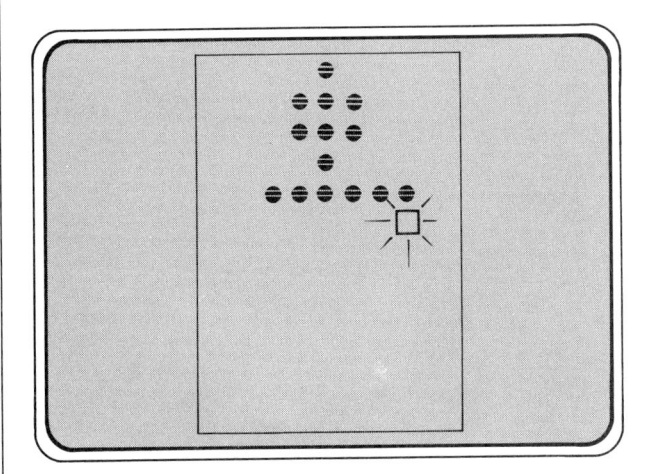

$6,6,6,6,6,6^S,2,4^S,4^S,2^S,6^S,6^S,$
$2^S,4,4^S,4,2,6^S,6^S,6^S,6^S,6^S,2^S$

Starting from the position shown in Figure 9.6, the player has begun to draw the picture shown in Figure 9.4. To the right of the screen appears the sequence of single-character commands that the player has entered. A superscript of "S" means "with the shift key."

The sequence begins with five 6's to move the picture cursor to the center of the top line. Then a 6^S causes the dot to be displayed and the picture cursor to be moved another position to the right.

A 2 moves the picture cursor to the position of the rightmost dot of the second line. 4^S twice followed by 2^S causes the second row of dots to be displayed and the picture cursor to be moved to the position of the leftmost dot of the third row. The subsequent commands continue this systematic back and forth drawing path, but notice that one mistake has been made.

Figure 9.8:   Starting to Draw the Alien

Unfortunately, your exuberance has led you to place an extra dot in the fifth row of the picture. To remove this dot, you must use the command that moves the picture cursor up one place, as shown in Figure 9.9. With the picture cursor positioned at the unwanted dot, any of the four "erase dot and move picture cursor" commands can be used to remove the dot. The "erase and move down" command is the one most nearly in keeping with the drawing strategy you have followed thus far.

Figure 9.10 shows how a bit-encoded version of the picture is obtained. Whenever the picture cursor is flashing, entry of the "E" command causes the encoded form of the picture to be displayed. Zeroes correspond to blank positions, ones to dot positions.

Figure 9.10 illustrates another feature of the Alien Life program. You have finished examining the encoded version of the picture and have pressed any key (except "Q") to cause the program to ask

LINES, COLUMNS:

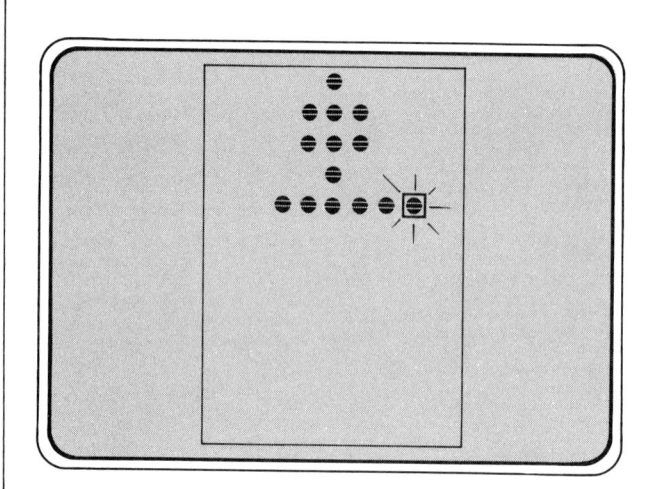

From the position shown in Figure 9.8, the player has entered an 8 to reposition the picture cursor at the last dot drawn. The picture cursor has a different appearance at a position at which there is a dot.

Now the player can remove the dot at which the cursor is positioned with any of the four commands 0, 7, 5 or 1. Each of these commands will leave the cursor positioned differently.

Figure 9.9: Removing a Dot

INSIDE BASIC GAMES

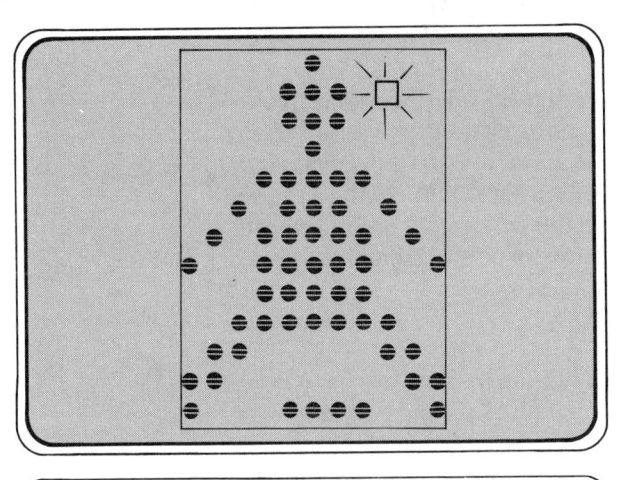

The picture of the Alien is complete, and the picture cursor has been moved to a position near the upper right corner.

The player then enters the "E" command.

```
00000100000000011100000000111000000000010
0000000111110000010111010001011111010100
11111001000111110000011111111000110000011
01100000001110001111001
```

The screen clears, and the encoding of the picture into a sequence of zeroes and ones is displayed. The player presses RETURN.

LINES, COLUMNS:

The program requests another frame size. The player presses RETURN to cause a return to the previous picture.

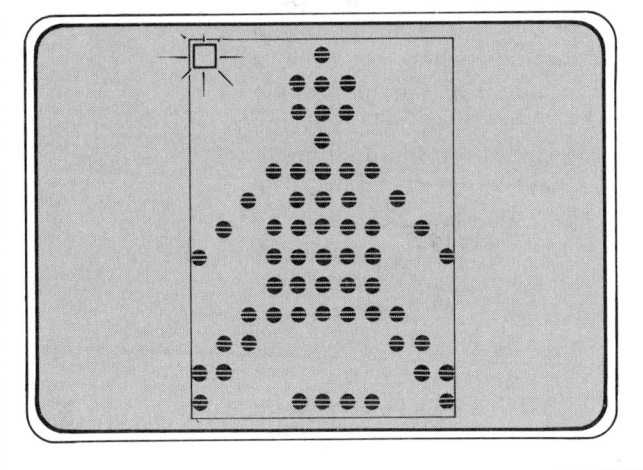

The program is now ready for more commands.

Figure 9.10: Encoding the Picture

Instead of entering a new set of dimensions, you have pressed RETURN, and the picture that you were previously working with has returned to the screen.

At this point, you would like to save the encoded version of your picture on cassette, disk or paper for transmission to another Alien Encounter player. You would also like to be able to enter encoded images received from other players into your home computer. The functions illustrated in Figure 9.10 are the actions of "stubs" whose ultimate purpose is the implementation of the cassette, disk and "manual" input/output of encoded pictures.

The Rules of Game of Life

Game of Life is a game in its own right, but we shall discuss it here only as an addition to the drawing capabilities of Alien Encounter.

Whenever the picture cursor is flashing, an "L" command can be entered. This command directs the program to transform the entire picture according to the rules of Game of Life. We shall state these rules without discussing their derivation, which was from observation of the actual behavior of populations of organisms.

The transformation of the picture according to the rules of Game of Life requires the computation of the number of neighbors of each position. Figure 9.11 illustrates what is meant by "neighboring" positions. The neighboring positions of a position are the "next" positions along the horizontal, vertical or 45-degree diagonal lines through the given position. In general, a position has eight, five or three neighboring positions, depending upon where the given position lies with respect to the "edges" of the area. The number of *neighbors* of a given position is the number of neighboring positions at which a dot appears. For example, in Figure 9.10, the dot position at the top of the alien's head (position (0,5)) has three neighbors: the three dots below it (at positions (1,4), (1,5) and (1,6)). The positions to the left and right of the top dot also have three neighbors. The neighbors of the position (0,4) are at (1,4), (0,5) and (1,5), and the neighbors of (0,6) are at (0,5), (1,5) and (1,6). The dot position directly below the top dot (position (1,5)), in the "middle" of the alien's head, has six neighbors: the dots at (0,5), (1,4), (1,6), (2,4), (2,5) and (2,6). The dots at the ends of the alien's arms have only one neighbor each.

Figure 9.12 shows the transformation rules of Game of Life. Figure 9.13 shows examples of the application of those rules. We do not need to make a thorough study of the many situations that can arise in Game of Life. This has been done elsewhere, and is only marginally relevant to our discussion.

Now, let's consider the effect of the "L" command on the alien's self-portrait. Figure 9.14 shows the stages through which the portrait passes until it reaches a constant configuration. Did the alien intend to convey additional information about itself by the behavior of its self-portrait under repeated application of Game of Life rules? Is there significance in the content of the final constant pattern or in the number of transformations required to reach that pattern? Perhaps the intermediate stages are also intended to convey information. These possibilities illustrate the added dimension that Game of Life has brought to Alien Encounter.

If you have used other Game of Life programs, you know that the drawing capabilities of Alien Encounter and the ability to alter the pattern between any two transformations are not usually available in Game of Life programs. Also, the development of Alien Life capabilities for storage and retrieval of pictures as bit patterns on external storage, as suggested at the end of the chapter, will provide another capability not usually available to Game of Life players.

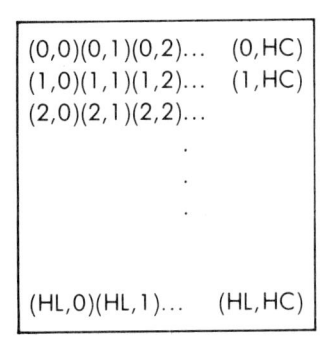

```
(0,0)(0,1)(0,2)...   (0,HC)
(1,0)(1,1)(1,2)...   (1,HC)
(2,0)(2,1)(2,2)...
         .
         .
         .
(HL,0)(HL,1)...      (HL,HC)
```

(1,1) has eight neighboring positions:

(0,0), (0,1), (0,2), (1,0), (1,2), (2,0), (2,1), (2,2).

(1,0) has five neighboring positions:

(0,0), (0,1), (1,1), (2,0), (2,1).

(0,0) has only three neighboring positions:
(0,1), (1,0), (1,1).

The positions of the picture area have been labeled with line and column numbers in order to illustrate the idea of "neighbors." In general, a position "away from the edges" has eight neighboring positions. A position "along the edge" but away from "the corner" has five neighboring positions. The positions "in the corners" have three neighboring positions. In the special cases in which the picture area has only one line or one column, the numbers of neighboring positions are either one or two.

Figure 9.11: Neighboring Positions

The Alien Life Program

Now that we have described the Alien Life game, let's look at the program that implements it. This program, shown in Figures 9.15 through 9.44, is similar in many ways to other programs in this book. In describing it we shall focus on the picture area and the commands. In discussing the picture area we shall study the programming that is concerned with the frame around the picture area, the dots that make up the picture, and the picture cursor. In discussing the commands, we shall describe the command encoding, and the provisions made for easy insertion of new commands.

The Picture Area

The picture area has three major components:

— the frame

— the picture dots

— the picture cursor.

The frame is drawn once, by routines called from *setframe* (Figure 9.16). The routine *center* (Figure 9.25) determines the position of the point at line zero, column zero in such a way that the frame will be centered on the screen. The routine *drawframe* (Figure 9.26) draws a rectangular box surrounding the picture area on the screen. The call

---

Number of Neighbors	New Contents
Less than 2	No dot
2	Old contents unchanged
3	Dot
Greater than 3	No dot

These are the rules that determine the transformation of the picture when the "L" (Game of Life) command is entered. The program first computes the number of neighbors of each position of the picture area, then determines the new contents of each position on the basis of the old contents and the number of neighbors, by using the rules shown above.

Figure 9.12: Game of Life Rules

INSIDE BASIC GAMES   **289**

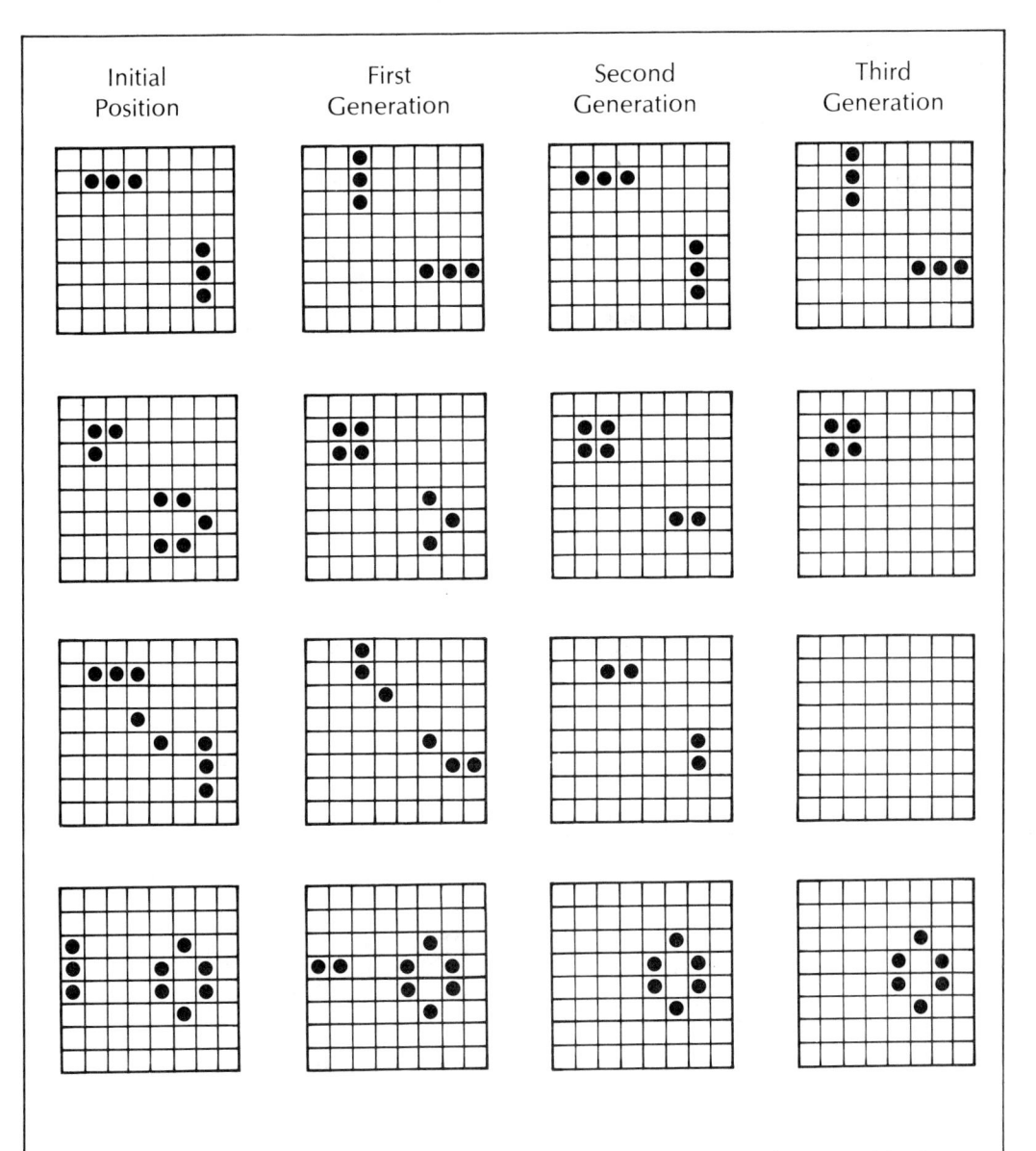

These examples illustrate what happens to dot patterns under repeated applications of the Game of Life rules. Some die out, some reach a constant configuration, while others pass repeatedly through a fixed sequence of configurations.

Figure 9.13: Game of Life Examples

290 ALIEN LIFE

Eleven applications of Game of Life rules to the alien's self-portrait transform it into a stable pattern consisting of three square clusters of four dots each.

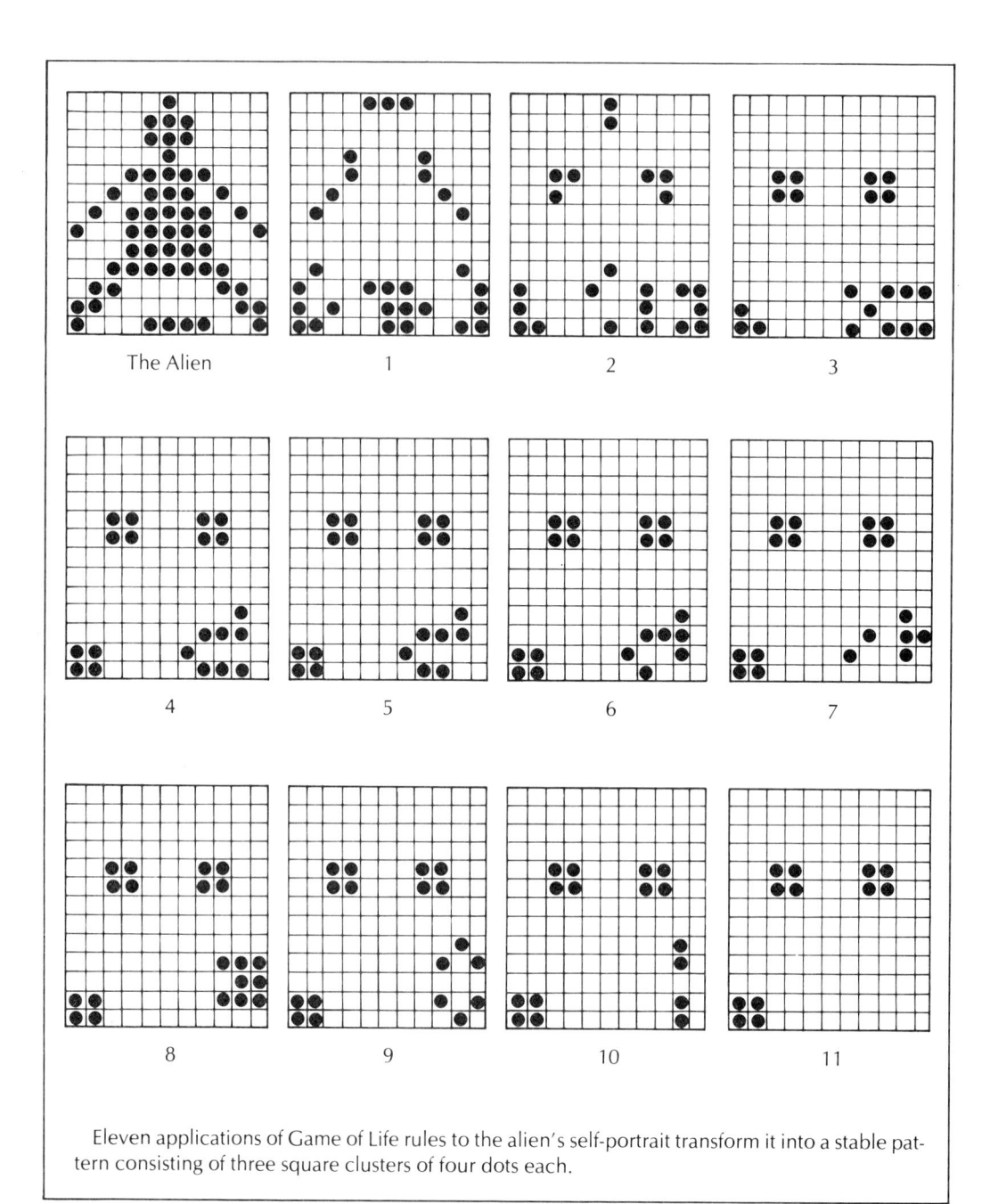

Figure 9.14: Alien Life Transformations

to *drawframe* in *setframe* is preceded by a call to *clearscreen*, so that the screen is clear when *drawframe* is called.

The picture dots are the image of the D array. The *drawdots* routine (Figure 9.27) assures this by placing a dot at each line-and-column position (LL,CC) for which D(LL,CC) <> 0, and a blank at each position (LL,CC) for which D(LL,CC) = 0. The order in which this is done affects the screen display. For example, when the "L" command is used to obtain a Game of Life transformation of the picture, the picture changes one line at a time.

```
#Alien Encounter and Game of Life
    GOSUB init                          #initialize arrays and strings
    repeat {
        GOSUB setframe                  #set up a new picture frame
        repeat {
            GOSUB drawcmd               #convert single-char input to
            IF C1 <> 0 THEN             #  C1,C2; C1 = 0 means done
                ON C1 GOSUB             #execute a drawing command:
                    space,              #  move picture cursor
                    mark,               #  make a mark (and move cursor)
                    unmark,             #  remove a mark (and move cursor)
                    miscmd              #  other commands
        } until (C1 = 0)
        GOSUB outstream                 #output picture as ones and zeroes
        GOSUB onech                     #signal to start another frame
    } until (X$ = "Q")                  #Q = exit
    END

100   GOSUB1190
110   GOSUB160
120   GOSUB700:IFC1=0THEN140
130   ONC1GOSUB720,730,740,750:GOTO120
140   GOSUB560:GOSUB990:IFX$<>"Q"THEN110
150   END
```

This is the main routine of the Alien Life program, which implements the Alien Encounter game and contains Game of Life as one of its features.

The structure of this routine is similar to that of other main routines in this book.

Figure 9.15: Alien Life

An alternative approach would be to blank the entire picture area, and then draw in the dots. One advantage of this approach, which can be implemented by a change in the *drawdots* routine, is that the time required for the drawing of the picture would be reduced in most cases. The choice between these alternatives is a matter of personal preference.

The *drawdots* routine is initially called from *setframe* (Figure 9.16).

```
#Set up picture frame
setframe    GOSUB clearscreen        #start on clear screen
            GOSUB primes             #ask for picture dimensions
            IF PX = -1 THEN {        #PX = -1 to load existing pattern
                GOSUB blockin        #get the pattern
                GOSUB choices        #get P1, P2, HL, HC
            }
            else {
                HL = P1 - 1          #highest line number
                HC = P2 - 1          #highest column number
                GOSUB cleardot       #clear HL × HC section of D
            }
            GOSUB center             #compute LZ and CZ for lower left corner
            GOSUB clearscreen        #clear the screen
            GOSUB drawframe          #draw a frame
            GOSUB drawdots           #draw in dots (from blockin pattern)
            CC = 0: LL = 0           #set the cursor
            GOSUB newcursor
            RETURN

160   GOSUB980:GOSUB200:IFPX<>-1THEN180
170   GOSUB540:GOSUB550:GOTO190
180   HL=P1-1:HC=P2-1:GOSUB600
190   GOSUB610:GOSUB980:GOSUB620:GOSUB660:CC=0:LL=0:GOSUB840:RETURN
```

The *setframe* routine acquires the picture frame dimensions and draws the picture frame in the center of the screen. As one option, it can accept "block input," that is, direct input of a picture as ones and zeroes. This input can come from the keyboard or from an external storage device on which it was previously saved.

Figure 9.16: Setting the Frame

Thereafter, it is only called from the *life* routine (Figure 9.44). The routines that implement the drawing commands simply display (or erase) one dot at a time using cursor control, rather than redrawing the entire picture. The drawing routines that display or erase dots are shown in Figure 9.32. They call subroutines *showdot* and *zapdot* (Figure 9.27) for the actual screen change.

The picture cursor is an entirely different element from the CRT cursor that is controlled by your home computer system and used to guide your input of programs and data. The picture cursor is maintained entirely by the Alien Life program. This maintenance requires control of the following:

— location of the picture cursor

— choice of cursor character

— flashing

— restoration of screen contents when cursor is removed.

The location of the picture cursor is controlled by the line and column variables LL and CC. These are set to zero in the *setframe* routine (Figure 9.16) and the *life* routine (Figure 9.44), thereby placing the picture cursor in the upper left corner of the picture area. The drawing commands cause the picture cursor to move from that location. The routines shown in Figure 9.33 carry out the change in the coordinate variables LL and CC. The actual motion on the screen occurs in the *cursin* routine (Figure 9.36).

The *cursin* routine is called from *drawcmd* (Figure 9.28) in order to obtain the next command from the player. *Cursin* first selects the cursor character by calling *newcursor* (Figure 9.32). It then executes a loop in which it alternately displays the chosen cursor character and a blank character while waiting for keyboard input. This alternating display is the "flashing" of the picture cursor. The frequency of flashing is determined by the variable FQ, which is set in the *init* routine (Figure 9.38).

The *cursin* routine shown in Figure 9.36 uses the Pet TI feature to control the alternation. For that reason, the variable FQ is specified in "jiffies" (sixtieths of a second), the units in which the Pet's TI clock is kept. For the TRS-80 version, the *repeat...until* loop is replaced by a FOR...NEXT loop using limits

1 TO FQ∗TN

#Get picture dimensions — returns PX = −1 for block input

primes   repeat {
        PRINT "LINES, COLUMNS: ";   #ask for dimensions
        GOSUB stringin   #get reply
        IF XX$ = "" THEN   #RETURN signals block input
          { PX = −1: RETURN }
        else {
          GOSUB getprimes   #parse input string
          IF PX = 0 THEN   #invalid syntax
            XX = −1
          else {
            GOSUB checkprimes   #check the numbers
            IF XX = 1 THEN {   #if non-prime,
              GOSUB askok   # ask if OK anyway
              IF OK = 1 THEN
                XX = 0
            }
          }
        }
        } until (XX = 0)
        P1 = PX: P2 = PY   #PX, PY become lines, cols
        RETURN

```
200   PRINT"LINES,COLUMNS: ";:GOSUB1010:IFXX$<>""THEN220
210   PX=−1:RETURN
220   GOSUB310:IFPX<>0THEN240
230   XX=−1:GOTO200
240   GOSUB420:IFXX<>1THEN270
250   GOSUB290:IFOK<>1THEN270
260   XX=0
270   IFXX<>0THEN200
280   P1=PX:P2=PY:RETURN
```

The *primes* routine interacts with the player to obtain the picture frame dimensions. For the Alien Encounter game, the dimensions should be prime numbers. If they are not, the routine requests confirmation from the player before accepting them.

Figure 9.17: Getting the Picture Dimensions

where TN has an empirically determined value that results in execution of the loop in FQ jiffies. Also, the TRS-80 version uses the instruction

$X\$ = INKEY\$$

where the Pet version uses

GET X$

The Apple version cannot use this technique, since the Apple

GET X$

instruction causes the Apple's CRT cursor to flash, and waits until the player has input a character. Thus, the Apple version of *cursin* consists entirely of a call to *cursor* with C and L set to CZ + CC and LZ + LL, followed by the instruction

GET X$

The final point to be discussed is the restoration of the screen contents. The call to *uncursor* (Figure 9.32), which occurs immediately before *cursin* returns, causes the appropriate character (as dictated by the contents of the D array) to be displayed at the location at which the cursor had been flashing.

#Ask whether non-primes are OK — return OK = 0 if no, 1 if yes

askok   PRINT "NON-PRIMES OK? ";   #ask the question
        GOSUB onech   #get the answer
        IF X$ = "N" THEN
            OK = 0   #"N" = no
        else
            OK = 1   #anything else = yes
        RETURN

290   PRINT"NON-PRIMES OK? ";:GOSUB990:OK=1:IFX$="N"THENOK=0
300   RETURN

The *askok* routine asks the player whether or not the non-prime dimensions just entered are actually acceptable. If the player has mistakenly entered a non-prime, an answer of "N" will cause the program to request input of new dimensions.

Figure 9.18: Asking for Confirmation of Non-Prime Input

#Extract a pair of dimensions from XX$ — PX = 0 on error exit
getprimes   LL = LEN(XX$)   #length of input
              IF LL < 3 THEN   #too short
                    errorexit
              else {
                    NP = 0   #no primes yet
                    FOR SX = 1 TO LL − 1
                          X$ = MID$(XX$,SX,1)   #next char of XX$
                        GOSUB sepcheck   #separator?
                        IF SP = 0 THEN   #no
                            NP = 1   #signal "on first"
                        else   #yes
                            IF NP = 0 THEN   #before first prime
                                IF X$ <> " " THEN   #leading space OK
                                    errorexit
                            else {
                                PX = VAL(LEFT$(XX$,SX − 1))   #first prime
                                PY = VAL(MID$(XX$,SX + 1))   #second prime
                                RETURN
                          }
                    NEXT SX
                errorexit   #didn't find a pair
              }

```
310   LL=LEN(XX$):IFLL>=3THEN330
320   PX=0:RETURN
330   NP=0:FORSX=1TOLL−1:X$=MID$(XX$,SX,1):GOSUB400:IFSP<>0THEN350
340   NP=1:GOTO390
350   IFNP<>0THEN380
360   IFX$=" "THEN390
370   PX=0:RETURN
380   PX=VAL(LEFT$(XX$,SX−1)):PY=VAL(MID$(XX$,SX+1)):RETURN
390   NEXTSX:PX=0:RETURN
```

The *getprimes* routine extracts a pair of dimensions from the string XX$ returned by the *stringin* routine. The input format is assumed to be: number, separator, number, where separator is either a comma or a blank. For example, "7,5" and "7 5" are acceptable formats.

Figure 9.19: Parsing the Dimension Input

This completes our discussion of the picture area. Now let's turn to a study of the Alien Life program's commands.

The Commands

The Alien Life program responds to a number of single-character commands. In this section we shall discuss:

— the encoding used to include primary and secondary command information in one number

— the built-in mechanism for accommodating new commands

— the association of specific characters with commands.

The *drawcmd* routine shown in Figure 9.28 illustrates the encoding used for commands. *Drawcmd* begins by calling *cursin* in order to obtain a single ASCII character, which *drawcmd* uses (in a way we shall soon describe) to reach the array element C(CX) containing the number that encodes the command information. The computation of C1 and C2 is based upon the assumption that the number is of the form

$$n_2 + 100 \times n_1$$

where n_2 is less than 100. The number n_2 becomes the value of C2 and n_1 becomes the value of C1. The encoding of these two command numbers into one number occurs in the *init* routine (Figures 9.38 through 9.40).

#Check for separator

sepcheck   IF X$ = " " OR X$ = "," THEN
            SP = 1
        else
            SP = 0
        RETURN

  400   SP=0:IFX$=" "ORX$=","THENSP=1
  410   RETURN

The *sepcheck* routine examines a character to determine whether or not the character is a separator. The separator characters recognized by this routine are comma and space, but others could be added easily.

Figure 9.20: Checking for a Separator

The primary and secondary command information is used as follows. First the contents of C1 are interpreted in the main routine (Figure 9.15). There are three primary command codes:

— the code for "exit"

— a drawing command code

— the "miscellaneous" code.

\#Check PX, PY — returns XX = −1 for reject, 0 for primes, 1 for non-primes

checkprimes IF PX > ML OR PY > MC THEN   #out of range
        XX = −1
    else {
        PP = PX: GOSUB primetest
        IF PR = 0 THEN
            XX = 1
        else {
            PP = PY: GOSUB primetest
            IF PR = 0 THEN
                XX = 1
            else
                XX = 0
        }
    }
    RETURN

```
420   IFPX<=MLANDPY<=MCTHEN440
430   XX=-1:RETURN
440   PP=PX:GOSUB480:IFPR<>0THEN460
450   XX=1:RETURN
460   PP=PY:GOSUB480:XX=0:IFPR=0THENXX=1
470   RETURN
```

The *checkprimes* routine examines the numbers passed to it in the variables PX and PY, and returns an indication in the variable XX. XX is set to −1 if either of the numbers exceeds the maximum allowed, to 0 if both numbers are primes, or to 1 if either of the numbers is a non-prime. The testing for primes is carried out by the *primetest* routine, which is called once for each of the numbers.

Figure 9.21: Testing the Dimensions for "Primeness"

In the case of "exit," the secondary command information is not used. For the drawing commands, the secondary information is interpreted in *movecursor* (Figure 9.31). For the "miscellaneous" commands, the secondary information is interpreted in *miscdo* (Figure 9.30).

The *miscdo* routine is the point at which new commands are integrated into the program. Game of Life provides a model for this process, since its integration into the program occurred after Alien Encounter was complete. The steps necessary to integrate Game of

---

#Test PP and return PR = 1 if PP is prime, PR = 0 otherwise

primetest   IF PP = 2 OR PP = 3 OR PP = 5 THEN
           primereturn
       else IF PP < 7 OR PP is even THEN
           notprimereturn
       else {
           FOR I = 3 TO INT(SQR(PP)) STEP 2   #try possible divisors
              IF I divides PP THEN
                 notprimereturn
           NEXT I
           primereturn
       }

```
480   IFPP=2ORPP=3ORPP=5THENPR=1:RETURN
490   IFPP<7ORPP/2=INT(PP/2)THENPR=0:RETURN
500   FORI=3TOINT(SQR(PP))STEP2
510   IFPP/I<>INT(PP/I)THEN530
520   PR=0:RETURN
530   NEXTI:PR=1:RETURN
```

The *primetest* routine examines the number passed to it in the variable PP and returns an indication in the variable PR. PR is set to 1 if the value of PP is prime, to 0 if not.

Figure 9.22: Checking PP's "Primeness"

300   ALIEN LIFE

```
#Stub for block input
blockin    RETURN                                    #keep current picture


#Stub for presentation of possible prime sets
choices    RETURN                                    #keep current dimensions


#Stub for output of the bit stream for the message
outstream  GOSUB clearscreen
           FOR LL = 0 TO HL
               FOR CC = 0 TO HC
                   IF D(LL,CC) = 1 THEN
                       PRINT "1";
                   else
                       PRINT "0";
               NEXT CC
           NEXT LL
           RETURN


   540  RETURN
   550  RETURN
   560  GOSUB980:FORLL=0TOHL:FORCC=0TOHC:IFD(LL,CC)<>1THEN580
   570  PRINT"1";:GOTO590
   580  PRINT"0";
   590  NEXTCC:NEXTLL:RETURN
```

The *blockin* routine is a stub that returns without doing anything. The effect of this inaction is to allow the previous pattern to be recalled. The intended use for the *blockin* routine is to load a string of zeroes and ones that were previously saved on external storage or entered at the keyboard by the player.

The *choices* routine is a stub intended for use with *blockin*. Its current inaction works with the *blockin* stub to allow the previous pattern to be recalled to the screen.

The *outstream* routine is a stub that simply prints a sequence of ones and zeroes corresponding to the picture. The intended use for *outstream* is to save the sequence of ones and zeroes on an external storage medium (e.g., disk, cassette) for later recall by *blockin*.

Figure 9.23: External Storage Stubs for Alien Life

Life into this program were:

— define the "L" command as "miscellaneous, number 1"

— cause *miscdo* to recognize a C2 value of 1 as a signal to call *life*

— provide the routine *life* (Figure 9.44) and its subroutines *neighbors* (Figures 9.41, 9.42) and *lifedeath* (Figure 9.43).

Let's look at these steps, since you may wish to add new commands of your own.

"L" was defined to be "miscellaneous, number 1" by the line

DATA a(L), misc, life

in Figure 9.39. The meanings of these symbols are explained in Figure 9.39. The code that interprets this DATA statement is shown in Figure 9.38. We shall discuss this code shortly.

Causing *miscdo* to recognize a C2 value of 1 as a signal to call *life* was done by including an explicit test for a value of 1. If additional commands are to be recognized, a structure similar to that of *movecursor* (Figure 9.31) should be adopted for *miscdo*. The ON...GOSUB construction is well suited to this kind of command dispatching.

Providing *life* and its subroutines was fairly simple to do. The subroutines *neighbors* and *lifedeath* implement the Game of Life

---

#Clear dot array

cleardot   FOR LX = 0 TO HL
              FOR CX = 0 TO HC
                D(LX,CX) = 0
              NEXT CX
            NEXT LX
          RETURN

  600   FORLX=0TOHL:FORCX=0TOHC:D(LX,CX)=0:NEXTCX:NEXTLX:RETURN

The *cleardot* routine initializes the array D to all zeroes. D will contain a 1 for each position in the picture at which a dot appears.

Figure 9.24: Clearing the Dot Array

rules in a straightforward manner, although the use of the D array for counting neighbors deserves study. A separate array for counting neighbors could have been used, but would have required substantially more memory space for the program.

The *life* routine makes calls on *drawdots* and *newcursor* and uses the picture dimensions established by a prior call to *setframe*. In all other respects, *life* is self-contained.

Finally, let's discuss the association of specific characters with commands. This is the function of the *repeat...until* loop in the *init* routine (Figure 9.38). This loop allows an unspecified number of DATA statements to be used. These statements are shown in Figure 9.39. Each contains an ASCII code, a primary command code and a secondary command code. In the loop, the ASCII code is converted into an array index, and the primary and secondary command codes are combined into a single number for storage in the C array at the index corresponding to the ASCII code.

If you wish to change the codes already assigned to commands (e.g., to adapt to a keyboard without a numeric pad), the DATA statements in Figure 9.39 must be altered. Furthermore, if you wish to use codes larger or smaller than those within the range defined by CM (character max) and CB (character base), you must use different values for *lochar* and *hichar* in Figure 9.38. These are the values used to limit the size of the C array.

#Compute upperleft corner for centered box

center   CZ = INT((SC − HC − 3)/2) + 1   #allow for frame
LZ = INT((SL − HL − 3)/2) + 1
RETURN

610   CZ=INT((SC−HC−3)/2)+1:LZ=INT((SL−HL−3)/2)+1:RETURN

The *center* routine determines the position at which the upper left corner of the picture frame must appear in order for the picture to be centered on the display screen.

Figure 9.25: Centering the Picture

#Draw the box that surrounds the picture area

drawframe	L = LZ − 1: C = CZ: GOSUB cursor	#upper left
	FOR CC = 0 TO HC	#top edge
	PRINT TH$;	
	NEXT CC	
	PRINT "down";	
	FOR LL = 0 TO HL	#right edge
	PRINT RV$; "cursor back"; "cursor down"	
	NEXT LL	
	L = LZ: C = CZ − 1: GOSUB cursor	#upper left
	FOR LL = 0 TO HL	#left edge
	PRINT LV$; "cursor back"; "cursor down";	
	NEXT LL	
	PRINT "right";	
	FOR CC = 0 TO HC	#bottom edge
	PRINT BH$;	
	NEXT CC	
	RETURN	

```
620* L=LZ−1:C=CZ:GOSUB940:FORCC=0TOHC:PRINTTH$;:NEXTCC:PRINTCHR$(17);
630* FORLL=0TOHL:PRINTRV$;CHR$(157);CHR$(17);:NEXTLL
640* L=LZ:C=CZ−1:GOSUB940:FORLL=0TOHL:PRINTLV$;CHR$(157);CHR$(17);:
     NEXTLL:PRINTCHR$(29);
650  FORCC=0TOHC:PRINTBH$;:NEXTCC:RETURN
```

The *drawframe* routine draws the box that surrounds the picture area on the screen. Pet or TRS-80 cursor-moving characters are used, but very similar routines can be used on the Apple. The Apple cursor-moving commands provide the necessary tools on those systems.

The actual BASIC instructions for the Apple are:

```
620  L=LZ−1:C=CZ:GOSUB940: FOR CC=0 TO HC: PRINT TH$;: NEXT CC
630  FOR LL=0 TO HL:L=LZ+LL:C=CZ−1:GOSUB940: PRINT LV$;
640  C=CZ+HC+1:GOSUB940: PRINT RV$;: NEXT LL
650  L=LZ+HL+1:C=CZ:GOSUB940: FOR CC=0 TO HC: PRINT BH$;: NEXT CC: RETURN
```

*In the TRS-80 version, the cursor back, down and right characters are represented by CHR$(24), CHR$(26) and CHR$(25).

Figure 9.26: Framing the Picture

ALIEN LIFE

#Draw dots as indicated by the D array

drawdots FOR LL = 0 TO HL
        L = LL + LZ: C = CZ: GOSUB cursor   #position cursor at start of line
        FOR CC = 0 TO HC
           IF D(LL,CC) <> 0 THEN
              PRINT DT$;   #print "dot"
           else
              PRINT " ";   #print blank
        NEXT CC
        NEXT LL

```
660   FORLL=0TOHL:L=LL+LZ:C=CZ:GOSUB940
670   FORCC=0TOHC:IFD(LL,CC)<>0THENPRINTDT$;:GOTO690
680   PRINT " ";
690   NEXTCC:NEXTLL:RETURN
```

The *drawdots* routine redraws the picture, placing either a dot or a space at each position, thereby erasing old dots. This routine is used to draw the new generation in the Game of Life transformation of the picture and to redraw a recalled previous pattern.

Figure 9.27: Drawing the Picture

#Get and decode a single-character drawing command

drawcmd GOSUB cursin   #flash picture cursor, wait for input
        CX = ASC(X$) − CB   #index to C array
        IF NOT 0 <= CX <= MX THEN
           CX = DX   #default value if out of range
        C1 = INT(C(CX)/100)   #primary command
        C2 = C(CX) − 100∗C1   #secondary command
        RETURN

```
700   GOSUB1080:CX=ASC(X$)−CB:IFCX>MXORCX<0THENCX=DX
710   C1=INT(C(CX)/100):C2=C(CX)−100∗C1:RETURN
```

The *drawcmd* routine accepts a single-character input from the player and translates it into a primary command C1 and a secondary command C2. The primary command is interpreted in the main routine (Figure 9.15). For most primary commands, the secondary command is interpreted in the *movecursor* routine (Figure 9.31). For the "miscellaneous" primary command, the secondary command is interpreted in the *miscdo* routine (Figure 9.30).

Figure 9.28: Decoding a Command

```
#Primary drawing commands (indexed by C1)

     space     GOSUB movecursor              #only move the cursor
               RETURN


     mark      GOSUB putmark                 #place a dot, then move cursor
               GOSUB movecursor
               RETURN


     unmark    GOSUB takemark                #remove a dot, then move cursor
               GOSUB movecursor
               RETURN


     miscmd    GOSUB miscdo                  # odd jobs
               RETURN



        720   GOSUB780:RETURN
        730   GOSUB800:GOSUB780:RETURN
        740   GOSUB810:GOSUB780:RETURN
        750   GOSUB760:RETURN
```

These are the routines called from the main routine (Figure 9.15) to process primary commands. Each routine first performs an action (or does nothing), then calls another routine to interpret the secondary command.

Figure 9.29: Primary Command Processing

306 ALIEN LIFE

#Miscellaneous command routine

miscdo   IF C2 = 1 THEN   #C2 = 1 encodes "Game of Life"
              GOSUB life
          RETURN

```
760   IFC2=1THENGOSUB1450
770   RETURN
```

The *miscdo* routine is the point in the program at which new commands can be added. For example, if a command to erase the entire picture were desired, an appropriate DATA statement could be added to the initialization routine (Figure 9.39) specifying a primary command of 4 (miscellaneous) and a secondary command of 2. Then *miscdo* could be altered to recognize a C2 value of 2 and call a subroutine that consisted of calls to *cleardots* and *drawdots*. Then, whenever the player entered the character specified in the DATA statement, the entire picture (but not the frame) would be erased.

Figure 9.30: Taking Miscellaneous Action

#Dispatch routine for positioning the picture cursor

movecursor   IF 1 <= C2 <= 4 THEN   #check C2 range
                 ON C2 GOSUB
                     cursleft,   #move cursor left
                     cursright,   #move cursor right
                     cursup,   #move cursor up
                     cursdown   #move cursor down
               RETURN

```
780   IFC2<1ORC2>4THENRETURN
790   ONC2GOSUB860,880,900,920:RETURN
```

The *movecursor* routine interprets the secondary command for the primary commands mark, unmark and space. The secondary command determines where the cursor will move after the primary command is executed.

Figure 9.31: Telling the Cursor Where to Go

#Routines to display or erase dot characters

putmark   GOSUB showdot   #place a dot in the picture
                D(LL,CC) = 1   # at LL,CC and remember
                RETURN   # it in D

takemark   GOSUB zapdot   #remove the dot at LL,CC
                D(LL,CC) = 0   # and zero its position in D
                RETURN

uncursor   IF D(LL,CC) = 1 THEN   #if cursor was obscuring a dot,
                      GOSUB showdot   # replace the dot
                else
                      GOSUB zapdot   #otherwise, remove all marks
                RETURN

newcursor   IF D(LL,CC) = 1 THEN   #if cursor will be obscuring a dot,
                      CS$ = OD$ # show "on dot" cursor
                else
                      CS$ = ND$ #otherwise, "no dot" cursor
                RETURN

```
800  GOSUB1170:D(LL,CC)=1:RETURN
810  GOSUB1180:D(LL,CC)=0:RETURN
820  IFD(LL,CC)=1THENGOSUB1170:RETURN
830  GOSUB1180:RETURN
840  CS$=ND$:IFD(LL,CC)=1THENCS$=OD$
850  RETURN
```

These routines display or erase dot characters. In addition to the ordinary dots that make up the picture, there are two characters used for the picture cursor. One is used when the cursor is at an unused position in the picture, another is used when the cursor is obscuring a dot.

Figure 9.32: Displaying Dots and Cursors

#Routines to move the picture cursor

cursleft   IF CC > 0 THEN
              CC = CC − 1
         else
              CC = HC
         RETURN

cursright   IF CC < HC THEN
              CC = CC + 1
         else
              CC = 0
         RETURN

cursup   IF LL > 0 THEN
              LL = LL − 1
         else
              LL = HL
         RETURN

cursdown IF LL < HL THEN
              LL = LL + 1
         else
              LL = 0
         RETURN

```
860  IFCC>0THENCC=CC-1:RETURN
870  CC=HC:RETURN
880  IFCC<HCTHENCC=CC+1:RETURN
890  CC=0:RETURN
900  IFLL>0THENLL=LL-1:RETURN
910  LL=HL:RETURN
920  IFLL<HLTHENLL=LL+1:RETURN
930  LL=0:RETURN
```

These routines compute the new position (LL,CC) of the picture cursor after a cursor-moving command. The cursor position values are kept in the ranges 0 <= LL <= HL and 0 <= CC <= HC. The values "wrap around," going from HC or HL to zero when increased past the limit, and going from 0 to HC or HL when decreased past zero.

Figure 9.33: Moving the Picture Cursor

#Position the actual cursor at column C, line L.

cursor   PRINT "home";   #move to upperleft corner
        C = C mod SC   #keep on the screen
        L = L mod SL
        IF C > 0 THEN   #if not far left,
           FOR ZZ = 1 TO C
              PRINT "right";   #PRINT "cursor right" characters
           NEXT ZZ
        IF L > 0 THEN   #if not top line,
           FOR ZZ = 1 TO L
              PRINT "down";   #PRINT "cursor down" characters
           NEXT ZZ
        RETURN

```
940*  PRINTCHR$(19);:C=C-SC*INT(C/SC):L=L-SL*INT(L/SL)
950*  IFC>0THENFORZZ=1TOC:PRINTCHR$(29);:NEXTZZ
960*  IFL>0THENFORZZ=1TOL:PRINTCHR$(17);:NEXTZZ
970   RETURN
```

   The *cursor* routine is similar to routines appearing in other programs in this book. Here, the CRT cursor is positioned at the column and line specified in the variables C and L; elsewhere in this book, the arguments are taken from variables CC and LL. In Alien Life, CC and LL are not available for use as arguments for *cursor*, because they are used for the picture cursor. This illustrates the central problem with the use of subroutines in BASIC: there are no dummy arguments. Arguments are passed to subroutines in global variables.
   The actual BASIC instructions for the Apple version are:

```
940   C=C-SC*INT(C/SC):L=L- SL*INT(L/SL)
950   HTAB C+1
960   LTAB L+1
970   RETURN
```

---

*In the TRS-80 versions of these lines, the home, cursor right and cursor down characters are represented by CHR$(28), CHR$(25), CHR$(26).

Figure 9.34: Positioning the CRT Cursor

310 ALIEN LIFE

```
#Clear the screen
clearscreen  PRINT "clr";
             RETURN
#Single character input
onech        repeat
                 GET X$
                 until (X$ <> "")
#String input
stringin     XX$ = ""
             repeat {
                 GOSUB onech
                 IF X$ <> "delete" THEN {
                     PRINT X$;
                     IF X$ <> "return" THEN
                         XX$ = XX$ + X$
                     else break
                     }
                 else IF LEN(XX$) <> 0 THEN {       #if there are any characters,
                     PRINT "delete string";         #  then remove the last one
                     XX$ = LEFT$(XX$, LEN(XX$) - 1)
                     }
                 }
             RETURN
 980  PRINTCHR$(147);:RETURN
 990  GETX$:IFX$="" THEN990
1000  RETURN
1010  XX$=""
1020  GOSUB990:IFX$=CHR$(20)THEN1050
1030  PRINTX$;:IFX$=CHR$(13)THEN1070
1040  XX$=XX$+X$:GOTO1020
1050  IFLEN(XX$)=0THEN1020
1060  PRINTX$;:XX$=LEFT$(XX$,LEN(XX$)-1):GOTO1020
1070  RETURN
```

These three routines are identical to routines that appear in other programs in this book. (See Figure 4.10 for Apple and TRS-80 versions.)

Figure 9.35: Familiar Utility Routines

```
#Input character while displaying cursor — Pet version
cursin      GOSUB newcursor                              #set cursor character
            CF = 1                                       #flashing loop
            repeat {
               IF CF = 1 THEN
                   GOSUB showcursor                      #cursor on
               else
                   GOSUB showblank                       #cursor off
               CF = -CF: TZ = TI                         #reverse "flag", get time
               repeat {                                  #wait for input
                   GET X$
                   IF X$ <> "" THEN
                      { GOSUB uncursor: RETURN }         #remove cursor
                   } until (TI - TZ >= FQ)               #time to change cursor
            }

showcursor  C = CZ + CC: L = LZ + LL: GOSUB cursor
            PRINT CS$;: RETURN

showblank   C = CZ + CC: L = LZ + LL: GOSUB cursor
            PRINT " ";: RETURN

1080  GOSUB 840:CF=1
1090  IFCF=1THENGOSUB1150:GOTO1110
1100  GOSUB1160
1110  CF=-CF:TZ=TI
1120  GETX$:IFX$<>""THENGOSUB820:RETURN
1130  IFTI-TZ<FQTHEN1120
1140  GOTO1090
1150  C=CZ+CC:L=LZ+LL:GOSUB940:PRINTCS$;:RETURN
1160  C=CZ+CC:L=LZ+LL:GOSUB940:PRINT" ";:RETURN
```

The *cursin* routine and its two subroutines *showcursor* and *showblank* provide a flashing picture cursor, using the Pet TI feature for timing. On TRS-80 and Apple systems, different approaches must be taken. The Apple GET command provides a flashing cursor (unlike Pet's GET). The TRS-80 requires the use of a "delay counter" or "timing loop." The TRS-80 version appears in Figure 9.36a. The actual BASIC instructions for the Apple version are:

```
1080   C=CZ+CC: L=LZ+LL: GOSUB940: GOSUB990: RETURN
```

Figure 9.36: *Flashing the Picture Cursor During Input*

```
1080 GOSUB 840:CF=1
1090 IFCF=1THENGOSUB1150:GOTO1110
1100 GOSUB1160
1110 CF=-CF
1120 FORQQ=1TOCJ*FQ:X$=INKEY$:IFX$<>""THENGOSUB820:RETURN
1130 NEXTQQ
1140 GOTO1090
1150 C=CZ+CC:L=LZ+LL:GOSUB940:PRINTCS$;:RETURN
1160 C=CZ+CC:L=LZ+LL:GOSUB940:PRINT" ";:RETURN
```

Figure 9.36a: TRS-80 BASIC for Cursin

#Show a dot at LL,CC

showdot   L = LZ + LL: C = CC + CZ: GOSUB cursor
              PRINT DT$;
              RETURN

#Erase the dot at LL,CC

zapdot   L = LZ + LL: C = CC + CZ: GOSUB cursor
              PRINT " ";
              RETURN


```
1170 C=CZ+CC:L=LZ+LL:GOSUB940:PRINTDT$;:RETURN
1180 C=CZ+CC:L=LZ+LL:GOSUB940:PRINT" ";:RETURN
```

The *showdot* and *zapdot* routines show or erase a dot at the position determined by the variables LL and CC.

Figure 9.37: Showing and Erasing Dots

```
#Initialization for Alien Life
init    READ SL, SC: DATA lines, columns            #screen size
        READ BH$, TH$, LV$, RV$                     #line segments for
        DATA "bottom", "top", "left", "right"       #  picture frame
        READ OD$, ND$: DATA "ondot", "nodot"        #picture cursor characters
        READ FQ: DATA frequency                     #cursor flash rate
        READ DT$: DATA "dot"                        #picture dot
        READ MC, ML, RL                             #ideal max picture size, and
        DATA maxcols, maxlines, memory limit        #  limit imposed by memory
        MC = min(MC, SC - 2, RL)                    #max columns in picture
        ML = min(ML, SL - 2, RL)                    #max lines in picture
        DIM D(ML, MC)                               #dot array for picture
        READ CB, CM: DATA lochar, hichar            #range of ASCII command chars
        MX = CM - CB: IF MX < 1 THEN STOP           #size of command array
        DIM C(MX)
        FOR CY = 0 TO MX
            C(CY) = misc*100 + no-op                #fill command array with no-op
            NEXT CY
        repeat {                                    #load the command array
            READ CH                                 #ASCII code for char
            IF CH <> -1 THEN {
                CY = CH - CB                        #index for this character
                IF NOT 0 <= CY <= MX THEN STOP
                READ C1, C2                         #primary and secondary
                C(CY) = 100*C1 + C2                 #command codes
            }
        } until (CH = -1)                           #-1 terminates
        #DATA statements for command table (see Figure 9.39)
        #Format: DATA ASCII code, primary command, secondary command
        READ CH: DATA defaultchar                   #for commands out of range
        DX = CH - CB                                #default index to C array
        RETURN
```

(Actual BASIC instructions appear in Figure 9.40.)

Figure 9.38: Initializing Alien Life

314 ALIEN LIFE

```
#DATA statements for command table (from initialization routine)
#Format is: a(character), primary command, secondary command.
#a(character) stands for the ASCII code for the character.
#-1 instead of an ASCII code terminates the list.
#Primary command takes the following values:
#     0          exit       (picture is output as zeroes and ones)
#     1          space      (move the picture cursor)
#     2          mark       (make a dot and move the picture cursor)
#     3          unmark     (remove a dot and move the picture cursor)
#     4          misc       (other commands)
#Secondary command for space, mark or unmark:
#     1          left       (picture cursor moves left)
#     2          right      (picture cursor moves right)
#     3          up         (picture cursor moves up)
#     4          down       (picture cursor moves down)
#Secondary command for miscellaneous:
#     0          no-op      (do nothing)
#     1          life       (transform picture by Game of Life rules)
DATA    a(4),       space,      left,   a(6),       space,      right
        a(8),       space,      up,     a(2),       space,      down
DATA    a(shift 4), mark,       left,   a(shift 6), mark,       right
        a(shift 8), mark,       up,     a(shift 2), mark,       down
DATA    a(0),       unmark,     left,   a(5),       unmark,     right
        a(7),       unmark,     up,     a(1),       unmark,     down
DATA    a(L),       misc,       life
DATA    a(E),       exit,       0,      -1
```

These DATA statements define the single-character commands recognized by the program. The C array has one entry for each ASCII code in the range of possible command codes defined by the variables CB and CM. The command corresponding to the value of CB is encoded in C(0). Each command is encoded as a 2-digit number to the base 100. That is, if C1 and C2 contain the primary and secondary codes to be encoded in C(CY), then the value stored in C(CY) is computed by 100*C1 + C2.

The use of a final −1 to terminate the list makes it possible to add or delete command codes easily.

Figure 9.39: Building the Commands

```
1190*      READSL,SC:DATA24,40
1210**     READBH$,TH$,LV$,RV$:DATA"—","—"," |","| "
1220**     READOD$,ND$:DATA"●","□"
1230**     READFQ:DATA18
1240**     READDT$:DATA"●"
1250***    READMC,ML,RL:DATA37,37,16
1260       IFMC>SC−2THENMC=SC−2
1270       IFMC>RLTHENMC=RL
1280       IFML>SL−2THENML=SL−2
1290       IFML>RLTHENML=RL
1300       DIMD(ML,MC)
1310**     READCB,CM:DATA48,185
1320       MX=CM−CB:IFMX<1THENSTOP
1330       DIMC(MX)
1340       FORCY=0TOMX:C(CY)=400:NEXTCY
1350       READCH:IFCH=−1THEN1380
1360       CY=CH−CB:IFCY<0ORCY>MXTHENSTOP
1370       READC1,C2:C(CY)=100*C1+C2:GOTO1350
1380       DATA52,1,1,54,1,2,56,1,3,50,1,4
1390****   DATA180,2,1,182,2,22,2,2,184,2,3,178,2,4
1400       DATA48,3,1,53,3,2,55,3,3,49,3,4
1410       DATA76,4,1
1420       DATA69,0,0,−1
1430       READCH:DATA52
1440       DX=CH−CB:RETURN
```

These are the actual BASIC instructions for the *init* routine shown in Figures 9.38 and 9.39. The *init* routine establishes values for many variables that could have been constants throughout the program. The use of variables facilitates the changes necessary to adapt the program to different home computer systems.

The techniques used for setting the values of the C array should be studied.

---

*In the TRS-80 version of this line, the screen size is represented by the values 16 and 64.

**The Apple and TRS-80 versions of these lines differ as follows:

```
1210  READ BH$,TH$,LV$,RV$: DATA "−","−","!","!"
1220  and 1230 omitted for Apple.
1220  READOD$,ND$:DATA"#", "*"       ⎫
1230  READFQ:DATA18:READCJ:DATA .25  ⎬ for TRS-80
1240  READ DT$: DATA "0"             ⎭
1310  READ CB,CM: DATA 0,127
```

***The value 16 for RL allows the program to run on an 8K Pet. For larger systems, 100 can be used, meaning that there is essentially no memory limit on the array size, since anything that can fit on the screen can be stored in memory.

****In the Apple and TRS-80 versions of this line, 180, 182, 184 and 178 are replaced by 36, 38, 40 and 34.

Figure 9.40: Initializing Alien Life in BASIC

```
#Count neighbors
neighbors  IF D(0,0) is odd THEN
               add 2 to D(0,1), D(1,0), D(1,1)
           IF D(0,HC) is odd THEN
               add 2 to D(0,HC – 1), D(1,HC), D(1,HC – 1)
           IF D(HL,0) is odd THEN
               add 2 to D(HL,1), D(HL – 1,0), D(HL – 1,1)
           IF D(HL,HC) is odd THEN
               add 2 to D(HL,HC – 1), D(HL – 1,HC), D(HL – 1,HC – 1)
           IF HC > 1 THEN
               FOR CC = 1 TO HC – 1
                   IF D(0,CC) is odd THEN
                       add 2 to D(0,CC – 1), D(0,CC + 1),
                           D(1,CC – 1), D(1,CC), D(1,CC + 1)
                   IF D(HL,CC) is odd THEN
                       add 2 to D(HL,CC – 1), D(HL,CC + 1),
                           D(HL – 1,CC – 1), D(HL – 1,CC), D(HL – 1,CC + 1)
                   NEXT CC
           IF HL > 1 THEN
               FOR LL = 1 TO HL – 1
                   IF D(LL,0) is odd THEN
                       add 2 to D(LL – 1,0), D(LL + 1,0),
                           D(LL – 1,1), D(LL,1), D(LL + 1,1)
                   IF D(LL,HC) is odd THEN
                       add 2 to D(LL – 1,HC), D(LL + 1,HC),
                           D(LL – 1,HC – 1), D(LL,HC – 1), D(LL + 1,HC – 1)
                   IF HC > 1 THEN
                       FOR CC = 1 TO HC – 1
                           IF D(LL,CC) is odd THEN
                               add 2 to D(LL,CC – 1), D(LL,CC + 1),
                                   D(LL – 1,CC – 1), D(LL – 1,CC), D(LL – 1,CC + 1),
                                   D(LL + 1,CC – 1), D(LL + 1,CC), D(LL + 1,CC + 1)
                           NEXT CC
                   NEXT LL
           RETURN
```

Figure 9.41: Counting Neighbors

```
1460  D=D(0,0)/2:IFD=INT(D)THEN1480
1470  D(0,1)=D(0,1)+2:D(1,0)=D(1,0)+2:D(1,1)=D(1,1)+2
1480  D=D(0,HC)/2:IFD=INT(D)THEN1500
1490  D(0,HC-1)=D(0,HC-1)+2:D(1,HC)=D(1,HC)+2:D(1,HC-1)=D(1,HC-1)+2
1500  D=D(HL,0)/2:IFD=INT(D)THEN1520
1510  D(HL,1)=D(HL,1)+2:D(HL-1,0)=D(HL-1,0)+2:D(HL-1,1)=D(HL-1,1)+2
1520  D=D(HL,HC)/2:IFD=INT(D)THEN1550
1530  D(HL,HC-1)=D(HL,HC-1)+2:D(HL-1,HC)=D(HL-1,HC)+2
1540  D(HL-1,HC-1)=D(HL-1,HC-1)+2
1545  IFHC<=1THEN1615
1550  FORCC=1TOHC-1:D=D(0,CC)/2:IFD=INT(D)THEN1580
1560  D(0,CC-1)=D(0,CC-1)+2:D(0,CC+1)=D(0,CC+1)+2:D(1,CC-1)=D(1,CC-1)+2
1570  D(1,CC)=D(1,CC)+2:D(1,CC+1)=D(1,CC+1)+2
1580  D=D(HL,CC)/2:IFD=INT(D)THEN1610
1590  D(HL,CC-1)=D(HL,CC-1)+2:D(HL,CC+1)=D(HL,CC+1)+2:
      D(HL-1,CC)=D(HL-1,CC)+2
1600  D(HL-1,CC-1)=D(HL-1,CC-1)+2:D(HL-1,CC+1)=D(HL-1,CC+1)+2
1610  NEXTCC
1615  IFHL<=1THENRETURN
1620  FORLL=1TOHL-1:D=D(LL,0)/2:IFD=INT(D)THEN1650
1630  D(LL-1,0)=D(LL-1,0)+2:D(LL+1,0)=D(LL+1,0)+2:D(LL-1,1)=D(LL-1,1)+2
1640  D(LL,1)=D(LL,1)+2:D(LL+1,1)=D(LL+1,1)+2
1650  D=D(LL,HC)/2:IFD=INT(D)THEN 1675
1660  D(LL-1,HC)=D(LL-1,HC)+2:D(LL+1,HC)=D(LL+1,HC)+2:
      D(LL,HC-1)=D(LL,HC-1)+2
1670  D(LL-1,HC-1)=D(LL-1,HC-1)+2:D(LL+1,HC-1)=D(LL+1,HC-1)+2
1675  IFHC<=1THENNEXTLL:RETURN
1680  FORCC=1TOHC-1:D=D(LL,CC)/2:IFD=INT(D)THEN1730
1690  D(LL,CC-1)=D(LL,CC-1)+2:D(LL,CC+1)=D(LL,CC+1)+2
1700  D(LL-1,CC-1)=D(LL-1,CC-1)+2:D(LL-1,CC+1)=D(LL-1,CC+1)+2
1710  D(LL+1,CC-1)=D(LL+1,CC-1)+2:D(LL+1,CC+1)=D(LL+1,CC+1)+2
1720  D(LL-1,CC)=D(LL-1,CC)+2:D(LL+1,CC)=D(LL+1,CC)+2
1730  NEXTCC:NEXTLL:RETURN
```

The *neighbors* routine encodes the number of neighbors of each point by adding twice the number to the value (0 or 1) of the D array entry. The routine scans the entire picture, and whenever a dot is found, the neighbor count of each of the neighboring positions is incremented. The routine is complicated, since positions near the edges of the picture don't have all eight neighboring positions. The routine could be greatly reduced through the use of a dummy "border" in the D array, since every "actual" point would then have eight neighboring positions. Positions in the "border" would never contain dots.

Figure 9.42: BASIC for Neighbors

#Make life or death decisions

```
lifedeath   FOR LL = 0 TO HL
               FOR CC = 0 TO HC
                   NB = INT(D(LL,CC)/2)          #2*neighbors is stored in D
                   PT = D(LL,CC) - 2*NB          #dot value is "low bit"
                   IF case
                       NB < 2 OR NB > 3 THEN
                           D(LL,CC) = 0          #no dot
                       NB = 3 THEN
                           D(LL,CC) = 1          #dot
                       NB = 2 THEN
                           D(LL,CC) = PT         #keep current contents
                   NEXT CC
               NEXT LL
            RETURN
```

```
1740  FORLL=0TOHL:FORCC=0TOHC:D=D(LL,CC):NB=INT(D/2):PT=D-2*NB
1750  IFNB<2ORNB>3THEND(LL,CC)=0:GOTO1780
1760  IFNB=3THEND(LL,CC)=1:GOTO1780
1770  D(LL,CC)=PT
1780  NEXTCC:NEXTLL:RETURN
```

The *lifedeath* routine decides whether or not a given position shall have a dot. The following algorithm is used:

— If the position has fewer than 2, or more than 3 neighbors, no dot shall appear.
— If the position has 2 neighbors, then the new generation shall have a dot there if and only if the old generation had a dot there.
— If the position has 3 neighbors, then a dot shall appear there.

As the new D array is constructed from the old, the neighbor counts disappear, and the array values are once again only ones and zeroes.

Figure 9.43: Deciding Where the Dots Shall Be

Improvements and Additions

Here are some of the many possible improvements and additions to Alien Life:

- Implement external storage for bit patterns. The *blockin*, *choices* and *outstream* routines (Figure 9.23) are the only ones that must be changed.

- Eliminate the call to *drawdots* in *setframe* (Figure 9.16) when a new frame has been specified. This change will speed up the program's initialization.

- Add new drawing commands:

    — Commands that move first, then perform the action.

    — Commands that move diagonally.

    — A command to "undo" the last command.

- Modify the *neighbors* routine to use "wraparound." That is, count (LL,HC) as a neighboring position to (LL,0) or (0,CC) as a neighboring position to (HL,CC). With this approach, every point has eight neighboring positions. The game becomes Life on a torus. (Torus is the mathematical name for a donut-shaped surface.) Do you see why?

- Speed up the Game of Life transformations.

#Transform pattern by Game of Life rules

life GOSUB neighbors:GOSUB lifedeath: GOSUB drawdots   #transform and draw
     CC = 0: LL = 0: GOSUB newcursor: RETURN   #upper left corner

1450 GOSUB1460:GOSUB1740:GOSUB660:CC=0:LL=0:GOSUB840:RETURN

The *life* routine is called from *miscdo* to implement a transformation of the picture according to Game of Life rules. First a call is made to *neighbors* to count the neighbors of each position in the picture. (A neighbor is a dot in one of the eight positions surrounding a position.) Then a call is made to *lifedeath* to transform the contents of the D array (i.e., the dots of the picture) according to the Game of Life rules, which depend upon the number of neighbors of each point. Finally, *drawdots* is called to display the picture, and the cursor is reset to the upper left corner by a call to *newcursor*.

Figure 9.44: *Applying Life Rules to the Picture*

Summary

The Alien Encounter game allows the player to design dot matrix picture messages. These messages are to be transmitted to aliens as bit streams containing numbers of bits that are products of two primes. Messages of this type have the best chance of being interpreted correctly by a receiver who has to guess the message format.

The well-known Game of Life has been "grafted" onto Alien Encounter, so that the picture messages being prepared can be transformed according to Game of Life rules.

The Alien Life program that implements the composite game has two important aspects: the picture area and the commands. The components of the picture area are the frame, the picture dots and the picture cursor. The *setframe, center* and *drawframe* routines participate in the drawing of the frame. The picture dots are the image of the D array, and are displayed by the *drawdots* routine. The picture cursor is under the control of the *cursin* routine, which determines the cursor character and location, implements the flashing, and restores the screen contents when the picture cursor is removed.

The single-character commands recognized by the Alien Life program are defined in the *init* routine. Specific characters are associated with specific commands through the C array, whose indices are derived from ASCII characters and whose contents are single numbers encoding primary and secondary command information. The primary information selects among "exit," "drawing" or "miscellaneous," and the secondary information selects within the latter two categories. Game of Life is encoded under the "miscellaneous" category.

Many improvements and additions are possible, primarily in the provision of external storage facilities for encoded images.

APPENDIX **A**

The ASCII Character Set

The table opposite shows the "standard" ASCII character set. The decimal numeric codes from 0 to 127 are listed in order, and the corresponding character is shown for each code.

All BASIC systems use slight variations of this code. The following program is helpful in determining the decimal code values assigned to special characters in your system:

    1 GOSUB 2: PRINT ASC(X$);" ";: GOTO 1

    2 (Single-character input subroutine, as shown for Apple, Pet and TRS-80 systems in Figure 2.11. Line numbers 720 and 730 in Figure 2.11 become 2 and 3 here.)

After entering this program and typing RUN, you can press any key and see an immediate display of the decimal value of the code corresponding to the key you pressed.

In addition to the codes 0 to 127, the Pet and TRS-80 systems use codes 128 to 255 for special purposes. On a TRS-80, codes 128 to 191 encode "graphics blocks," and codes 192 to 255 encode strings of blanks. On a Pet, codes 128 to 255 correspond to special "graphics" characters. The following program will allow you to display the characters corresponding to a range of codes:

    1 INPUT "RANGE ";N1,N2

    2 FOR XX=N1 TO N2: PRINT "(";CHR$(XX);") ";: NEXT XX: PRINT: GOTO1

The program will first request a pair of numbers specifying the desired range and then display the characters corresponding to the numbers in the range. Each character will be enclosed within parentheses.

ASCII Character Set

CODE	CHAR	CODE	CHAR	CODE	CHAR	CODE	CHAR
0	NUL	32[1]		64	@	96[5]	`
1	SOH	33	!	65	A	97	a
2	STX	34	"	66	B	98	b
3	ETX	35	#	67	C	99	c
4	EOT	36	$	68	D	100	d
5	ENQ	37	%	69	E	101	e
6	ACK	38	&	70	F	102	f
7	BEL	39[2]	'	71	G	103	g
8	BS	40	(	72	H	104	h
9	TAB	41	)	73	I	105	i
10	LF	42	*	74	J	106	j
11	VT	43	+	75	K	107	k
12	FF	44[3]	,	76	L	108	l
13	CR	45	−	77	M	109	m
14	SO	46	.	78	N	110	n
15	SI	47	/	79	O	111	o
16	DLE	48	0	80	P	112	p
17	DC1	49	1	81	Q	113	q
18	DC2	50	2	82	R	114	r
19	DC3	51	3	83	S	115	s
20	DC4	52	4	84	T	116	t
21	NAK	53	5	85	U	117	u
22	SYN	54	6	86	V	118	v
23	ETB	55	7	87	W	119	w
24	CAN	56	8	88	X	120	x
25	EM	57	9	89	Y	121	y
26	SUB	58	:	90	Z	122	z
27	ESC	59	;	91	[	123	{
28	FS	60	<	92	\	124	|
29	GS	61	=	93	]	125[6]	}
30	RS	62	>	94	↑	126	~
31	US	63	?	95[4]	←	127[7]	RUBOUT

[1] space
[2] single quote
[3] comma
[4] or underline
[5] accent mark
[6] or ALT MODE
[7] or DEL

APPENDIX B
TRS-80 BASIC Listings

Addition Drill

```
100 N1=INT(RND(0)*10)
110 N2=INT(RND(0)*10)
120 PRINT "WHAT IS";N1;"+";N2;
130 INPUT A
140 PRINT
150 IF A<>N1+N2 THEN PRINT "I'M SORRY, THAT'S WRONG--TRY AGAIN":GOTO 120
160 PRINT "THAT'S RIGHT--NOW TRY ANOTHER ONE"
170 GOTO 100
```

Arithmetic Drill

```
100 INPUT "HOW MANY PLACES";D
110 D=INT(D):IF D<=0 OR D>3 THEN 100
120 DEF FNR(X)=INT(RND(0)*10[X)
130 OP=INT(RND(0)*4)+1
140 ON OP GOSUB 190,220,260,290
150 PRINT "WHAT IS";N1;OP$;N2;
160 INPUT A:PRINT
170 IF A<>EX THEN PRINT "I'M SORRY, THAT'S WRONG--TRY AGAIN":GOTO 150
180 PRINT "THAT'S RIGHT -- NOW TRY ANOTHER ONE":GOTO 130
190 OP$="+"
200 N1=FNR(D):N2=FNR(D)
210 EX=N1+N2:RETURN
220 OP$="-"
230 N1=FNR(D):NN=FNR(D):IF NN<=N1 THEN N2=NN:GOTO 250
240 N2=N1:N1=NN
250 EX=N1-N2:RETURN
260 OP$="x"
270 N1=FNR(D):N2=FNR(D)
280 EX=N1*N2:RETURN
290 OP$="/"
300 N2=FNR(D):IF N2=0 THEN 300
310 NN=FNR(D):N1=NN*N2
320 EX=NN:RETURN
```

Guess

```
90 CLEAR100
100 GOSUB 670
110 GOSUB 590
120 GOSUB 200:PRINT:PRINT IN$:PRINT
130 G=0
140 GOSUB 360:G=G+1:GOSUB 410:GOSUB 480
150 IF IP<>N AND IP<>-1 THEN 140
160 GOSUB 520:GOSUB 710
170 IF NX$="N" THEN 110
180 IF NX$="E" THEN END
190 GOTO 120
200 GOSUB 740
210 FOR ZZ=BH TO TH:HT(ZZ)=0:NEXT ZZ
220 ON GM GOSUB 310,240
230 RETURN
240 INPUT "HIDDEN WORD";H$:N=LEN(H$)
250 IF N=0 OR N>MX THEN 240
260 IN$="GUESS WORDS OF"+STR$(N)+" LETTERS.":GL=N
270 FOR ZZ=1 TO N:HH=ASC(MID$(H$,ZZ,1))
280 HT(HH)=HT(HH)+2[(ZZ-1):NEXT ZZ
290 VG=N:MG=20
```

```
300 GOSUB 740:RETURN
310 FOR ZZ=1 TO N
320 HH=INT(RND(0)*10)+L:IF HT(HH)<>0 THEN 320
330 HT(HH)=HT(HH)+2[(ZZ-1):NEXT ZZ
340 IN$="GUESS NUMBERS OF "+STR$(N)+" DIGITS.":GL=N
350 RETURN
360 INPUT "WHAT DO YOU GUESS"; G$
370 IF LEN(G$)<>GL THEN PRINT IN$:GOTO 360
380 FOR ZZ=1 TO GL
390 GC(ZZ)=ASC(MID$(G$,ZZ,1)):NEXT ZZ
400 RETURN
410 RG=0:IP=0
420 FOR ZZ=1 TO GL:HH=HT(GC(ZZ))
430 IF HH=0 THEN 460
440 RG=RG+1:HH=INT(HH/2[(ZZ-1))
450 IF HH<>2*INT(HH/2) THEN IP=IP+1
460 NEXT ZZ
470 RETURN
480 IF IP=N THEN PRINT "THAT´S RIGHT! ";:GOTO 510
490 IF G=MG THEN PRINT "TOO MANY GUESSES -- YOU LOSE!":IP=-1:GOTO 510
500 PRINT "RIGHT:";RG,"IN PLACE:"; IP
510 RETURN
520 IF IP<>N THEN 570
530 IF G=1 THEN PRINT "YOU MUST BE PSYCHIC.":GOTO 580
540 PRINT "YOU GOT IT IN";G;"GUESSES."
550 IF G<=VG THEN PRINT:PRINT "VERY GOOD."
560 GOTO 580
570 IF IP=-1 THEN PRINT "YOU COULDN´T GET IT IN";MG;"TRIES."
580 RETURN
590 GOSUB 740:PRINT "GAME: ";:GOSUB 720
600 N=VAL(X$):IF N<>0 THEN GM=1:GOTO 630
610 IF X$="W" THEN GM=2:GOTO 630
620 N=4:GM=1
630 ON GM GOSUB 650,660
640 RETURN
650 VG=N+2:MG=2*VG:BH=L:TH=L+9:RETURN
660 VG=0:MG=99:BH=0:TH=127:RETURN
670 DIM HT(127)
680 MX=20:DIM GC(MX)
690 L=ASC("0")
700 RETURN
710 GOSUB 720:NX$=X$:RETURN
720 X$=INKEY$:IF X$="" THEN 720
730 RETURN
740 CLS:RETURN
```

Hangman

```
100 GOSUB 670
110 GOSUB 590
120 GOSUB 200:PRINT:PRINT IN$:PRINT
130 G=0
140 GOSUB 360:G=G+1:GOSUB 410:GOSUB 480
150 IF IP<>N AND IP<>-1 THEN 140
160 GOSUB 520:GOSUB 710
170 IF NX$="N" THEN 110
180 IF NX$="E" THEN END
190 GOTO 120
200 GOSUB 740
210 INPUT"HIDDEN WORD";H$:N=LEN(H$)
220 IF N=0 OR N>MX THEN 210
230 INPUT "HOW MANY WRONG GUESSES ARE ALLOWED";WG
240 IN$="GUESS ONE LETTER OR THE WHOLE WORD"
```

```
250 GL=1:IP=0
260 MG=99:VG=5
270 GOSUB 740
280 FOR ZZ=0 TO 127:HT(ZZ)=0:NEXT ZZ
290 FOR ZZ=1 TO N:HH=ASC(MID$(H$,ZZ,1))
300 GS(ZZ)=ASC("-")
310 HT(HH)=HT(HH)+2[(ZZ-1):NEXT ZZ
320 RETURN
360 PRINT:FOR ZZ=1 TO N:PRINT CHR$(GS(ZZ));:NEXT ZZ
365 PRINT "    WRONG GUESSES LEFT:";WG
370 INPUT "WHAT DO YOU GUESS";G$
375 IF LEN(G$)=N THEN RETURN
380 IF LEN(G$)<>GL THEN PRINT IN$:GOTO 370
390 FOR ZZ=1 TO GL
395 GC(ZZ)=ASC(MID$(G$,ZZ,1)):NEXT ZZ
400 RETURN
410 IF LEN(G$)<>N THEN 440
420 IF G$=H$ THEN IP=N:RETURN
430 WG=WG-1:RETURN
440 IX=0:FOR ZZ=1 TO GL:HH=HT(GC(ZZ))
445 IF HH=0 THEN 460
450 FOR YY=1 TO N:IF HH<>2*INT(HH/2) THEN IP=IP+1:IX=IX+1:GS(YY)=GC(ZZ)
455 HH=INT(HH/2):NEXT YY
460 NEXT ZZ
465 IF IX=0 THEN WG=WG-1
470 RETURN
480 IF IP=N THEN PRINT "THAT´S RIGHT! ";:GOTO 500
490 IF G=MG OR WG<0 THEN PRINT "TOO MANY GUESSES -- YOU LOSE!":IP=-1
500 RETURN
520 IF IP<>N THEN 570
530 IF G=1 THEN PRINT "YOU MUST BE PSYCHIC.":GOTO 580
540 PRINT "YOU GOT ";H$;" IN";G;"GUESSES."
550 IF G<=VG THEN PRINT:PRINT "VERY GOOD."
560 GOTO 580
570 IF IP=-1 THEN PRINT "YOU COULDN´T GET ";H$
580 RETURN
590 GOSUB 740:RETURN
670 DIM HT(127)
680 MX=20:DIM GC(MX)
690 L=ASC("0")
700 RETURN
710 GOSUB 720:NX$=X$:RETURN
720 X$=INKEY$:IF X$="" THEN 720
730 RETURN
740 CLS:RETURN
```

Clock

```
100 GOSUB 1110:GOSUB 920
110 GOSUB 340:GOSUB 240:GOSUB 250:GOSUB 120:GOTO 110
120 CM$=INKEY$:IF CM$="" THEN 230
130 GOSUB 920
140 IF CM$="F" THEN GOSUB 450:GOTO 220
150 IF CM$="S" THEN GOSUB 480:GOTO 220
160 IF CM$="T" THEN GOSUB 520:GOTO 220
170 IF CM$="A" THEN GOSUB 550:GOTO 220
180 IF CM$="Q" THEN GOSUB 590:GOTO 220
190 IF CM$="R" THEN GOSUB 600:GOTO 220
200 IF CM$="Z" THEN GOSUB 690:GOTO 220
210 IF CM$="C" THEN GOSUB 720:GOTO 220
220 GOSUB 920
230 RETURN
240 LL=TL:CC=TC:GOSUB 930:RETURN
```

APPENDIX B

```
250 GOSUB 1050:HR=VAL(HR$)
260 IF HR>11 THEN AP$="PM":GOTO 280
270 AP$="AM"
280 IF HR=0 THEN HR$="12":GOTO 310
290 IF HR>12 THEN HR$=STR$(HR-12):GOTO 310
300 HR$=STR$(HR)
310 HR$=RIGHT$(HR$,2)
320 PRINT HR$;":";MN$;":";SC$;" ";AP$;
330 RETURN
340 GOSUB 830:ET=JT
350 IF ET=>OE THEN 380
360 IF AJ<>-1 THEN AJ=AJ-86400
370 IF AL<>-1 THEN AL=AL-86400
380 IF OE=ET THEN 440
390 OE=ET
400 IF AL=-1 OR AL>ET THEN 420
410 GOSUB 610:AL=AL+86400
420 IF AJ=-1 OR AJ>ET THEN 440
430 DT=AI:GOSUB 810:AJ=AJ+AF
440 RETURN
450 INPUT "ADJUSTMENT INTERVAL (SEC)";FT
460 IF FT<=0 OR FT<>INT(FT) THEN 450
470 IN=1:GOSUB 510:RETURN
480 INPUT "ADJUSTMENT INTERVAL (SEC)";FT
490 IF FT<=0 OR FT<>INT(FT) THEN 480
500 IN=-1:GOSUB 510:RETURN
510 AI=IN:AF=FT:GOSUB 830:AJ=JT+AF:RETURN
520 PRINT "SET TIME"
530 GOSUB 640:PRINT "PRESS RETURN TO START CLOCK"
540 GOSUB 1030:GOSUB 1080:RETURN
550 PRINT "SET ALARM"
560 GOSUB 640:AL=TM:GOSUB 830
570 IF AL<JT THEN AL=AL+86400
580 RETURN
590 AL=-1:GOSUB 630:RETURN
600 GOSUB 630:RETURN
610 LL=AR:CC=AC:GOSUB 930
620 PRINT AL$:RETURN
630 GOSUB 920:RETURN
640 INPUT "TIME (HHMMSS)";TM$
650 AH=VAL(LEFT$(TM$,2)):IF AH<0 OR AH=>24 THEN 640
660 AM=VAL(MID$(TM$,3,2)):IF AM<0 OR AM=>60 THEN 640
670 AS=VAL(RIGHT$(TM$,2)):IF AS<0 OR AS=>60 THEN 640
680 TM=3600*AH+60*AM+AS:RETURN
690 INPUT "CHANGE AMOUNT (HOURS)";CH
700 IF CH<>INT(CH) OR CH<=-24 OR CH=>24 THEN 690
710 CS=CH*3600:GOSUB 750:RETURN
720 INPUT "CHANGE AMOUNT (SECONDS)";CS
730 IF CS<>INT(CS) OR CS<=-60 OR CS=>60 THEN 720
740 GOSUB 750:RETURN
750 GOSUB 830:CT=JT
760 DT=CS:GOSUB 810:GOSUB 830
770 IF CS=>0 OR JT<=CT THEN 785
775 IF AJ<>-1 THEN AJ=AJ+86400
780 IF AL<>-1 THEN AL=AL+86400
785 IF CS<=0 OR JT=>CT THEN 800
790 IF AJ<>-1 THEN AJ=AJ-86400
795 IF AL<>-1 THEN AL=AL-86400
800 RETURN
810 GOSUB 830:JT=JT+DT+1:GOSUB 850
820 GOSUB 1090:GOSUB 1080:RETURN
830 GOSUB 1050
840 JT=VAL(SC$)+60*VAL(MN$)+3600*VAL(HR$):RETURN
850 IF JT<0 THEN JT=JT+86400:GOTO 850
860 IF JT=>86400 THEN JT=JT-86400:GOTO 860
```

```
870 SC=JT-60*INT(JT/60)
880 MN=(JT-SC)/60:MN=MN-60*INT(MN/60)
890 HR=(JT-SC-60*MN)/3600
900 TX=SC+100*MN+10000*HR+1000000
910 TM$=RIGHT$(STR$(TX),6):RETURN
920 CLS:RETURN
930 IF LL<0 THEN LL=LL+16:GOTO 930
940 IF LL>15 THEN LL=LL-16:GOTO 940
950 IF CC<0 THEN CC=CC+64:GOTO 950
960 IF CC>63 THEN CC=CC-64:GOTO 960
970 PRINT CHR$(28);
980 IF LL=0 THEN 1000
990 FOR XX=1 TO LL:PRINT CHR$(26);:NEXT XX
1000 IF CC=0 THEN 1020
1010 FOR XX=1 TO CC:PRINT CHR$(25);:NEXT XX
1020 RETURN
1030 X$=INKEY$:IF X$="" THEN 1030
1040 RETURN
1050 TT$=TIME$
1060 HR$=MID$(TT$,10,2):MN$=MID$(TT$,13,2):SC$=RIGHT$(TT$,2)
1070 RETURN
1080 RETURN
1090 IF TT$=TIME$ THEN 1090
1100 RETURN
1110 OE =-1:AL=-1:AJ=-1
1120 AL$=" ALARM "
1130 READ TL,TC:DATA 14,52
1140 READ AR,AC:DATA 11,17
1150 READ FT,IN:DATA 2900,-1:GOSUB 510
1160 RETURN
```

Card Memory

```
100 GOSUB 960
110 GOSUB 860:GOSUB 920
120 GOSUB 860:GOSUB 910:GOSUB 180:GOSUB 240
130 GOSUB 390:GOSUB 660
140 GOSUB 830:IF NX$="R" THEN GOSUB 270:GOTO 140
150 IF NX$="N" THEN 110
160 IF NX$="E" THEN END
170 GOTO 120
180 FOR XX=1 TO 52:DD(XX)=XX:NEXT XX
190 FOR XX=1 TO NC:YY=53-XX
200 ZZ=INT(RND(0)*YY)+1:DK(XX)=DD(ZZ)
210 IF XX=52 OR YY<=ZZ THEN 230
220 FOR WW=ZZ TO YY-1:DD(WW)=DD(WW+1):NEXT WW
230 NEXT XX:RETURN
240 FOR I=1 TO NC
250 CD=DK(I):GOSUB 290:DL=DC:GOSUB 870:GOSUB 860:NEXT I
260 RETURN
270 GOSUB 860:FOR I=1 TO NC:CD=DK(I):GOSUB 290
280 NEXT I:RETURN
290 GOSUB 310:PRINT CD$
300 RETURN
310 FV=CD-INT(CD/13)*13
320 SU=(CD-FV)/13+1
330 IF FV=0 THEN FV=13:SU=SU-1
340 FV$=FV$(FV):SU$=SU$(SU)
350 CD$=SU$+" "+FV$:RETURN
360 IF SU<1 OR SU>4 OR FV<1 OR FV>13 THEN CD=-1:RETURN
370 CD=13*(SU-1)+FV
380 RETURN
```

APPENDIX B

```
390 PRINT "REPEAT CARDS IN ORDER"
400 FOR I=1 TO NC
410 GOSUB 490:GOSUB 500:IF CD=-1 THEN 410
420 IF CD=-2 THEN GOSUB 270:RETURN
430 IF CD=DK(I) THEN GOSUB 470:GOTO 450
440 GOSUB 480:CD=DK(I)
450 GOSUB 290:DL=SC:GOSUB 870:GOSUB 860:NEXT I
460 RETURN
470 RG=RG+1:RETURN
480 RETURN
490 INPUT "CARD";CD$:RETURN
500 IF LEN(CD$)=>3 THEN 530
510 IF CD$="Q" THEN CD=-2:RETURN
520 CD=-1:RETURN
530 L$=LEFT$(CD$,1):R$=MID$(CD$,3):VR=VAL(R$)
540 IF L$="C" THEN SU=1:GOTO 590
550 IF L$="D" THEN SU=2:GOTO 590
560 IF L$="H" THEN SU=3:GOTO 590
570 IF L$="S" THEN SU=4:GOTO 590
580 SU=-1
590 IF VR>0 THEN FV=VR:GOTO 650
600 IF R$="A" THEN FV=1:GOTO 650
610 IF R$="K" THEN FV=13:GOTO 650
620 IF R$="Q" THEN FV=12:GOTO 650
630 IF R$="J" THEN FV=11:GOTO 650
640 FV=-1
650 GOSUB 360:RETURN
660 PRINT:IF RG<>NC THEN 700
670 IF NC=1 THEN PRINT "YOU GOT IT!":GOTO 740
680 IF NC=2 THEN PRINT "YOU GOT BOTH OF THEM!":GOTO 740
690 PRINT "YOU GOT ALL";NC;"OF THEM!":GOTO 740
700 IF RG<>0 THEN 730
710 IF NC=1 THEN PRINT "YOU MISSED IT!":GOTO 740
720 PRINT "YOU DIDN'T GET ANY!":GOTO 740
730 PRINT "YOU GOT";RG;"OUT OF";NC
740 PRINT
750 IF RG<>NC OR NC<=BC THEN 780
760 PRINT "A NEW RECORD LENGTH!"
770 BC=NC:BD=DC
780 IF RG>BR THEN BR=RG
790 PRINT:GOSUB 800:RETURN
800 PRINT "BEST FOR THIS SET:";BR;"OUT OF";NC
810 IF BC>0 THEN PRINT:PRINT "BEST LENGTH:";BC;" (DELAY:";BD;"SEC)"
820 RETURN
830 GOSUB 840:NX$=X$:RETURN
840 X$=CHR$(RND(0)):X$=INKEY$:IF X$="" THEN 840
850 RETURN
860 CLS:RETURN
870 FOR DX=1TODL*SS:NEXT DX:RETURN
910 RG=0:RETURN
920 INPUT "NUMBER OF CARDS";NC:IF NC<1 OR NC>52 THEN 920
930 INPUT "INTERVAL (SEC)";DC
940 SC=1
950 BR=0:RETURN
960 DIM SU$(4),FV$(13),DK(52),DD(52)
965 READ SS:DATA 200
970 FOR I=1 TO 4:READ SU$(I)
980 DATA "CLUB","DIAMOND","HEART","SPADE"
990 NEXT I
1000 FOR I=1 TO 13:READ FV$(I)
1010 DATA "ACE","2","3","4","5","6","7","8","9","10","JACK","QUEEN","KING"
1020 NEXT I
1030 BC=0:BD=0:RETURN
```

Ten-key Flicker

```
90 CLEAR 100
100 GOSUB 430
110 GOSUB 420
120 GOSUB 340:GOSUB 200
130 GOSUB 230:IF XX=1 THEN PRINT "THAT'S RIGHT!":GOTO 160
140 IF XX=-1 THEN PRINT "YOU'VE GOT TO BE FASTER! -- IT WAS ";EX$;:GOTO 160
150 PRINT "THAT'S WRONG -- THE ANSWER IS ";EX$;
160 DL=2:GOSUB 300
170 GOSUB 200:GOSUB 390:IF NX$="E" THEN END
180 IF NX$="N" THEN 110
190 GOTO 120
200 CLS
210 PRINT A(7);L$;A(8);L$;A(9);B$;A(4);L$;A(5);L$;A(6);B$;A(1);L$;A(2);L$;A(3);
220 RETURN
230 DL=DC:TX$=TIME$
240 X$=INKEY$:IF X$<>"" THEN 280
250 IF DL<=0 THEN XX=-1:GOTO 290
260 IF TX$=TIME$ THEN 240
270 TX$=TIME$:DL=DL-1:GOTO 240
280 XX=0:IF X$=EX$ THEN XX=1
290 CLS:RETURN
300 IF DL=0 THEN RETURN
310 FOR I=1 TO DL:TX$=TIME$
320 IF TX$=TIME$ THEN 320
330 NEXT I:RETURN
340 N1=FNR9(1):N2=FNR9(1):N3=FNR9(1)
350 FOR I=1 TO 9:A(I)=N1:NEXT I:A(N2)=N3
360 IF N1=N3 THEN EX$="0":GOTO 380
370 EX$=RIGHT$(STR$(N2),1)
380 RETURN
390 GOSUB 400:NX$=X$:RETURN
400 X$=STR$(RND(0)):X$=INKEY$:IF X$="" THEN 400
410 RETURN
420 INPUT "DELAY INTERVAL";DC:RETURN
430 DIM A(9)
440 L$="":FOR LL=1 TO 27:L$=L$+CHR$(25):NEXT LL
450 B$=CHR$(24):FOR LL=1 TO 6:B$=B$+CHR$(26):NEXT LL
460 DEF FNR9(X)=INT(RND(0)*9)+1
470 RETURN
```

Timer

```
100 TX=0:GOSUB 190
110 X$=INKEY$:IFX$=""THEN110
120 IF X$="R" THEN TX=0:GOSUB 190:GOTO 110
130 IF X$=" " THEN GOSUB 150:GOTO 110
140 GOTO 110
150 TX$=TIME$
160 X$=INKEY$:IF X$<>"" THEN RETURN
170 IF TX$=TIME$ THEN 160
180 TX=TX+1:GOSUB 190:GOTO 150
190 IF TX=0 THEN CLS:GOTO 200
195 PRINT CHR$(28);
200 PRINT TX:RETURN
```

Birthday

```
100 GOSUB610
110 GOSUB510:DP$="BIRTHDATE: ":GOSUB120:GOSUB300:GOSUB400:GOSUB520:GOTO110
120 PRINT DP$;:GOSUB540:GOSUB140:IF XX<>1 THEN120
130 RETURN
140 L=LEN(XX$):IF L<5 THEN210
150 GOSUB220:MO=INT(VAL(MO$)):DA=INT(VAL(DA$)):YR=INT(VAL(YR$))
160 IF LEN(YR$)=2 THEN YR=YR+1900
170 GOSUB480:ML(2)=28:IF XX=1 THEN ML(2)=29
180 IF NOT (1<=MO AND MO<=12 AND YR=>1) THEN210
190 IF NOT (1<=DA AND DA<=ML(MO)) THEN210
200 IF NOT (YR=1582 AND MO=10 AND 4<DA AND DA<15) THEN XX=1:RETURN
210 XX=0:RETURN
220 XS=0:FOR XL=2 TO L-1:ZZ$=MID$(XX$,XL,1)
230 IF ZZ$<>"/" AND ZZ$<>" " AND ZZ$<>"-" THEN260
240 XS=XS+1:IF XS=1 OR XS=2 THEN SP(XS)=XL:GOTO260
250 MO$="":RETURN
260 NEXT XL:IF XS<>2 THEN MO$="":GOTO290
270 MO$=LEFT$(XX$,SP(1)-1):DA$=MID$(XX$,SP(1)+1,SP(2)-SP(1)-1)
280 YR$=MID$(XX$,SP(2)+1)
290 RETURN
300 YZ=YR-400*INT(YR/400):DW=DZ
310 IF YZ<101 THEN330
320 YZ=YZ-100:DW=DW+5:GOTO310
330 IF YZ<5 THEN350
340 YZ=YZ-4:DW=DW+5:GOTO330
350 IF YZ=0 THEN DW=DW-2:GOTO356
352 DW=DW+YZ-1
356 GOSUB450:DW=DW+JD-1:IF YR>1582 THEN390
360 IF YR=1582 AND MO>10 THEN390
370 IF YR=1582 AND MO=10 AND DA=>15 THEN 390
380 DW=DW+3
390 DW=DW-7*INT(DW/7):RETURN
400 PRINT:PRINT "YOU WERE BORN ON"
410 PRINT DW$(DW);", ";MO$(MO);STR$(DA);",";YR
420 PRINT "HOW NICE!"
430 PRINT
440 PRINT PM$(DW):RETURN
450 JD=J(MO)+DA:IF MO<=2 THEN RETURN
460 GOSUB480:IF XX=1 THEN JD=JD+1
470 RETURN
480 XX=0:IF YR<>4*INT(YR/4)THEN RETURN
490 IF YR=100*INT(YR/100)AND YR<>400*INT(YR/400)THEN RETURN
500 XX=1:RETURN
510 CLS:RETURN
520 X$=INKEY$:IF X$="" THEN 520
530 RETURN
540 XX$=""
550 GOSUB520:IF X$=CHR$(8) THEN 580
560 PRINT X$;:IF X$=CHR$(13) THEN600
570 XX$=XX$+X$:GOTO550
580 IF LEN(XX$)=0 THEN550
590 PRINT X$;:XX$=LEFT$(XX$,LEN(XX$)-1):GOTO550
600 RETURN
610 READ DZ:DATA 1
620 DIM DW$(6),J(12),MO$(12),ML(12),PM$(6)
630 FOR XX=0 TO 6:READ DW$(XX),PM$(XX):NEXT XX
640 DATA SUNDAY, "YOU´RE FAIR, AND WISE, AND GOOD, AND GAY"
650 DATA MONDAY,"MONDAY´S CHILD IS FAIR OF FACE"
660 DATA TUESDAY,"TUESDAY´S CHILD IS FULL OF GRACE"
670 DATA WEDNESDAY,"WEDNESDAY´S CHILD IS FULL OF WOE"
680 DATA THURSDAY,"THURSDAY´S CHILD HAS FAR TO GO"
690 DATA FRIDAY,"FRIDAY´S CHILD IS LOVING AND GIVING"
```

INSIDE BASIC GAMES

```
700 DATA SATURDAY,"SATURDAY´S CHILD MUST WORK FOR A LIVING"
710 FOR XX=1 TO 12:READ MO$(XX):NEXT XX
720 DATA JANUARY,FEBRUARY,MARCH,APRIL,MAY,JUNE,JULY,AUGUST,SEPTEMBER
730 DATA OCTOBER,NOVEMBER,DECEMBER
740 DT=0:FOR XX=1 TO 12:J(XX)=DT:READ ML(XX):DT=DT+ML(XX):NEXT XX
745 IF DT<>365 THEN STOP
750 DATA 31,28,31,30,31,30,31,31,30,31,30,31
760 RETURN
```

Calendar

```
 90 CLEAR500
100 GOSUB610:GOSUB900
105 GOSUB510
110 PRINT "MONTH: ";:GOSUB540:XX$="1/"+XX$:L=LEN(XX$):GOSUB220
112 XX$=DA$+" "+MO$+" "+YR$:GOSUB140:IF XX<>1 THEN110
115 GOSUB300:SZ=DW:NL=1:DL=ML(MO)+SZ-7
120 IF DL>0 THEN DL=DL-7:NL=NL+1:GOTO120
125 GOSUB510:GOSUB130:ND=ML(MO):GOSUB400
128 GOSUB520:GOTO105
130 PRINT:PRINT HD$:PRINT TL$:FOR LX=1 TO NL:FOR VV=1 TO VS:PRINT VL$:NEXT VV
132 PRINT BL$:NEXT LX
135 CC=TC:LL=LL(NL)+BH:GOSUB800:PRINT MO$(MO);YR;:RETURN
140 L=LEN(XX$):IF L<5 THEN210
150 GOSUB220:MO=INT(VAL(MO$)):DA=INT(VAL(DA$)):YR=INT(VAL(YR$))
160 IF LEN(YR$)=2 THEN YR=YR+1900
170 GOSUB480:ML(2)=28:IF XX=1 THEN ML(2)=29
180 IF NOT (1<=MO AND MO<=12 AND YR=>1) THEN210
190 IF NOT (1<=DA AND DA<=ML(MO)) THEN210
200 IF NOT (YR=1582 AND MO=10 AND 4<DA AND DA<15) THEN XX=1:RETURN
210 XX=0:RETURN
220 XS=0:FOR XL=2 TO L-1:ZZ$=MID$(XX$,XL,1)
230 IF ZZ$<>"/" AND ZZ$<>" " AND ZZ$<>"-" THEN260
240 XS=XS+1:IF XS=1 OR XS=2 THEN SP(XS)=XL:GOTO260
250 MO$="":RETURN
260 NEXT XL:IF XS<>2 THEN MO$="":GOTO290
270 MO$=LEFT$(XX$,SP(1)-1):DA$=MID$(XX$,SP(1)+1,SP(2)-SP(1)-1)
280 YR$=MID$(XX$,SP(2)+1)
290 RETURN
300 YZ=YR-400*INT(YR/400):DW=DZ
310 IF YZ<101 THEN330
320 YZ=YZ-100:DW=DW+5:GOTO310
330 IF YZ<5 THEN350
340 YZ=YZ-4:DW=DW+5:GOTO330
350 IF YZ=0 THEN DW=DW-2:GOTO356
352 DW=DW+YZ-1
356 GOSUB450:DW=DW+JD-1:IF YR>1582 THEN390
360 IF YR=1582 AND MO>10 THEN390
370 IF YR=1582 AND MO=10 AND DA=>15 THEN 390
375 IF LEN(G$)=N THEN RETURN
380 DW=DW+3
390 DW=DW-7*INT(DW/7):RETURN
395 GC(ZZ)=ASC(MID$(G$,ZZ,1)):NEXT ZZ
400 DY=1:FOR LX=1 TO NL:LL=LL(LX):ZC=1:IF LX=1 THEN ZC=SZ+1
405 FOR CX=ZC TO 7:CC=CC(CX):GOSUB800:PRINT RIGHT$(" "+STR$(DY),2);
410 DY=DY+1:IF DY>ND THEN RETURN
415 NEXT CX:NEXT LX:RETURN
450 JD=J(MO)+DA:IF MO<=2 THEN RETURN
460 GOSUB480:IF XX=1 THEN JD=JD+1
470 RETURN
480 XX=0:IF YR<>4*INT(YR/4) THEN RETURN
490 IF YR=100*INT(YR/100) AND YR<>400*INT(YR/400) THEN RETURN
500 XX=1:RETURN
```

```
510 CLS:RETURN
520 X$=INKEY$:IF X$="" THEN520
530 RETURN
540 XX$=""
550 GOSUB520:IF X$=CHR$(8) THEN 580
560 PRINT X$;:IF X$=CHR$(13) THEN600
570 XX$=XX$+X$:GOTO550
580 IF LEN(XX$)=0 THEN550
590 PRINT X$;:XX$=LEFT$(XX$,LEN(XX$)-1):GOTO550
600 RETURN
610 READ DZ:DATA 1
620 DIM DW$(6),J(12),MO$(12),ML(12),PM$(6)
630 FOR XX=0 TO 6:READ DW$(XX),PM$(XX):NEXT XX
640 DATA SUNDAY, "YOU´RE FAIR, AND WISE, AND GOOD, AND GAY"
650 DATA MONDAY,"MONDAY´S CHILD IS FAIR OF FACE"
660 DATA TUESDAY,"TUESDAY´S CHILD IS FULL OF GRACE"
670 DATA WEDNESDAY,"WEDNESDAY´S CHILD IS FULL OF WOE"
680 DATA THURSDAY,"THURSDAY´S CHILD HAS FAR TO GO"
690 DATA FRIDAY,"FRIDAY´S CHILD IS LOVING AND GIVING"
700 DATA SATURDAY,"SATURDAY´S CHILD MUST WORK FOR A LIVING"
710 FOR XX=1 TO 12:READ MO$(XX):NEXT XX
720 DATA JANUARY,FEBRUARY,MARCH,APRIL,MAY,JUNE,JULY,AUGUST,SEPTEMBER
730 DATA OCTOBER,NOVEMBER,DECEMBER
740 DT=0:FOR XX=1 TO 12:J(XX)=DT:READ ML(XX):DT=DT+ML(XX):NEXT XX
745 IF DT<>365 THEN STOP
750 DATA 31,28,31,30,31,30,31,31,30,31,30,31
760 RETURN
800 PRINT CHR$(28);:CC=CC-64*INT(CC/64):LL=LL-16*INT(LL/16)
805 IF CC>0 THEN FOR ZZ=1 TO CC:PRINT CHR$(25);:NEXT ZZ
810 IF LL>0 THEN FOR ZZ=1 TO LL:PRINT CHR$(26);:NEXT ZZ
815 RETURN
900 READ BW,BH:DATA4,2
905 READ LZ,CZ:DATA3,1
910 FOR XX=1 TO 6:LL(XX)=LZ+(XX-1)*BH:NEXT XX
915 FOR XX=1 TO 7:CC(XX)=CZ+(XX-1)*BW:NEXT XX
920 TL$="":BL$="":VL$=""
925 FOR XX=1TO7:TL$=TL$+"----":VL$=VL$+"!   ":BL$=BL$+"----":NEXT XX
930 VL$=VL$+"!":BL$=BL$+"-"
935 HD$=" SUN MON TUE WED THU FRI SAT"
940 VS=BH-1:TC=CZ+7:RETURN
```

Taxman

```
100 GOSUB530
110 GOSUB420:IF N=0 THEN END
120 GOSUB150:IF CH>0 THEN INPUT "YOUR SLICE";S
130 GOSUB260:IF PL>0 THEN120
140 GOSUB360:GOTO110
150 GOSUB520:CH=0:FOR YY=1 TO N STEP LS:TP=YY+LS-1:IF TP>N THEN TP=N
160 FOR XX=YY TO TP:IF HT(XX)<0 THEN NM$=L2$+SZ$:GOTO190
170 NM$=STR$(XX):IF XX<10 THEN NM$=L2$+NM$:GOTO190
180 IF XX<100 THEN NM$=L1$+NM$
190 RV=0:IF HT(XX)>0 THEN RV=1:CH=CH+1
200 GOSUB480:NEXT XX:PRINT:NEXT YY:GOSUB210:RETURN
210 PK=FNP(KP/PZ):PK$=LB$+STR$(PK)
220 PT=FNP(PY/PZ):PT$=LB$+STR$(PT)
230 PP=FNP(PL/PZ):PP$=LB$+STR$(PP)
240 PRINT:PRINT "TOTALS:":PRINT "YOU:";PK$;"%   TAXMAN:";PT$;"%   LEFT:";PP$;"%"
250 RETURN
260 IF CH=0 THEN300
270 S=INT(S):IF S<2 OR S>N THEN300
280 IF HT(S)<1 THEN300
290 GOSUB310:RETURN
300 PRINT "TAXMAN GETS THE REST.":PY=PY+PL:PL=0:RETURN
```

INSIDE BASIC GAMES

```
310 TX=0:KP=KP+S:PL=PL-S:HT(S)=-1:IF INT(N/S)<2 THEN330
320 FOR YY=2 TO INT(N/S):HT(YY*S)=HT(YY*S)-1:NEXT YY
330 FOR XX=INT(S/2) TO 1 STEP -1:IF HT(XX)<0 OR S<>XX*INT(S/XX) THEN350
340 HT(XX)=-1:TX=TX+XX:FOR YY=2 TO INT(N/XX):HT(YY*XX)=HT(YY*XX)-1:NEXT YY
350 NEXT XX:PY=PY+TX:PL=PL-TX:RETURN
360 PK=FNP(KP/PZ):PT=100-PK:NR=0:IF PK>BK THEN BK=PK:BZ=N:NR=1
370 PK$=LB$+STR$(PK):PT$=LB$+STR$(PT)
380 PRINT:PRINT "YOU KEPT";PK$;"%      TAXMAN GOT";PT$;"%"
390 IF NR=1 THEN PRINT "THAT´S A NEW RECORD!":GOTO410
400 BK$=LB$+STR$(BK):BZ$=LB$+STR$(BZ):PRINT "THE RECORD IS";BK$"% OF";BZ$
410 GOSUB500:RETURN
420 GOSUB520
430 INPUT "PIE SIZE";N:IF N=0 THEN RETURN
440 N=INT(N):IF N<4 OR N>MX THEN430
450 PL=N*(N+1)/2:PZ=PL:FOR XX=1 TO N:HT(XX)=0:NEXT XX
460 FOR XX=1 TO INT(N/2):FOR YY=2 TO INT(N/XX):HT(YY*XX)=HT(YY*XX)+1:NEXT YY:NEXT XX
470 KP=0:PY=0:RETURN
480 REM NO REVERSE VIDEO AVAILABLE
490 PRINT NM$;:RETURN
500 X$=INKEY$:IF X$="" THEN500
510 RETURN
520 CLS:RETURN
530 READ LS:DATA 15:READ DL:DATA 16
540 MX=LS*(DL-5):DIM HT(MX)
550 L=LEN(STR$(1)):SZ$="":FOR LL=1 TO L:SZ$=SZ$+" ":NEXT LL
560 LB$=" ":IF L>1 THEN LB$=""
570 L1$=LB$+" ":L2$=L1$+" "
580 DEF FNP(X) =INT(1000*X+.5)/10
590 RETURN
```

Matchup

```
 90 CLEAR1000
100 GOSUB2260
110 GOSUB1420
120 GOSUB2200:PRINT"()";:GOSUB2180:PRINTX$
130 IFX$="E"THENGOSUB210:GOTO120
140 IFX$="N"THENGOSUB260:GOTO120
150 IFX$="P"THENGOSUB970:GOTO120
160 IFX$="L"THENGOSUB2230:GOTO120
170 IFX$="S"THENGOSUB2240:GOTO120
180 IFX$<>"Q"THEN120
190 PRINT "REALLY? ";:GOSUB2180
200 IFX$="Y"THEN110
205 GOTO120
210 GOSUB1310:IFXX=0THENRETURN
220 GOSUB1290:GOSUB340
230 GOSUB460:GOSUB510:IFNOT(OK=1OROK=-1)THEN230
240 IFOK=1THENGOSUB380
250 RETURN
260 GOSUB1260:GOSUB1290:GOSUB300
270 GOSUB460:GOSUB510:IFNOT(OK=1OROK=-1)THEN270
280 IFOK=1THENGOSUB390
290 RETURN
300 WN$=""
310 FORQX=1TONQ(GP):WA(QX)=0:NEXTQX
320 FORQX=1TONQ(GC):WW(QX)=0:NEXTQX
330 RETURN
340 WN$=NM$(PP,GP)
350 FORQX=1TONQ(GP):WA(QX)=A(QX,PP,GP):NEXTQX
360 FORQX=1TONQ(GC):WW(QX)=W(QX,PP,GP):NEXTQX
370 RETURN
380 UP=PP:UG=GP:UC=GC:GOSUB420:RETURN
390 UG=GP:UC=GC
```

APPENDIX B

```
400 IFNP(GP)<MPTHENNP(GP)=NP(GP)+1:UP=NP(GP):GOSUB420:RETURN
410 PRINT"SORRY,NO ROOM.":RETURN
420 NM$(UP,UG)=WN$
430 FORQX=1TONQ(UG):A(QX,UP,UG)=WA(QX):NEXTQX
440 FORQX=1TONQ(UC):W(QX,UP,UG)=WW(QX):NEXTQX
450 RETURN
460 GOSUB550
470 FORQX=1TONQ(GP):GOSUB680:IFX$<>"Q"THENNEXTQX
480 IFX$="Q"THENRETURN
485 GOSUB2200:PRINT"NOW GIVE YOUR PREFERENCES FOR ANSWERS BY ";GN$(GC)
490 GOSUB2180:FORQX=1TONQ(GC):GOSUB880:IFX$<>"Q"THENNEXTQX
500 RETURN
510 GOSUB2200:PRINT"MORE? ";:GOSUB2180:OK=1
520 IFX$="Q"THENOK=-1
530 IFX$="Y"THENOK=0
540 RETURN
550 GOSUB2200:IFWN$<>""THEN590
560 PRINT"NAME* ";
570 GOSUB610:WN$=SS$:IFWN$:IFWN$=""THEN570
580 RETURN
590 PRINTWN$:PRINT"NAME: ";:GOSUB610:IFSS$<>""THENWN$=SS$
600 RETURN
610 SS$=""
620 GOSUB2180:IFX$=CHR$(13)THEN660
630 IFX$<>CHR$(20)THENPRINTX$;:SS$=SS$+X$:GOTO620
640 IFLEN(SS$)<>0THENPRINTX$;:SS$=LEFT$(SS$,LEN(SS$)-1)
650 GOTO620
660 IFSS$="Q"THENPRINTCHR$(8);:RETURN
665 IF SS$<>""THENPRINT
670 RETURN
680 FORQX=1TONQ(GP):GOSUB2200
690 GQ=GP:GOSUB1240:PRINT"CHOICE:";:IFWA(QX)<>0THENPRINTWA(QX);":      ";
700 GOSUB2180
710 IFX$="Q"ORX$=CHR$(13)THEN750
720 IFX$="D"THENWA(QX)=0:GOTO750
730 NN=VAL(X$):IF1<=NNANDNN<=NC(QX,GP)THENWA(QX)=NN:GOTO750
740 GOTO700
750 IFX$="Q"THENRETURN
760 NEXTQX:RETURN
770 FORFX=1TONF:PRINT:PRINT"VALUE";FV(FX);"    CHOICE";
780 IFFT(FX)<>0THENPRINTFT(FX);
790 PRINT":   ";
800 GOSUB2180:IFX$=CHR$(13)ORX$="Q"ORX$="N"THEN850
810 IFX$="D"THENPRINTX$;:FT(FX)=0:GOTO850
820 NN=VAL(X$)
830 IF1<=NNANDNN<=MCTHENPRINTX$;:FT(FX)=NN:GOTO850
840 GOTO800
850 IFX$="Q"ORX$="N"THEN870
860 NEXTFX
870 RETURN
880 FORQX=1TONQ(GC):GOSUB2200
890 GQ=GC:GOSUB1240:WC=WW(QX):GOSUB920:GOSUB970:GOSUB950:WW(QX)=WC
900 IFX$<>"Q"THENNEXTQX
910 RETURN
920 FORFX=1TONF:FT(FX)=WC-(MC+1)*INT(WC/(MC+1))
930 WC=(WC-FT(FX))/(MC+1):NEXTFX
940 RETURN
950 WC=0:FM=1:FORFX=1TONF:WC=WC+FT(FX)*FM:FM=FM*(MC+1):NEXT FX
960 RETURN
970 GOSUB1310:GOSUB1290:GOSUB2200:GOSUB1110
980 PW=PP:GW=GP:GA=GC
990 FORPX=1TONP(GC):PA=PX:GOSUB1050:GOSUB1140:NEXT PX
1000 PRINT GN$(GC);" BEST FOR ";NM$(PP,GP):GS=GC:GOSUB1120
1010 GOSUB1110:PA=PP:GA=GP:GW=GC
1020 FORPX=1TONP(GC):PW=PX:GOSUB1050:GOSUB1140:NEXT PX
1030 PRINT:PRINTGN$(GC);" THAT ";NM$(PP,GP);" IS BEST FOR":GS=GC:GOSUB1120
```

```
1040 GOSUB2180:RETURN
1050 SC=0:FORQX=1TONQ(GA)
1060 WC=W(QX,PW,GW):AV=A(QX,PA,GA):GOSUB1080
1070 SC=SC+FT*WT(QX,GA):NEXTQX:RETURN
1080 FT=0:GOSUB920
1090 FOR FX=1TONF:IFFT(FX)=AVTHENFT=FV(FX):RETURN
1100 NEXTFX:RETURN
1110 SX=0:RETURN
1120 IFSX<=0THENPRINT"NONE":RETURN
1130 FORSZ=1TOSX:PRINTTS(SZ),NM$(TI(SZ),GS):NEXTSZ:RETURN
1140 IFSX=0THENGOSUB1230:RETURN
1150 FORSZ=1TOSX:IFSC<=TS(SZ)THEN1210
1160 IFSX<MTTHENSX=SX+1
1170 SE=SX-1:IFSE<SZTHEN1200
1180 FORSY=SETOSZSTEP-1
1190 TS(SY+1)=TS(SY):TI(SY+1)=TI(SY):NEXTSY
1200 TS(SZ)=SC:TI(SZ)=PX:RETURN
1210 NEXTSZ:IFSX<MTTHENGOSUB1230
1220 RETURN
1230 SX=SX+1:TS(SX)=SC:TI(SX)=PX:RETURN
1240 PRINTQ$(QX,GQ)
1250 FORCX=1TONC(QX,GQ):PRINTCX;" ";C$(CS,QX,GQ):NEXTCX:RETURN
1260 PRINT:FORGX=1TO2:PRINTGX;" ";GN$(GX):NEXTGX:PRINT"GROUP:   ";
1270 GOSUB2180:GP=VAL(X$):IFNOT(GP=1ORGP=2)THEN1270
1280 PRINTX$;:RETURN
1290 IFGP=1THENGC=2:RETURN
1300 GC=1:RETURN
1310 GOSUB1260:GOSUB2200:IFNP(GP)>0THEN1330
1320 PRINT"NO ";GN$(GP);" ON FILE.":XX=0:RETURN
1330 XX=1:FORPX=1TONP(GP)STEP10:GOSUB2200
1340 PT=PX+9:IFPT>NP(GP)THENPT=NP(GP)
1350 FORPY=PXTOPT:PRINTPY;" ";NM$(PY,GP):NEXTPY
1360 PRINT"NUMBER: ";:GOSUB610:PP=VAL(SS$)
1370 IF1<=PPANDPP<=NP(GP)THENRETURN
1380 IFPP<>0THEN1360
1390 IFX$="Q"THENXX=0:RETURN
1400 NEXTPX
1410 PRINT"THAT'S ALL THE ";GN$(GP):XX=0:RETURN
1420 GOSUB2200:PRINT"::";:GOSUB2180:PRINTX$
1430 IFX$="I"THENGOSUB1510:GOTO1420
1440 IFX$="N"THENGOSUB1700:GOTO1420
1450 IFX$="1"THENGOSUB1740:GOTO1420
1460 IFX$="2"THENGOSUB1750:GOTO1420
1470 IFX$="L"THENGOSUB2210:GOTO1420
1480 IFX$="S"THENGOSUB2220:GOTO1420
1490 IFX$<>"Q"THEN1420
1500 RETURN
1510 FORGX=1TO2
1520 PRINT:PRINT"GROUP";GX;":   ";
1530 GOSUB610:IFSS$=""THEN1530
1540 GN$(GX)=SS$:N1(GX)=0
1550 IFNQ(GX)=>MQTHEN1590
1560 GOSUB2200:QX=NQ(GX)+1:GOSUB1600
1570 IFX$="Q"THEN1590
1580 NQ(GX)=QX:GOTO1550
1590 NEXTGX:RETURN
1600 PRINT:PRINT"Q";QX;"FOR ";GN$(GX)
1610 GOSUB610:IFSS$="Q"THENX$="Q":RETURN
1620 IFSS$=""THEN1610
1630 Q$(QX,GX)=SS$:NC(QX,GX)=0
1640 IFNC(QX,GX)=>MCTHEN1690
1650 CX=NC(QX,GX)+1:PRINTCX;" ";
1660 GOSUB610:IFSS$="Q"THEN1690
1670 IFSS$=""THEN1660
1680 C$(CX,QX,GX)=SS$:NC(QX,GX)=CX:GOTO1640
1690 PRINT:PRINT"WEIGHT:  ";:GOSUB610:WT(QX,GX)=VAL(SS$):RETURN
```

APPENDIX B

```
1700 FORGX=1TO2:PRINTGX;" ";GN$(GX)
1710 GOSUB610:IFSS$<>""THENGN$(GX)=SS$
1720 NEXTGX:RETURN
1740 GX=1:GOSUB1760:RETURN
1750 GX=2:GOSUB1760:RETURN
1760 GOSUB2200:PRINT"*";:GOSUB2180:PRINTX$:IFX$="q"THENRETURN
1770 IFX$="E"THENGOSUB1990
1780 IFX$="D"THENGOSUB1850
1790 IFX$="A"THENGOSUB1820
1800 IFX$="B"THENGOSUB1910
1810 GOTO1760
1820 IFNQ(GX)=>MQTHENPRINT"NO ROOM":RETURN
1830 QX=NQ(GX)+1:GOSUB1600:IFX$<>"Q"THENNQ(GX)=QX
1840 RETURN
1850 GOSUB2160:IFQQ<>0THENGOSUB1870
1860 RETURN
1870 NQ(GX)=NQ(GX)-1:IFQQ>NQ(GX)THENRETURN
1880 FORQX=QQTONQ(GX):Q$(QX,GX)=Q$(QX+1,GX):IFNC(QX+1,GX)<=0THEN1900
1890 FORCX=1TONC(QX+1,GX):C$(CX,QX,GX)=C$(CX,QX+1,GX):NEXTCX
1900 NC(QX,GX)=NC(QX+1,GX):WT(QX,GX)=WT(QX+1,GX):NEXTQX:RETURN
1910 IFNQ(GX)=>MQTHENPRINT"NO ROOM":RETURN
1920 PRINT"AHEAD OF ";:GOSUB2160:IFQQ=0THENRETURN
1930 FORQX=NQ(GX)TOQQSTEP-1:Q$(QX+1,GX)=Q$(QX,GX)
1940 IFNC(QX,GX)<=0THEN1960
1950 FORCX=1TONC(QX,GX):C$(CX,QX+1,GX)=C$(CX,QX,GX):NEXTCX
1960 NC(QX+1,GX)=NC(QX,GX):WT(QX+1,GX)=WT(QX,GX):NEXTQX
1970 NQ(GX)=NQ(GX)+1:QX=QQ:GOSUB1600:IFX$="Q"THENGOSUB1870
1980 RETURN
1990 GOSUB2160:IFQQ=0THENRETURN
2000 PRINTQ$(QQ,GX):GOSUB610:IFSS$="Q"THENRETURN
2010 IFSS$<>""THENQ$(QQ,GX)=SS$
2020 GOSUB2050:PRINT:PRINT"WEIGHT:";WT(QQ,GX);":    ";:GOSUB610
2030 IFVAL(SS$)<>0THENWT(QQ,GX)=VAL(SS$)
2040 RETURN
2050 CX=1
2060 IFCX>NC(QQ,GX)THEN2130
2070 PRINTCX;" ";C$(CX,QQ,GX):GOSUB610:IFSS$="Q"THENRETURN
2080 IFSS$<>"D"THEN2110
2090 NC(QQ,GX)=NC(QQ,GX)-1:IFCX>NC(QQ,GX)THEN2060
2100 FORCY=CXTONC(QQ,GX):C$(CY,QQ,GX)=C$(CY+1,QQ,GX):NEXTCY:GOTO2060
2110 IFSS$<>""THENC$(CX,QQ,GX)=SS$
2120 CX=CX+1:GOTO2060
2130 IFNC(QQ,GX)=>MCTHENRETURN
2140 CX=NC(QQ,GX)+1:PRINX;" ";:GOSUB610:IFSS$=""ORSS$="Q"THENRETURN
2150 C$(CX,QQ,GX)=SS$:NC(QQ,GX)=CX:TO2130
2160 PRINT"QUESTION NUMBER: ";:GOSUB610:QQ=VAL(SS$):IFQQ<0ORQQ>NQ(GX)THEQ=0
2170 RETURN
2180 X$=INKEY$:IFX$=""THEN2180
2190 RETURN
2200 CLS:RETURN
2210 GOSUB2250:RETURN
2220 GOSUB2250:RETURN
2230 GOSUB2250:RETURN
2240 GOSUB2250:RETURN
2250 PRINT GETOT IMPLEMENTED":RETURN
2260 GM=2:READMQ,MC,MP:DATA10,5,30
2270 DIMNQ(GM),NM$(MP,GM),Q$(MQ,GM)
2280 DIMWT(MQ,GM),NC(MQ,GM),C$(MC,MQ,GM)
2290 DIMA(MQ,MP,GM),W(MQ,MP,GM)
2300 READMT:DATA3
2310 DIMTS(MT),TI(MT)
2320 READNF:DATA4
2330 DIMFV(NF),FT(NF)
2340 FORFX=1TONF:READFV(FX):NEXTFX
2DATA 2,1,-1,-2
2360 RETURN
```

Craps

```
90 CLEAR 200
100 GOSUB 1950
130 GOSUB 960:GOSUB 2660:GOSUB 160
140 ON IC GOSUB 560,230,1400,1690,1820,1560
150 GOTO 130
160 REM ANALYSE INPUT CODES
170 IF X$=" " THEN IC=2:RETURN
180 IF X$="N" THEN IC=3:RETURN
190 IF X$="Q" THEN IC=4:RETURN
200 IF X$="S" THEN IC=5:RETURN
210 IF X$="P" THEN IC=6:RETURN
220 IC = 1:RETURN
230 REM PLAY ONE GAME
240 GOSUB 370
250 IF D=2 OR D=3 OR D=12 THEN GOSUB 490:RETURN
260 IF D=7 OR D=11 THEN GOSUB 420:RETURN
270 PT=D:TH=1
280 RS$="POINT IS "+STR$(PT):GOSUB 1020
290 GOSUB 980
300 GOSUB 2660:GOSUB 370
310 TH=TH+1:GOSUB 980
320 IF D=PT THEN GOSUB 420:GOTO 360
330 IF D=7 THEN GOSUB 490:GOTO 360
340 RS$=STR$(D)+" -- THROW AGAIN":GOSUB 1020
350 GOTO 300
360 RETURN
370 REM THROW ROUTINE
380 D1=INT(6*RND(X)+1)
390 D2=INT(6*RND(X)+1)
400 GOSUB 2500
410 D=D1+D2:RETURN
420 REM WIN ROUTINE
430 WN=WN+BT
440 RD=2*BT:PS=PS+1
450 IF PS>BR THEN BR=PS
460 ZP=CP:GOSUB 1890
470 RS$="YOU WIN!":GOSUB 1020
480 GOSUB 1380:RETURN
490 REM LOSE ROUTINE
500 WN=WN-BT
510 RD=OB:PS=0
520 ZP=CP:GOSUB 1890
530 RS$="SORRY, YOU LOSE":GOSUB 1020
540 GOSUB 1380:GOSUB 1560
550 RETURN
560 REM ASK FOR BET
570 GOSUB 700
580 ON B GOSUB 860,880,900,920,770
590 IF 0<BT AND BT<=MX THEN 690
600 IF BT<=MX THEN 650
610 LL=BL:CC=BC:GOSUB 1270
620 PRINT "HOUSE LIMIT IS";MX;"   ";
630 BT=MX:IF OB>MX THEN OB=MX
640 GOSUB 1380:GOTO 690
650 LL=BL:CC=BC:GOSUB 1270
660 PRINT "SORRY, PLEASE ENTER A NEW BET CODE";
670 GOSUB 2660
680 GOTO 570
690 RETURN
700 REM BETCODE
710 IF X$="D" THEN B=1:RETURN
720 IF X$="R" THEN B=2:RETURN
730 IF X$="W" OR X$="L" THEN B=3:RETURN
```

```
740 IF X$="O" THEN B=4:RETURN
750 IF X$="B" THEN B=5:RETURN
760 B=4:RETURN
770 REM NUMBER
780 AA=0:GOSUB 940
790 GOSUB 2660
800 NN=ASC(X$)-ASC("0")
810 IF NN<0 OR NN>9 THEN 840
820 PRINT X$;:AA=10*AA+NN
830 GOTO 790
840 BT=AA:OB=BT
850 RETURN
860 REM DOUBLE
870 BT=2*BT:RETURN
880 REM RIDE
890 BT=RD:RETURN
900 REM WON
910 BT=ABS(WN):RETURN
920 REM OLD
930 BT=OB:RETURN
940 REM BETPROMPT
950 GOSUB 1180:PRINT "BET: ";:RETURN
960 REM START ROUTINE
970 GOSUB 1250:GOSUB 1050:RETURN
980 REM PRINT GAME LINE
990 LL=GL:CC=GC:GOSUB 1270
1000 PRINT "POINT:";PT;" THROWS:";TH;
1010 RETURN
1020 REM PRINT RESULT LINE
1030 GOSUB 1150
1040 PRINT RS$;:GOSUB 1380:GOSUB 2580:RETURN
1050 REM PRINT STATISTICS LINE
1060 GOSUB 1210
1070 IF WN<0 THEN PRINT "LOST",:GOTO 1090
1080 PRINT "WON",
1090 PRINT "BET","THIS RUN","BEST RUN"
1100 PRINT ABS(WN),BT,PS,BR;"   ";
1110 IF CP=0 THEN 1140
1120 CC=NC:LL=NL:GOSUB 1270
1130 PRINT NM$(CP);
1140 RETURN
1150 REM CLEAR RESULT LINE
1160 CC=RC:LL=RL:GOSUB 1270:PRINT RC$;
1170 CC=RC: LL=RL:GOSUB 1270:RETURN
1180 REM CLEAR FOR BET
1190 GOSUB 960
1200 CC=BC:LL=BL:GOSUB 1270:RETURN
1210 REM CLEAR STATISTICS LINE
1220 CC=0:LL=SL:GOSUB 1270
1230 PRINT BL$:PRINT BL$;
1240 CC=0:LL=SL:GOSUB 1270:RETURN
1250 REM CLEAR THE SCREEN
1260 CLS:RETURN
1270 REM CURSOR ROUTINE
1280 PRINT CHR$(28);
1290 IF CC<0 THEN CC=CC+64:GOTO 1290
1300 IF CC>63 THEN CC=CC-64:GOTO 1300
1310 IF LL<0 THEN LL=LL+16:GOTO 1310
1320 IF LL>15 THEN LL=LL-16:GOTO 1320
1330 IF CC=0 THEN 1350
1340 FOR ZZ=1 TO CC:PRINT CHR$(25);:NEXT ZZ
1350 IF LL=0 THEN 1370
1360 FOR ZZ=1 TO LL:PRINT CHR$(26);:NEXT ZZ
1370 RETURN
1380 REM DELAY
```

INSIDE BASIC GAMES 341

```
1390 FOR ZZ=1 TO DC:NEXT ZZ:RETURN
1400 REM ADD A NEW PLAYER
1410 IF NP<MP THEN 1450
1420 CC=BC:LL=BL:GOSUB 1270
1440 GOSUB 1380:RETURN
1450 NP=NP+1
1460 GOSUB 1250
1470 INPUT "MAY I HAVE YOUR NAME PLEASE";NM$(NP)
1480 PRINT:PRINT "THANK YOU.   ";
1490 IF NP<=1 THEN 1540
1500 PRINT "YOU WILL FOLLOW ";NM$(NP-1)
1510 ZP=CP:GOSUB 1890
1520 GOSUB 1860:ZP=NP:GOSUB 1890
1530 ZP=CP:GOSUB 1920
1540 CP=1:ZP=1:GOSUB 1890
1550 GOSUB 1380:GOSUB 1380:RETURN
1560 REM PASS DICE TO THE NEXT PLAYER
1570 IF NP=0 THEN RETURN
1580 ZP=CP:GOSUB 1890
1590 CP=CP+1:IF CP>NP THEN CP=1
1600 ZP=CP:GOSUB 1920:PS=0
1610 RETURN
1620 REM DISPLAY THE PLAYERS
1630 GOSUB 1250
1640 PRINT:PRINT "NUMBER","WON/LOST","NAME":PRINT
1650 FOR ZZ=1 TO NP
1660 PRINT ZZ,ST(1,ZZ),NM$(ZZ):NEXT ZZ
1670 PRINT
1680 RETURN
1690 REM PLAYER QUITS
1700 GOSUB 1620
1710 INPUT "WHAT IS YOUR NUMBER, PLEASE";QP
1720 IF QP=0 THEN RETURN
1730 IF QP<=0 OR QP>NP THEN PRINT "PLEASE REFER TO THE LIST":GOTO 1710
1740 IF QP=CP THEN GOSUB 1560
1750 IF QP=NP THEN 1790
1760 FOR ZZ=QP TO NP-1:FOR YY=1 TO 4
1770 ST(YY,ZZ)=ST(YY,ZZ+1):NEXT YY
1780 NM$(ZZ)=NM$(ZZ+1):NEXT ZZ
1790 NP=NP-1:IF CP=>QP THEN CP=CP-1
1800 IF NP=0 THEN GOSUB 1860
1810 RETURN
1820 REM SIDE BETS
1830 CC=BC:LL=BL:GOSUB 1270
1840 PRINT "SIDEBETS NOT YET IMPLEMENTED";
1850 GOSUB 1380:RETURN
1860 REM ZAP THE STATS
1870 WN=0:BT=0:BR=0:OB=0
1880 RETURN
1890 REM SAVE STATS
1900 ST(1,ZP)=WN:ST(2,ZP)=BT:ST(3,ZP)=BR:ST(4,ZP)=OB
1910 RETURN
1920 REM FETCH STATS
1930 WN=ST(1,ZP):BT=ST(2,ZP):BR=ST(3,ZP):OB=ST(4,ZP)
1940 RETURN
1950 REM INITIALIZE
1960 READ SL:DATA 13
1970 READ GL,GC:DATA 11,21
1980 READ RL,RC:DATA 10,22
1990 READ L1,C1:DATA 1,10
2000 READ L2,C2:DATA 3,25
2010 READ BL,BC:DATA 0,3
2020 READ NL,NC:DATA 15,26
2030 READ DC: DATA 100
2040 READ MX: DATA 500
```

APPENDIX B

```
2050 FOR ZZ=0 TO 6: FOR XX=1 TO 5
2060 READ D$(ZZ,XX)
2070 NEXT XX: NEXT ZZ
2080 DATA "   "," ","   "," ","   "," ","   "," ","   "
2130 DATA " --- ",":   :",":   :",":   :","   --- "
2180 DATA " --- ",":*  :",":   :",":  *:","   --- "
2230 DATA " --- ",":*  :",":  * :",": *:","   --- "
2280 DATA " --- ",":* *:",":   :",":* *:","   --- "
2330 DATA " --- ",":* *:",":  * :",":* *:","   --- "
2380 DATA " --- ",":* *:",":* *:",":* *:","   --- "
2430 READ BF$: DATA " "
2440 READ RC$: DATA "                          "
2450 READ BL$: DATA "  "
2460 GOSUB 1860: PS=0
2470 NP=0:MP=9:CP=0
2480 DIM NM$(9),ST(4,9)
2490 RETURN
2500 REM DRAW DICE
2510 FOR LX=L1 TO L1+4
2520 LL=LX:CC=C1:GOSUB 1270
2530 PRINT D$(D1,LX-L1+1);:NEXT LX
2540 FOR LX=L2 TO L2+4
2550 LL=LX:CC=C2:GOSUB 1270
2560 PRINT D$(D2,LX-L2+1);:NEXT LX
2570 RETURN
2580 REM CLEAR OFF THE DICE
2590 FOR LX=L1 TO L1+4
2600 LL=LX:CC=C1:GOSUB 1270
2610 PRINT D$(0,LX-L1+1);:NEXT LX
2620 FOR LX=L2 TO L2+4
2630 LL=LX:CC=C2:GOSUB 1270
2640 PRINT D$(0,LX-L2+1);:NEXT LX
2650 RETURN
2660 X$=INKEY$:IF X$="" THEN 2660
2670 RETURN
```

Alien Life

```
90 CLEAR100
100 GOSUB1190
110 GOSUB160
120 GOSUB700:IFC1=0THEN140
130 ONC1GOSUB720,730,740,750:GOTO120
140 GOSUB560:GOSUB990:IFX$<>"Q"THEN110
150 END
160 GOSUB980:GOSUB200:IFPX<>-1THEN180
170 GOSUB540:GOSUB550:GOTO190
180 HL=P1-1:HC=P2-1:GOSUB600
190 GOSUB610:GOSUB980:GOSUB620:GOSUB660:CC=0:LL=0:GOSUB840:RETURN
200 PRINT"LINES,COLUMNS: ";:GOSUB1010:IFXX$<>""THEN220
210 PX=-1:RETURN
220 GOSUB310:IFPX<>0THEN240
230 XX=-1:GOTO200
240 GOSUB420:IFXX<>1THEN270
250 GOSUB290:IFOK<>1THEN270
260 XX=0
265 STOP
270 IFXX<>0THEN200
280 P1=PX:P2=PY:RETURN
290 PRINT"NON-PRIMES OK? ";:GOSUB990:OK=1:IFX$="N"THENOK=0
300 RETURN
310 LL=LEN(XX$):IFLL=>3THEN330
320 PX=0:RETURN
```

INSIDE BASIC GAMES 343

```
330 NP=0:FORSX=1TOLL-1:X$=MID$(XX$,SX,1):GOSUB400:IFSP<>0THEN350
340 NP=1:GOTO390
350 IFNP<>0THEN380
360 IFX$=" "THEN390
370 PX=0:RETURN
380 PX=VAL(LEFT$(XX$,SX-1)):PY=VAL(MID$(XX$,SX+1)):RETURN
390 NEXTSX:PX=0:RETURN
400 SP=0:IFX$=" "ORX$=","THENSP=1
410 RETURN
420 IFPX<=MLANDPY<=MCTHEN440
430 XX=-1:RETURN
440 PP=PX:GOSUB480:IFPR<>0THEN460
450 XX=1:RETURN
460 PP=PY:GOSUB480:XX=0:IFPR=0THENXX=1
470 RETURN
480 IFPP=2ORPP=3ORPP=5THENPR=1:RETURN
490 IFPP<7ORPP/2=INT(PP/2)THENPR=0:RETURN
500 FORI=3TOINT(SQR(PP))STEP2
510 IFPP/I<>INT(PP/I)THEN530
520 PR=0:RETURN
530 NEXTI:PR=1:RETURN
540 RETURN
550 RETURN
560 GOSUB980:FORLL=0TOHL:FORCC=0TOHC:IFD(LL,CC)<>1THEN580
570 PRINT"1";:GOTO590
580 PRINT"0";
590 NEXTCC:NEXTLL:RETURN
600 FORLX=0TOHL:FORCX=0TOHC:D(LX,CX)=0:NEXTCX:NEXTLX:RETURN
610 CZ=LZ-1:C=CZ:GOSUB940:FORCC=0TOHC:PRINTTH$;:NEXTCC:PRINTCHR$(26);
620 L=LZ-1:C=CZ:GOSUB940:FORCC=0TOHC:PRINTTH$;:NEXTCC:PRINTCHR$(26);
630 FORLL=0TOHL:PRINTRV$;CHR$(24);CHR$(26);:NEXTLL
640 L=LZ:C=CZ-1:GOSUB940:FORLL=0TOHL:PRINTLV$;CHR$(24);CHR$(26);:NEXTLL:PRINTCHR$(25);
650 FORCC=0TOHC:PRINTBH$;:NEXTCC:RETURN
660 FORLL=0TOHL:FORCC=0TOHC:L=LL+LZ:C=CZ:GOSUB940
670 FORCC=0TOHC:IFD(LL,CC)<>0THENPRINTDT$;:GOTO690
680 PRINT " ";
690 NEXTCC:NEXTLL:RETURN
700 GOSUB1080:CX=ASC(X$)-CB:IFCX>MXORCX<0THENCX=DX
710 C1=INT(C(CX)/100):C2=C(CX)-100*C1:RETURN
720 GOSUB780:RETURN
730 GOSUB800:GOSUB780:RETURN
740 GOSUB810:GOSUB780:RETURN
750 GOSUB760:RETURN
760 IFC2=1THENGOSUB1450
770 RETURN
780 IFC2<1ORC2>4THENRETURN
790 ONC2GOSUB860,880,900,920:RETURN
800 GOSUB1170:D(LL,CC)=1:RETURN
810 GOSUB1180:D(LL,CC)=0:RETURN
820 IFD(LL,CC)=1THENGOSUB1170:RETURN
830 GOSUB1180:RETURN
840 CS$=ND$:IFD(LL,CC)=1THENCS$=OD$
850 RETURN
860 IFCC>0THENCC=CC-1:RETURN
870 CC=HC:RETURN
880 IFCC<HCTHENCC=CC+1:RETURN
890 CC=0:RETURN
900 IFLL>0THENLL=LL-1:RETURN
910 LL=HL:RETURN
920 IFLL<HLTHENLL=LL+1:RETURN
930 LL=0:RETURN
940 PRINTCHR$(28);:C=C-SC*INT(C/SC):L=L-SL*INT(L/SL)
950 IFC>0THENFORZZ=1TOC:PRINTCHR$(25);:NEXTZZ
960 IFL>0THENFORZZ=1TOL:PRINTCHR$(26);:NEXTZZ
970 RETURN
```

```
 980 CLS:RETURN
 990 X$=INKEY$:IFX$=""THEN990
1000 RETURN
1010 XX$=""
1020 GOSUB990:IFX$=CHR$(20)THEN1050
1030 PRINTX$;:IFX$=CHR$(13)THEN1070
1040 XX$=XX$+X$:GOTO1020
1050 IFLEN(XX$)=0THEN1020
1060 PRINTX$;:XX$=LEFT$(XX$,LEN(XX$)-1):GOTO1020
1070 RETURN
1080 GOSUB 840:CF=1
1090 IFCF=1THENGOSUB1150:GOTO1110
1100 GOSUB1160
1110 CF=-CF
1120 FORQQ=1TOCJ*FQ:X$=INKEY$:IFX$<>""THENGOSUB820:RETURN
1130 NEXTQQ
1140 GOTO1090
1150 C=CZ+CC:L=LZ+LL:GOSUB940:PRINTCS$;:RETURN
1160 C=CZ+CC:L=LZ+LL:GOSUB940:PRINT" ";:RETURN
1170 C=CZ+CC:L=LZ+LL:GOSUB940:PRINTDT$;:RETURN
1180 C=CZ+CC:L=LZ+LL:GOSUB940:PRINT" ";:RETURN
1190 READSL,SC:DATA16,64
1210 READBH$,TH$,LV$,RV$:DATA"-","-","!","!"
1220 READOD$,ND$:DATA"#","*"
1230 READFQ:DATA18:READCJ:DATA.25
1240 READDT$:DATA"0"
1250 READMC,ML,RL:DATA37,37,37
1260 IFMC>SC-2THENMC=SC-2
1270 IFMC>RLTHENMC=RL
1280 IFML>SL-2THENML=SL-2
1290 IFML>RLTHENML=RL
1300 DIMD(ML,MC)
1310 READCB,CM:DATA0,127
1320 MX=CM-CB:IFMX<1THENSTOP
1330 DIMC(MX)
1340 FORCY=0TOMX:C(CY)=400:NEXTCY
1350 READCH:IFCH=-1THEN1380
1360 CY=CH-CB:IFCY<0ORCY>MXTHENSTOP
1370 READC1,C2:C(CY)=100*C1+C2:GOTO1350
1380 DATA52,1,1,54,1,2,56,1,3,50,1,4
1390 DATA36,2,1,38,2,2,40,2,3,34,2,4
1400 DATA48,3,1,53,3,2,55,3,3,49,3,4
1410 DATA76,4,1
1420 DATA69,0,0,-1
1430 READCH:DATA52
1440 DX=CH-CB:RETURN
1450 GOSUB1460:GOSUB1740:GOSUB660:CC=0:LL=0:GOSUB840:RETURN
1460 D=D(0,0)/2:IFD=INT(D)THEN1480
1470 D(0,1)=D(0,1)+2:D(1,0)=D(1,0)+2:D(1,1)=D(1,1)+2
1480 D=D(0,HC)/2:IFD=INT(D)THEN1500
1490 D(0,HC-1)=D(0,HC-1)+2:D(1,HC)=D(1,HC)+2:D(1,HC-1)=D(1,HC-1)+2
1500 D=D(HL,0)/2:IFD=INT(D)THEN1520
1510 D(HL,1)=D(HL,1)+2:D(HL-1,0)=D(HL-1,0)+2:D(HL-1,1)=D(HL-1,1)+2
1520 D=D(HL,HC)/2:IFD=INT(D)THEN1550
1530 D(HL,HC-1)=D(HL,HC-1)+2:D(HL-1,HC)=D(HL-1,HC)+2
1540 D(HL-1,HC-1)=D(HL-1,HC-1)+2
1545 IFHC<=1THEN1615
1550 FORCC=1TOHC-1:D=D(1,CC)/2:IFD=INT(D)THEN1580
1560 D(0,CC-1)=D(0,CC-1)+2:D(0,CC+1)=D(0,CC+1)+2:D(1,CC-1)=D(1,CC-1)+2
1570 D(1,CC)=D(1,CC)+2:D(1,CC+1)=D(1,CC+1)+2
1580 D=D(HL,CC)/2:IFD=INT(D)THEN1610
1590 D(HL,CC-1)=D(HL,CC-1)+2:D(HL,CC+1)=D(HL,CC+1)+2:D(HL-1,CC)=D,CC)+2
1600 D(HL-1,CC-1)=D(HL-1,CC-1)+2:D(HL-1,CC+1)=D(HL-1,CC+1)+2
1610 NEXTC1THENRETURN
1620 FORLL=1TOHL-1:D=D(LL,0)/2:IFD=INT(D)THEN1650
1630 D(LL-1,0)=D+1,0)=D(LL+1,0)+2:D(LL-1,1)=D(LL-1,1)+2
```

```
1640 D(LL,1)=D(LL,1)+2:D(LL+1,1)=D(LL+1
1650 D=D(LL,HC)/2:IFD=INT(D)THEN1675
1660 D(LL-1,HC)=D(LL-1,HC)+2:D(LL+1,HC)=:D(LL,HC-1)=D(LL,HC-1)+2
1670 D(LL-1,HC-1)=D(LL-1,HC-1)+2:D(LL+1,HC-1)=D(LL+1,H5  IFHC<=1THENNEXTLL:RETURN
1680 FORCC=1TOHC-1:D=D(LL,CC)/2:IFD=INT(D)THEN1730
D(LL,CC-1)+2:D(LL,CC+1)=D(LL,CC+1)+2
1700 D(LL-1,CC-1)=D(LL-1,CC-1)+2:D(LL-1,CC(LL-1,CC+1)+2
1710 D(LL+1,CC-1)=D(LL+1,CC-1)+2:D(LL+1,CC+1)=D(LL+1,CC+1)+2
172L-1,CC)+2:D(LL+1,CC)=D(LL+1,CC)+2
1730 NEXTCC:NEXTLL:RETURN
1740 FORLL=0TOHL:F(LL,CC):NB=INT(D/2):PT=D-2*NB
1750 IFNB<2ORNB>3THEND(LL,CC)=0:GOTO1780
1760 IFTHEND(LL,CC)=1:GOTO1780
1770 D(LL,CC)=PT
1780 NEXTCC:NEXTLL:RETURN
```

Index

Abstraction, levels of, 155-57, 184
Action, continuity and speed of, 35, 197-99, 205, 244
Algorithm, 145-46
Apple
  differences from other systems of. *See* System-dependent features
  game paddle, 64, 249
Argument, 180
Aristotle, 151, 152
Array, size of in bytes, 217, 220
ASC, 41
ASCII, 38, 40. *See also* ASC
Bits. *See* Powers of two
Blanks (in BASIC listings), 212-13. *See also* Space saving
Boldface, 8
Braces, 149, 150-51
Break (in Free BASIC), 151
Calendar, Gregorian, 112
Cannibalization, 23-24, 59, 121
Cards, shuffling of, 84, 86
Case. *See* IF case
Change, provision for, 186-89, 301-2. *See also* Generalization
Chaliapin, Fyodor, 245
CHR$, 50
CLR key, 50. *See also* Screen clearing
CLS, 63. *See also* Screen clearing
Colon, 7, 14
Comma, 6
Comment, 149. *See also* REM
Condition, 7, 150, 151, 152, 153
Constant, 5, 187
Context, switching of, 245
Control-C, recognition of as Apple break character, 50
Counting, techniques for, 32-33, 42, 44, 53
Coupling (of programs), 56
CTRL key, 283
Cursor
  characters for moving of, 102, 125, 242
  positioning of, 123, 125, 242
Dates, parsing of, 112
DATA. *See* READ...DATA
Day (of the week), 112
Days of the Week (anon.), 111
Default, 46. *See also* IF; IF case
Delay. *See* Timing loop

Deletion, algorithm for, 84-86, 205-8, 241
Dijkstra, Edsger, 155, 185
Documentation, 192-94
Double precision, 220
Drake, Frank, 277
Editing, 172, 200-205, 210
Else. *See* IF...THEN; IF case
  vs. ELSE, 152
Encoding, 24, 33, 38, 86-88, 134, 217-20, 297-98
END, 31
Errors, handling of, 197, 199-200
Events, assuring desired frequency of, 69
Feedback. *See* Score
Files, maintenance of, 208-10
Flowchart, 1, 10, 11-14, 145
Free BASIC, 9, 144-58
  translation to BASIC of, 149-50, 153-55, 193-94
Function
  isolation of, 81-82, 86-88, 185-86, 241
  separation of, 66, 78, 123-24, 241-42
Function (BASIC), definition of, 14-15
Game of Life, transformation rules of, 286
Game paddle. *See* Apple game paddle
Generalization, 11
GET, 50, 64, 295. *See also* Input, single character
Hits, array of. *See* Encoding
HOME, 50
How to Manage Structured Programming (Yourdon), 185
IF case, 152
IF...THEN, 7, 8, 151-52
Indentation, 149, 152
Initialization, 31, 188, 302
INKEY$, 50, 64, 295
INPUT, 4, 6, 7, 14
Input, single character, 35, 50
Insertion, algorithm for, 205-8
INT, 4, 44
Intelligent Life in the Universe (Shklovskii and Sagan), 275
Interaction, 197-99
Jiffy, 293. *See also* Pet clock
Law of the Excluded Middle, 151
Leap year, 117

LEN, 41
LINE INPUT, 119, 162, 199
Line number, 2, 147, 153, 193
Loop. *See* Program structure
   limit of. *See* Constant; Parameter
Lowercase/uppercase, distinction between, 148-49, 152
Memory, limitations imposed by, 162, 213, 217
Menu, 197, 241
Midnight, 69
MID$, 41
Mnemonics, selection of, 171
Modularity, 181, 241
Murphy's Law, 186-87
NOT, 150
Number, decomposition of, 220.
   See also Powers of two
Number pad, 98, 281
Odd/even, determining which of a number is, 44
ON...GOSUB, 15-16, 48, 301
Orthogonality, 281-83
Parameter 149, 187
Pascal, 157
Pet
   clock for, 63-64, 66, 107, 283
   differences from other systems of. *See* System-dependent features
Pigeonhole. *See* Variable
Powers of two, 29, 39, 44
Prime number, 275
PRINT, 4-7
Processes, 245
   concurrent, 78
Program
   design of, 48, 50, 66, 74-76, 121, 124, 145-46, 292, 301
   structure of, 29-31, 44, 56, 74-76, 146, 149, 151, 189-90, 301.
      See also Stubs; Top-down
Programming
   GOTO-less, 146, 153, 155-57
   structured, 155-57
Prompts, 197-99, 200
Pseudocode, 146
Question mark, 4
Radio, transmission by, 275
Random numbers, 3, 39-40, 104
Randomizing, 248-49
READ...DATA, 188
REM, 213
Repeat...until, 150-51
Reverse video, 131
RND, 3, 248
Round-off, 220
Score, 34, 45, 92

Screen
   clearing of, 41, 50, 199
   management of, 241-42
   scrolling of, 41
Semicolon, 5-6, 123
SHIFT key, 283
Shift, logical right, 44
Shuffling. *See* Card shuffling
Space saving, 212-28
Strings, input of. *See* LINE INPUT
Structured Programming (Dahl et al.), 155
Stub, 74-76, 189-90, 286
Subroutine, 16, 153, 180
Symbolic address. *See* Variable
Synchronization, 78, 98
System-dependent features, 4-5, 7, 14, 38, 50, 63-64, 102, 125, 152, 212-13, 217, 220, 242, 249, 293-95
System independence, 136
Tab, 6
Taxman (Harrison), 131
TI. *See* Pet clock
Time, Julian, 66-69
TIME$, 63, 78. *See also* Pet clock
Timing loop, 94, 102, 293-95
Top-down, 74-76, 157, 181-85, 241.
   See also Program structure
Trade-off, 59, 217
Traditions of Devon (Bray), 111
TRS-80, differences from other systems of. *See* System-dependent features
Until. *See* Repeat...until
Uppercase. *See* Lowercase/uppercase distinction
VAL, 48
Value, 5
Variable, 3, 5
   dummy, 15
   global, 180, 181
While, 153
Wirth, Nicklaus, 185
Working copy, 210
Wraparound, 281
Yourdon, Ed, 31, 185
Zeroes, leading, 81
<> (not equal), 34
#. *See* Comment

The SYBEX Library

INSIDE BASIC GAMES
by Richard Mateosian 350 pp., 240 Illustr., Ref. B245
Teaches interactive BASIC programming through games. Games are written in Microsoft BASIC and can run on the TRS-80, APPLE II and PET/CBM.

FIFTY BASIC EXERCISES
by J.P. Lamoitier 250 pp., 195 Illustr., Ref. B250
Teaches BASIC by actual practice using graduated exercises drawn from everyday applications. All programs written in Microsoft BASIC.

YOUR FIRST COMPUTER
by Rodnay Zaks 280 pp., 150 Illustr., Ref. C200A
The most popular introduction to small computers and their peripherals: what they do and how to buy one.

MICROPROCESSORS: FROM CHIPS TO SYSTEMS
by Rodnay Zaks 420 pp., 255 Illustr., Ref. C201
A simple and comprehensive introduction to microprocessors from both a hardware and software standpoint: what they are, how they operate, how to assemble them into a complete system.

MICROPROCESSOR INTERFACING TECHNIQUES
by Rodnay Zaks and Austin Lesea 460 pp., 400 Illustr., Ref. C207
Complete hardware and software interconnect techniques including D to A conversion, peripherals, standard buses and troubleshooting.

PROGRAMMING THE 6502
by Rodnay Zaks 390 pp., 160 Illustr., Ref. C202
Assembly language programming for the 6502, from basic concepts to advanced data structures.

6502 APPLICATIONS BOOK
by Rodnay Zaks 280 pp., 205 Illustr., Ref. D302
Real life application techniques: the input/output book for the 6502.

6502 GAMES
by Rodnay Zaks 300 pp., 140 Illustr., Ref. G402
Third in the 6502 series. Teaches more advanced programming techniques, using games as a framework for learning.

PROGRAMMING THE Z80
by Rodnay Zaks 620 pp., 200 Illustr., Ref. C280
A complete course in programming the Z80 microprocessor and a thorough introduction to assembly language.

PROGRAMMING THE Z8000
by Richard Mateosian 310 pp., 125 Illustr., Ref. C281
How to program the Z8000 16-bit microprocessor. Includes a description of the architecture and function of the Z8000 and its family of support chips.

THE CP/M HANDBOOK (with MP/M)
by Rodnay Zaks 330 pp., 100 Illustr., Ref. C300
An indispensable reference and guide to CP/M — the most widely used operating system for small computers.

DON'T (or How to Care for Your Computer)
by Rodnay Zaks 200 pp., 100 Illustr., Ref. C400
The correct way to handle and care for all elements of a computer system including what to do when something doesn't work.

INTRODUCTION TO PASCAL (Including UCSD PASCAL)
by Rodnay Zaks 440 pp., 130 Illustr., Ref. P310
A step-by-step introduction for anyone wanting to learn the Pascal language. Describes UCSD and Standard Pascals. No technical background is assumed.

THE PASCAL HANDBOOK
by Jacques Tiberghien 490 pp., 350 Illustr., Ref. P320
A dictionary of the Pascal language, defining every reserved word, operator, procedure and function found in all major versions of Pascal.

PASCAL PROGRAMS FOR SCIENTISTS AND ENGINEERS
by Alan Miller 250 pp., 80 Illustr., Ref. P340
A comprehensive collection of frequently used algorithms for scientific and technical applications, programmed in Pascal. Includes such programs as curve-fitting, integrals and statistical techniques.

FIFTY PASCAL PROGRAMS
by Rodnay Zaks and Rudolph Langer 275 pp., 90 Illustr., Ref. P350
A collection of 50 Pascal programs ranging from mathematics to business and games programs. Explains programming techniques and provides actual practice.

MICROPROCESSOR LEXICON (with Ten Language Vocabulary)
140 pp., Ref. X2
All the definitions and acronyms of microcomputer jargon defined in a handy pocket-size edition. Includes translations of the most popular terms into ten languages.

MICROPROGRAMMED APL IMPLEMENTATION
by Rodnay Zaks 350 pp., Ref. Z10
An expert-level text presenting the complete conceptual analysis and design of an APL interpreter, and actual listings of the microcode.

SELF STUDY COURSES

Recorded live at seminars given by recognized professionals in the microprocessor field.

INTRODUCTORY SHORT COURSES:
Each includes two cassettes plus special coordinated workbook (2½ hours).

S10—INTRODUCTION TO PERSONAL AND BUSINESS COMPUTING
A comprehensive introduction to small computer systems for those planning to use or buy one, including peripherals and pitfalls.

S1—INTRODUCTION TO MICROPROCESSORS
How microprocessors work, including basic concepts, applications, advantages and disadvantages.

S2—PROGRAMMING MICROPROCESSORS
The companion to S1. How to program any standard microprocessor, and how it operates internally. Requires a basic understanding of microprocessors.

S3—DESIGNING A MICROPROCESSOR SYSTEM
Learn how to interconnect a complete system, wire by wire. Techniques discussed are applicable to all standard microprocessors.

INTRODUCTORY COMPREHENSIVE COURSES:
Each includes a 300-500 page seminar book and seven or eight C90 cassettes.

SB1—MICROPROCESSORS
This seminar teaches all aspects of microprocessors: from the operation of an MPU to the complete interconnect of a system. The basic hardware course. (12 hours)

SB2—MICROPROCESSOR PROGRAMMING
The basic software course: step by step through all the important aspects of microcomputer programming. (10 hours)

ADVANCED COURSES:
Each includes a 300-500 page workbook and three or four C90 cassettes.

SB3—SEVERE ENVIRONMENT/MILITARY MICROPROCESSOR SYSTEMS
Complete discussion of constraints, techniques and systems for severe environment applications, including Hughes, Raytheon, Actron and other militarized systems. (6 hours)

SB5—BIT-SLICE
Learn how to build a complete system with bit slices. Also examines innovative applications of bit slice techniques. (6 hours)

SB6—INDUSTRIAL MICROPROCESSOR SYSTEMS
Seminar examines actual industrial hardware and software techniques, components, programs and cost. (4½ hours)

SB7—MICROPROCESSOR INTERFACING
Explains how to assemble, interface and interconnect a system. (6 hours)

SOFTWARE

BAS 65™ CROSS-ASSEMBLER IN BASIC
8" diskette, Ref. BAS 65
A complete assembler for the 6502, written in standard Microsoft BASIC under CP/M®.

8080 SIMULATORS
Turns any 6502 into an 8080. Two versions are available for APPLE II.
APPLE II cassette, Ref. S6580-APL(T)
APPLE II diskette, Ref. S6580-APL(D)

APPLICATION PROGRAMS
Actual programs described in SYBEX books are available in computer-readable form as follows:

BOOK	SOFTWARE	REF
INSIDE BASIC GAMES (B245)	Diskette for APPLE	A45D
INSIDE BASIC GAMES (B245)	Cassette for APPLE	A45T
INSIDE BASIC GAMES (B245)	Cassette for PET	P45T
INSIDE BASIC GAMES (B245)	Diskette for TRS-80	R45D
INSIDE BASIC GAMES (B245)	Cassette for TRS-80	R45T
FIFTY BASIC EXERCISES (B250)	Cassette for TRS-80	R50T
6502 APPLICATIONS (D302)	Cassette for SYM	302T
6502 GAMES (G402)	Cassette for SYM	402T

FOR A COMPLETE CATALOG OF OUR PUBLICATIONS

SYBEX

U.S.A.
2344 Sixth Street
Berkeley,
California 94710
Tel: (415) 848-8233
Telex: 336311

SYBEX-EUROPE
Centre Paris Daumesnil
4 Place Felix Eboue
75012 Paris, France
Tel: (1) 341-7110
Telex: 211801

SYBEX-VERLAG
Heyestr. 22
4000 Düsseldorf 12
Germany
Tel: (0211) 287066
Telex: 08 588 163